AP Chemistry For Dumm...

P9-ECN-683

Cheat Sheet

Atomic Structure

$E = h\nu$

$c = \lambda\nu$

$\lambda = h/(m\upsilon)$

$p = m\upsilon$

$E = (-2.178 \times 10^{-18}/n^2)$ joule

E = energy

ν = frequency

υ = velocity

λ = wavelength

p = momentum

m = mass

n = principal quantum number

Speed of light, $c = 3.0 \times 10^8$ m s^{-1}

Avogadro's number = 6.022×10^{23} mol^{-1}

Planck's constant, $h = 6.63 \times 10^{-34}$ J s

Electron charge, $e = -1.602 \times 10^{-19}$ coulomb

Boltzmann's constant, $k = 1.38 \times 10^{-23}$ J K

1 electron volt per atom = 96.5 kJ mol^{-1}

Equilibrium

$K_a = ([H^+][A^-])/[HA]$

$K_b = ([OH^-][HB^+])/[B]$

$K_w = K_a \times K_b = 1.0 \times 10^{-14}$ (25°C)

$pH = -\log[H^+]$

$pOH = -\log[OH^-]$

$14 = pH + pOH$

$pH = pK_a + \log([A^-]/[HA])$

$pOH = pK_b + \log([HB^+]/[B])$

$pK_a = -\log K_a$

$pK_b = -\log K_b$

$K_p = K_c(RT)^{\Delta n}$

Thermochemistry/Kinetics

$\Delta S° = \Sigma S°$ products $- \Sigma S°$ reactants

$\Delta H° = \Sigma \Delta H_f°$ products $- \Sigma \Delta H_f°$ reactants

$\Delta G° = \Sigma \Delta G_f°$ products $- \Sigma \Delta G_f°$ reactants

$\Delta G° = \Delta H° - T\Delta S° = -RT \ln K = -nFE°$

$\Delta G = \Delta G° + RT \ln Q = \Delta G° + 2.303\, RT \log Q$

$q = mC\Delta T$

$C_p = \Delta H/\Delta T$

$\ln[A]_t - \ln[A]_0 = -kt$

$(1/[A]_t) - (1/[A]_0) = kt$

$\ln k = (-E_a/R)(1/T) + \ln A$

Oxidation–Reduction/Electrochemistry

$Q = ([C]^c [D]^d) / ([A]^a [B]^b)$ where aA + bB $\to$ cC + dD

Current, $I = $ charge / time

1 ampere = 1 coulomb second^{-1}

$E_{cell} = E°_{cell} - (RT \ln Q) / (nF) = E°_{cell} - (0.0592 \log Q) / n$ (at 25°C)

$\log K = nE° / 0.0592$

AP Chemistry For Dummies®

Cheat Sheet

Gases, Liquids and Solutions

$PV = nRT$

$(P + n^2a/V^2)(V − nb) = nRT$

$P_A = P_{total} \times X_A$ (where X_A = moles A / total moles)

$P_{total} = P_A + P_B + P_C + \ldots$

STP = 0.000 °C and 1.000 atm

1 atm = 760 mm Hg = 760 torr

$moles = mass / molar\ mass$

K = °C + 273

$(P_1V_1)/T_1 = (P_2V_2)/T_2$

$density = mass / volume$

$u_{rms} = (3kT/mass)^{1/2} = (3RT/molar\ mass)^{1/2}$

Kinetic energy per molecule = $\frac{1}{2}\ mu^2$

Kinetic energy per mole = 1.5 RT

$rate_1 / rate_2 = (molar\ mass_2 / molar\ mass_1)^{1/2}$

Molarity, M = moles solute / liters solution

Absorbance, A = (molar absorbitivity)(pathlength)(concentration)

Molality, m = moles solute / kilograms solvent

$\Delta T_f = iK_f \times$ molality

$\Delta T_b = iK_b \times$ molality

(i = van't Hoff factor, # particles)

Osmotic pressure, π = (molar mass)(RT)

Abbreviations

$K = K_{eq}$ = equilibrium constant

u_{rms} = root mean squared speed

K_a = weak acid dissociation constant

K_b = weak base dissociation constant

$K_p = K_{eq}$ using gas pressures

$K_c = K_{eq}$ using molar concentrations

P = pressure

V = volume

n = moles

T = temperature

S° = standard entropy

H° = standard enthalpy

G° = standard free energy

E° = standard reduction potential

q = heat

C = specific heat capacity

E_a = activation energy

k = rate constant

A = frequency factor

K_f = molal freezing point depression constant

K_b = molal boiling point elevation constant

Gas constant, R = 8.31 J mol⁻¹ K⁻¹ = 0.0821 L atm mol⁻¹ K⁻¹ = 8.31 volt coulomb mol⁻¹ K⁻¹

Copyright © 2009 Wiley Publishing, Inc.
All rights reserved.

Item 8976-8.

For more information about Wiley Publishing, call 1-800-762-2974.

For Dummies: Bestselling Book Series for Beginners

AP Chemistry

FOR

DUMMIES®

AP Chemistry

FOR

DUMMIES®

by Peter Mikulecky, Michelle Rose Gilman, and Kate Brutlag

WILEY

Wiley Publishing, Inc.

AP Chemistry For Dummies®

Published by
Wiley Publishing, Inc.
111 River St.
Hoboken, NJ 07030-5774
www.wiley.com

For general information on our other products and services, please contact our Customer Care Department within the U.S. at 800-762-2974, outside the U.S. at 317-572-3993, or fax 317-572-4002.

For technical support, please visit www.wiley.com/techsupport.

Wiley also publishes its books in a variety of electronic formats. Some content that appears in print may not be available in electronic books.

Library of Congress Control Number: 2008938864

ISBN: 978-0-470-38976-8

Manufactured in the United States of America

10 9 8 7 6 5 4 3 2 1

WILEY

About the Authors

Peter Mikulecky grew up in Milwaukee, an area of Wisconsin unique for its high human-to-cow ratio. After a breezy four-year tour in the Army, Peter earned a bachelor of science degree in biochemistry and molecular biology from the University of Wisconsin–Eau Claire, and a PhD in biological chemistry from Indiana University. With science seething in his DNA, he sought to infect others with a sense of molecular wonderment. Having taught, tutored, and mentored in classroom and laboratory environments, Peter was happy to find a home at Fusion Learning Center and Fusion Academy. There, he enjoys convincing students that biology and chemistry are, in fact, fascinating journeys, not entirely designed to inflict pain on hapless teenagers. His military training occasionally aids him in this effort. He is the author of *AP Biology For Dummies* and *Chemistry Workbook For Dummies*.

Michelle Rose Gilman is most proud to be known as Noah's mom. A graduate of the University of South Florida, Michelle found her niche early, and at 19, she was working with emotionally disturbed and learning-disabled students in hospital settings. At 21, she made the trek to California, where she found her passion for helping teenagers become more successful in school and life. What started as a small tutoring business in the garage of her California home quickly expanded and grew to the point where traffic control was necessary on her residential street.

Today, Michelle is the founder and CEO of Fusion Learning Center and Fusion Academy, a private school and tutoring/test prep facility in Solana Beach, California, serving more than 2,000 students per year. She is the author of *ACT For Dummies, Pre-Calculus For Dummies, AP Biology For Dummies, Chemistry Workbook For Dummies, GRE For Dummies, Pre-Calculus Workbook for Dummies*, and other books on self-esteem, writing, and motivational topics. Michelle has overseen dozens of programs over the last 20 years, focusing on helping kids become healthy adults. She currently specializes in motivating the unmotivatable adolescent, comforting their shell-shocked parents, and assisting her staff of 35 teachers.

Michelle lives by the following motto: There are people content with longing; I am not one of them.

Kate Brutlag has been a full-fledged science dork since she picked up her first book on dinosaurs as a child. A native of Minnesota, Kate enjoys typical regional activities such as snow sports and cheese eating. Kate left Minnesota as a teen to study at Middlebury College in Vermont and graduated with a double major in physics and Japanese. Seeking to unite these two highly unrelated passions, she spent a year in Kyoto, Japan, on a Fulbright scholarship researching Japanese constellation lore. Kate was quickly drawn back to the pure sciences, however, and she discovered her love for education through her work at Fusion Academy, where she currently teaches upper-level sciences and Japanese. She is the author of *Chemistry Workbook for Dummies*.

Dedication

We would like to dedicate this book to our families and friends who supported us during the writing process. Also, to all our students who motivate us to be better teachers by pushing us to find unique and fresh ways to reach them.

Authors' Acknowledgments

Thanks to Bill Gladstone from Waterside Productions for being an amazing agent and friend. Thanks to Jennifer Connolly, our project editor, for her clear feedback, support, and abundant patience. A special shout-out to our acquisitions editor, Stacey Kennedy, who, for reasons unclear, seems to keep wanting to work with us.

Publisher's Acknowledgments

We're proud of this book; please send us your comments through our Dummies online registration form located at www.dummies.com/register/.

Some of the people who helped bring this book to market include the following:

Acquisitions, Editorial, and Media Development

Project Editor: Jennifer Connolly

Acquisitions Editor: Stacy Kennedy

Copy Editor: Jennifer Connolly

Technical Editor: Phil Dyer

Senior Editorial Manager: Jennifer Ehrlich

Editorial Supervisor: Carmen Krikorian

Editorial Assistants: Erin Calligan Mooney, Joe Niesen, Jennette ElNagger, and David Lutton

Cover photo: © Thinkstock Images

Cartoons: Rich Tennant (www.the5thwave.com)

Composition Services

Project Coordinator: Katie Key

Layout and Graphics: Carrie A. Cesavice, Reuben W. Davis

Proofreaders: Melanie Hoffman, Harry Lazarek

Indexer: Joan K. Griffitts

Publishing and Editorial for Consumer Dummies

> **Diane Graves Steele,** Vice President and Publisher, Consumer Dummies

> **Joyce Pepple,** Acquisitions Director, Consumer Dummies

> **Kristin A. Cocks,** Product Development Director, Consumer Dummies

> **Michael Spring,** Vice President and Publisher, Travel

> **Kelly Regan,** Editorial Director, Travel

Publishing for Technology Dummies

> **Andy Cummings,** Vice President and Publisher, Dummies Technology/General User

Composition Services

> **Gerry Fahey,** Vice President of Production Services

> **Debbie Stailey,** Director of Composition Services

Contents at a Glance

Table of Contents

Introduction

· ·

One of the central lessons of chemistry is that the starting and ending points of a reaction don't have much to do with the business in between. Your journey through AP chemistry is much the same. In the beginning, you decided to take AP chemistry. In the end, you'll take the AP chemistry exam. These moments are fixed points on the horizon behind you and the horizon in front of you. Time will move you from one point to the other. What you make of the intervening landscape is mostly up to you.

But along the way, we'd like to help.

About This Book

The College Board is very clear about the chemistry concepts and problems it wants you to master. The clear guidelines provided by the College Board lie at the heart of this book. We emphasize what the Board emphasizes, and we suspect that those are the very things that *you'd* like to emphasize as you prepare for the exam.

Chemistry is the practical science, which is one reason why a practical attitude permeates this book. We also realize that you're not simply reading this book — you're *using* it for a specific purpose. Your simple, clear goal: to ace the AP Chemistry exam. You want your grasp of chemistry to be just as clear as your goal for the AP exam. This book is here to support your quest for chemistry clarity; think of it as window cleaner for your brain. (Window cleaner is largely an aqueous solution of ammonia, by the way.)

The structure of this book largely follows the sequence you might follow in an AP chemistry course, so it lends itself to use as companion text as you go through the course. On the other hand, we have written the book to be quite modular, so you can pick and choose the chapters that interest you in any sequence. Cross-references are provided in any instances where there is must-know information discussed in another chapter. You don't have to read this thing cover to cover.

What You're Not to Read

Of course, you can read anything you want to, up to and including the publishing legalese on the opening page. But there are certain bits of the book that are less vital than others, and we alert you to them so you can make efficient decisions about how to spend your reading time.

Any text highlighted by the technical stuff icon (see "Icons Used in This Book," later in this Introduction) is, well, really technical. Now, in a book about chemistry, there's bound to be an abundance of technical material. But information accompanied by this icon is stuff that you won't need for the AP exam — though it certainly won't hurt you to know it!

In addition, any material placed within a sidebar is something of an "aside," items that are relevant and which you may find interesting or helpful, but definitely not required reading. These are items you might peruse as you sip tea and take a short break from hard-core study.

Foolish Assumptions

You're unique, and so is everyone else who uses this book. But to make it the most useful book possible, we've made some assumptions about you. And you should know what those assumptions are. We assume that

✔ **You have a basic facility with algebra and arithmetic.** You should know how to solve simple equations for an unknown variable, including the occasional quadratic equation. You should know how to work with exponents and logarithms. That's about it for the math. At no point will we ask you to, say, consider the contradictions between the Schrödinger equation and stochastic wavefunction collapse.

✔ **You have access to a college-level textbook in chemistry,** such as the one you use in your AP chemistry course. If you're not enrolled in an AP hemistry course or for some other reason don't have a good textbook, you can still benefit from this book. Every set of practice problems in this book is preceded by a chapter that gives a concise but thorough review of the topic at hand.

✔ **You don't like to waste time.** Neither do we. Chemists in general aren't too fond of time-wasting. That said, let's move straightaway to the organization of the book.

How This Book Is Organized

This book is divided into eight thematic parts. Some of the shorter parts serve very specific functions, such as familiarizing you with the AP exam and providing study tips, or giving you the opportunity to take full-length, model AP chemistry exams. The larger parts at the heart of the book are a comprehensive review of the material chosen by the College Board for emphasis in AP chemistry courses and for inclusion on AP chemistry exams. Chapters that extensively review chemistry content alternate with chapters that present pithy summaries of the *most* important points and test you against an arsenal of practice questions.

Part I: Thriving Inside the Test Tube: How to Prepare for the Exam

Like any big standardized exam, the AP chemistry exam tests more than chemistry — it tests your ability to take a test. In addition to knowing your chemistry cold, on test day you'll want the confidence that comes from knowing exactly what's coming. This part demystifies the exam, the exam-taking process, and the scoring process. In addition, it gives sage advice about how to tackle the long-term challenge of preparing for the exam without incurring permanent mental and emotional damage along the way.

Part II: Building Matter from the Ground Up: Atoms and Bonding

Chemistry is about building things with atoms and then breaking the things you built. Then, you build new things with the same building blocks as before. This part introduces you to the building blocks (atoms) and reviews the basic rules of construction (bonding). A surprisingly small number of guidelines give rise to an impressive array of molecular shapes, and we show you how.

Part III: Colliding Particles to Produce States

In addition to ridiculously tiny things like atoms and bonds, chemistry is about bigger, easier-to-notice things, like whether something is a liquid or not. But explaining *why* something is a liquid and *why* it turns to a solid or a gas requires you to squint once more at the tiny particles that compose that something. This part explains the "big" properties of matter (its state, temperature, pressure, volume) in terms of the behavior of the "little" particles that make up the matter.

Part IV: Transforming Matter in Reactions

Like any venerable and worthy science, chemistry extracts simplicity from an intimidating mountain of complexity. To do so, chemistry uses two basic tricks. First, chemistry keeps careful tabs on matter. By keeping a close count of each atom, ion, or electron involved in a reaction, it becomes possible to place any of a million chemical reactions into one of a few basic categories. Although they wear different clothes, there are really only a few kinds of reactions. In addition to accounting for all its material bits, chemistry uses a second trick to make sense of things: it keeps track of energy. By paying attention to different forms and amounts of energy, chemistry can predict whether or not reactions will go forward or backward and even calculate the rate at which they'll get to where they're going. This part delves into chemistry's two tricks: following the matter and following the energy.

Part V: Describing Patterns and Predicting Properties

Sometimes you can tell a lot about the molecular details of a reaction by observing some obvious things with your naked eyes. What color is the sample? Did a solid appear as if from nowhere from within a liquid? Is some sort of gas bubbling out of the beaker? Connecting observations like these to the details of a reaction is the task of descriptive chemistry, the heart of Part V. Connecting properties to structures is also a major focus of organic chemistry, a topic we also cover succinctly in this part.

Part VI: Getting Wet and Dirty: Labs

Labwork is a major portion of the AP chemistry experience, and the College Board includes "lab knowledge" questions on the AP chemistry exam. You should know your way around a general chemistry lab, including how to select the right piece of glassware or equipment for a particular task. We cover lab equipment, common lab tasks, and safety practices in this part. Then, we give you an overview of each of the 22 labs officially recommended by the College Board, emphasizing the key concepts and data analyses with which you need to be familiar.

Part VII: Measuring Yields: Practice Tests

There's no substitute for experience. The best way to assess your readiness for the AP chemistry exam is to take an AP chemistry exam. Taking a full-length, timed exam is also a great way to train your brain for the test-taking rhythms you'll experience on test day. Because taking practice tests is such a good idea, we've put two of them in this part. And because it's just possible that you don't know *every* answer, we've included complete answers with explanations for all the answers.

Part VIII: The Part of Tens

It would be nice if life came with a package: "Little Lists of Helpful Reminders." This part of your life *does* come with such lists. This part combs through the entire book, pulling together top-ten lists to help you review the most important ideas and equations.

Icons Used in This Book

Nicely nestled along the margins of this workbook you'll find a selection of helpful icons. Think of them as landmarks, familiar signposts to guide you as you cruise the highways of chemistry.

 Within already-concise summaries of chemical concepts, passages marked by this icon represent the choicest, must-know bits of information. You will need to know this stuff in order to solve problems.

 Sometimes there's an easy way and a hard way. This icon alerts you to passages intended to highlight an easier way. It's worth your while to linger for a moment. You may find yourself nodding quietly as you jot down a grateful note or two.

 This icon gives you technical info that you don't *have* to know to understand the rest of the section, but we sure think it's interesting to read about!

 This icon is just what it says. It flags the kind of information that you ignore at your own peril. If we could make it *our* own peril, we would. But we can't, so we use this icon.

Where to Go from Here

This book can help you in different ways, depending on your position along the AP chemistry path. If you have yet to enter the AP chemistry classroom, you can use the book to smooth your rough edges and build confidence so that when the course hits you, you're ready.

If you're just beginning the AP chemistry course, then smile. You've just discovered a lovely traveling companion. Use the book to supplement your schoolwork and as a way to get extra help and practice with weak spots. And be sure to take advantage of the full-length practice exams to help prepare yourself as test day approaches.

If you're nearing the middle or the end of the course, then use the book as first aid, shoring up your understanding of those areas you already *know* give you trouble. In parallel, use the book as a helpful secondary source for the remaining part of the course.

Finally, if you're not actually taking the AP course but could use some help with a general chemistry course you're taking in college, you'll probably find this book a helpful, friendly, and lightweight addition to your backpack.

Part I

Thriving Inside the Test Tube: How to Prepare for the Exam

The 5th Wave   By Rich Tennant

"I always get a good night's sleep the day before a test so I'm relaxed and alert the next morning. Then I grab my pen, eat a banana and I'm on my way."

In this part . . .

The AP Chemistry exam is mostly about chemistry, but it's also about *exams*. Knowing the ins and outs of the exam itself is just plain good sense, and is part of your training for exam day. By knowing the structure and idiosyncrasies of the test, you turn them to your advantage. This part deconstructs the test to let you see it for its parts. In addition, we offer you tips on how to pace yourself for the long-term project of preparing to show the AP Chemistry who's boss. (By the way, you're the boss.)

Chapter 1

Knowing What You're Up Against: The AP Chemistry Exam

*J*ust the words "Advanced Placement" can bring about trembling fits of fear in even the best of Chemistry students. The good 'ol folks at the Educational Testing Service (ETS) — the company hired by the College Board to write and grade the AP exams — have been disseminating this fear for over 50 years, and they've had the time to get quite good at it. Consider this chapter your best bet at releasing your fear, getting some early college credit, and planning what to do with all the free time you'll have in college by not having to take an intro chemistry course.

As you get intimate with the AP test, you will become familiar with why you should (or shouldn't) take this test, the test's structure, how you will be scored, and how to plan your time effectively. In this chapter we uncover the often-overlooked strategies for taking the exam itself.

You already know a whole bunch about chemistry. After all, you've probably spent the best part of this year listening eagerly to your high school chemistry teacher (in between texting your friend and downloading the latest Modest Mouse tune). Of course, nothing takes the place of full comprehension of the material, but knowing chemistry is only part of the puzzle (albeit a really big part). We can show you some other things that you can do to help raise your score that have little to do with chemistry. So sit back, relax, and let us show you the way.

Seeing If You Have the Right Stuff

Taking the AP exam sends out a beacon to the folks at any college that you believe you can think and perform on the college level, but can you? Although taking an AP chemistry course is not a prerequisite to taking the AP chemistry exam, we highly recommend that you do. Plus, you must be honest with yourself. If you received poor grades in chemistry in high school and could care less about stoichiometry, or if you accidentally blew up your high school chem lab by combining various chemicals, you might want to consider spending your time in an area you truly enjoy (perhaps a money-making vocation where you can pay for that lab to be rebuilt).

However, check out what your high score on the AP exam (see "Getting the Skinny on the Scoring," later in this chapter) can get you:

- ✔ You prove to the college that you understand advanced material and already possess what it takes to be successful in college. Well, go you!

- ✔ You have the opportunity to receive credit or advanced standings at most universities around the country. In plain English, this means you can sleep in a bit, because you won't have to take an intro chemistry class in college because you already proved that you know this stuff! Even better, if you aren't planning on studying science in college, you probably won't ever have to take another science class again! Yippee!

Those smart AP creators have AP tests in a variety of different subjects, so if you're not sure about your dedication to this particular test, you might want to visit their Web site and glance at the other subjects they offer. If you're just taking the AP chemistry class and don't plan on taking the exam at the end, then that's cool too. Skip this chapter and use the rest of the book for succeeding in the AP chemistry class.

Knowing the Breakdown — So You Don't Have One!

Considering that the AP folks don't provide anti-anxiety salt licks as you walk into the exam, knowing the breakdown of the exam ahead of time prepares you for what you're up against and makes you more confident so that you're better relaxed to take the exam.

The AP Chemistry exam takes three hours and includes two sections; a 90-minute, 75-question, multiple-choice section and a 95-minute, 6-question, free-response section. We explain these sections further in the sections that follow.

Making your way through multiple-choice questions: Section 1

You can't use a calculator on the multiple-choice section. With the recent capabilities of graphing calculators, it could probably take the test for you! The AP higher-ups figured this one out a long time ago.

You have to answer (or try to answer) 75 questions on the multiple-choice section. Each question has five possible answer choices. We have a feeling that the AP creators couldn't agree on what to put on the exam, so they just tossed everything into it. The questions you will encounter cover a large amount of material. Not all students will be exposed to all the material that the test may cover, so don't be surprised to see topics that may be unfamiliar to you. You can expect basic factual questions as well as heavy-duty, thought-provoking problems, including 10 to 12 math problems to be done without a calculator. This section is worth 50 percent of your score.

You will only have about 1.2 minutes to answer each question. If you don't know the immediate answer, skip the problem and move on.

Taking on free-response questions: Section 2

The second section of the exam consists of free-response questions. You will find a total of six questions:

- Three quantitative problems (one on chemical equilibrium) lasting 55 minutes.

- One of the questions will be on chemical reactions, which will require you to write balanced net equations for chemical reactions.

- Reaction question and two essay questions lasting 40 minutes.

- One of the six questions will be a laboratory-based question that could be located in either the quantitative section or the essay section, so pay close attention when we cover labs throughout this book because the lab question could be on anything.

The free-response questions are long, make no bones about it. They will require you to demonstrate your problem-solving skills, show knowledge of chemical reactions, and question your ability to reason and explain ideas logically and clearly. They are broken down into multiple parts. Basically, the test will ask you multiple questions about the concept and require you to show concept mastery.

You won't be given an incomprehensible topic for any of the free-response questions. You will be asked to solve a fairly basic chemistry concept, but they will want you to go into detail and ask you several different questions about it. There will be formulas to use and numbers to work with. Although the free-response questions are long and daunting, they're not there to trick you. The multiple parts often help to lead you to the right way to tackle the problem. The AP people want to see your true chemistry acumen, and this is the best way to assess that. We will discuss how to tackle free-response problems later in this chapter.

You can use certain tools during this portion of the test. The following list shows you what you can bring in and what the AP folks provide, as well as the restrictions on these tools:

- **Calculators:** You can use your calculator for the first 55 minutes of the free-response section, but you can't share your calculator with another student. The AP folks also restrict which calculators you can use on the AP test. Because technology changes so rapidly, you should check the College Board Web site (www.collegeboard.com) for an up-to-the-minute list of acceptable calculators.

- **Tables containing commonly used equations:** You can use these tables only during the free-response section. You can't use them on the multiple-choice section. The free-response section requires you to solve in-depth problems and to write essays where the knowledge of the concepts and how to apply the principles are the most important parts of solving the problem. The College Board gives everybody the table of equations to make it fair for the students who may not have the equations stored in their calculators. The AP folks have some heart, after all! But remember, because the test gives you the equations, you will receive no credit for answers simply written down in equation form without supported explanations or some type of logical development.

Tackling the Topics Covered

Before the AP exam was written, a few geeky chemists studied the chemistry curricula of many of the nation's best colleges. They combined their reports and came away with a clear idea of the stuff being taught to chemistry college students around the country. The end result culminated in the AP chemistry exam covering five key areas, and because we're obsessive about making sure you're the most informed you can be (and because we're control freaks), we've outlined each area for you below:

✔ **Structure of matter** (20 percent of test; see Chapters 3 through 8):

- **Atomic theory and atomic structure:** Evidence for the atomic theory; atomic masses; atomic number and mass numbers; electron energy levels; periodic relationships

- **Chemical bonding:** Binding forces; molecular models, geometry of molecules and ions, structural isomerism of simple organic molecules and coordination complexes

- **Nuclear chemistry: nuclear equations, half-lives and radioactivity**

✔ **States of matter** (20 percent of test; see Chapters 9 through 12):

- **Gases:** Laws of ideal gases; kinetic molecular theory

- **Liquids and solids:** Liquids and solids from the kinetic-molecular viewpoint; phase diagrams of one component systems; changes of state, including critical points and triple points; structure of solids

- **Solutions:** Types of solutions and factors affecting solubility; methods of expressing concentration; Raoult's law and colligative properties; nonideal behavior

✔ **Reactions** (35 to 40 percent of test; see Chapters 13 through 22):

- **Reaction types:** Acid-base reactions; precipitation reactions; oxidation-reduction reactions

- **Stoichiometry:** Ionic and molecular species present in chemical systems; balancing of equations including those for redox reactions; mass and volume relations

- **Equilibrium:** Concept of dynamic equilibrium, physical and chemical; quantitative treatment

- **Kinetics:** Concept of rate of reaction; use of experimental data and graphical analysis to determine reactant order, rate constants, and reaction rate laws; effect of temperature change on rates; energy of activation; relationship between the rate-determining step and a mechanism

- **Thermodynamics:** State functions; first law; second law; relationship of change in free energy to equilibrium constants and electrode potentials

✔ **Descriptive chemistry** (10 to 15 percent of test; see Chapters 23 to 27): Relationships in the periodic table (periodicity); chemical reactivity and products of chemical reactions; introduction to organic chemistry

✔ **Laboratory** (5 to 10 percent of test; see Chapters 27 and 28): Making observations of chemical reactions and substances; recording data; interpreting results; communicating the results

Understanding the AP Test Questions

The AP chemistry test offers no surprises. Every question covers one or more basic fundamental chemistry concept, tests you on your knowledge of the five areas described in the section above, "Tackling the Topics Covered." Following the basic guidelines in this section helps you tackle the multiple-choice and free-response sections of the exam.

Making sense of multiple choice

Of the two parts of this test, the multiple-choice part is easiest because the correct answer is staring you right in your face; you just need to find it. In the following sections, we describe the different setups of multiple-choice questions, and we've also included a section that shows you some educated-guessing techniques to help you whittle down the obvious wrong answers.

Clearing out the "crowded" questions

Basically, there are three types of multiple-choice questions. The first type of questions you encounter we call "crowded together" questions. These are the easiest because they take very limited time to complete. With crowded together questions, you get some information upfront lettered A through E. After this information, you get two to three questions pertaining to the initial A through E information. You just pick the correct letter choice to solve the questions.

Standing out from the crowd: Dealing with "loner" questions

Another type of question that makes up much of the multiple-choice questions we like to call "loner questions." Each loner question covers only one lonely topic at a time. You're presented with a question providing you with the information you need to complete the problem, usually five possible choices, again lettered A through E.

Putting a spin on interpretative questions

The interpretation questions do not occur very often, but they're important to know about. Basically, the interpretation questions take two somewhat related loner questions and stuff them together and then ask you two or more questions about the information. You are given a graph, diagram, or data table, and are then asked two questions about the presented visual.

Getting through the questions . . . with as many right answers as possible

The AP test creators throw everything at you at once, so don't think that the test starts easy and gets more difficult — it doesn't. Manage your time on the text using these tips:

- ✔ **Short questions:** Answer the questions that are the shortest first, leaving more time for you to put a little more time into the longer, more intense questions.

- ✔ **Easy questions:** Answer the questions that make you want to smile first (the concepts you know the most about), and leave the questions that make you want to vomit for last.

- ✔ **Educated guessing:** Educated guessing is a good thing. Eliminating even one possible answer increases your choices of hitting the right one. You don't get penalized for leaving an answer blank, but you do get a fraction ($\frac{1}{4}$) removed from your score for a wrong answer. On the other hand, guessing correctly earns you a full point. In other words, completely random guessing neither hurts nor harms you. Educated guessing helps you. If you don't know how to solve the problem but want to make an educated guess, keep these ideas in mind to make the best possible educated guess:

  - Take a quick glance through the possible answers.

  - The AP folks are not trying to trick you, so if something looks blatantly wrong, it probably is. So you're probably safe to eliminate it.

  - Eliminate the answers that don't seem to match up to the question.

  - Eliminate answers that appear too close to the actual question.

Finagling the free-response questions

The free-response section takes 95 minutes and has six questions divided into two parts:

- ✔ **Part A:** Part A takes 55 minutes, in which you answer three questions for which you're allowed to use your calculator. Part A covers one equilibrium problem and two other problems.

- ✔ **Part B:** Part B takes 40 minutes and covers the last three questions, one being a reactions question and the other two being essay questions. A lab question is thrown in the mix somewhere. Calculators are *not* permitted in Part B.

Part A counts for 60 percent of your score in the free-response section with each question counting 20 percent. Spending a little extra time in Part A is worth it.

The test giveth . . .

Before you begin the free-response section, you are given a plethora of information:

- ✔ You will be bombarded with four, highly coveted pages of chemistry-related material, consisting of a whole bunch of equations and constants covering atomic structure, equilibrium, thermochemistry, liquids, gases, solutions, oxidation-reduction, and electrochemistry.

- ✔ You will also receive a limited periodic table and a reduction potential table.

Right now, you might be thinking, "Cool, I'll be getting all the info I need to solve the problems." But, when you get to the test site and look at the questions, you'll soon realize that the good AP pros didn't just give you *only* the info you need, but they also gave you a ton more than you'll ever use on the test. So, even though you have information to take from, you still need to be chem smart to know when and where to use the info given to you.

Breathing easy by knowing what to expect

On the free-response portion of the exam, you need to answer the questions in your own logical words and commit to your answers. The topics on the AP exam refer to general chemistry topics, so rest assured that the AP pros won't toss some rarely taught concept your way to see if they can throw you a curve ball and screw up your game.

Most questions have multiple parts. You'll first encounter some type of initial chemistry information (it could be a figure, a graph or a concept) and then see questions labeled a, b, c, and so on. Each of these subquestions requires you to write an answer of at least a sentence and sometimes a paragraph or two or to give a multistep equation as your answer.

You have approximately 18 minutes (three problems in 55 minutes) and 13 minutes (three problems in 40 minutes) to complete each free-response problem.

Putting partial answers into practice

Don't be stingy with your answers! Partial credit is given for saying at least *something* right in the answer. You receive points for writing certain correct portions of the answer, so take your time, write all you know, and do not rush. However, be sure what you write is relevant to the question. No points are given for writing about chemistry in general when the subject is sodium chloride.

If you come across a problem that freaks you out, remind yourself that there must be some part of the subquestion that you're more familiar with. Read the entire problem before starting it. Do not skip a problem simply because the vomit feeling is welling up inside of you. Take a deep breath, scan for any part of the question that you might be able to write about, and begin writing. You can still get a fairly high score even when you receive only partial credit for certain subquestions.

More tips directly from the source

The College Board is nice enough to offer some tips of its own. In the following list, we offer you a condensed version of these tips to make your life easier:

- **Show all your work:** Partial credit is given for partial solutions.

- **If you do work that is incorrect, simply put an X through it:** Don't take the time to erase.

- **Organize your answers clearly:** If the scorers cannot follow your reasoning, they might not give you credit.

- **You do not need to simplify, but if you're asked to calculate, you must simplify all numerical expressions or carry out all numerical calculations to get credit.**

- **Do not use what they call a "scattershot" approach:** Avoid writing a whole bunch of equations or nonsense, hoping that that one among them will be correct.

Getting the Skinny on the Scoring

While you wait patiently to find out if you're a genius, the secret scorers are busy at work squinting their eyes, rubbing their chins, and contemplating your entire chemistry future. You ultimately receive a score between 1 and 5. We know, after all the hard work you did, the hours of studying, and the sweating at the exam, the highest score you can receive is a measly 5! The nerve of those people!

While those AP pros might not be the most creative bunch of folk, they did do a lot of researching to figure out how to score you. They periodically compare the performance of AP students with that of college students tested on the same material. It comes down to this:

- **A score of 5** on the AP test is comparable to a college student's earning an A in his college-level chemistry course.

- **A score of 4** equates to B in a college course, and so on.

- **A score of 3 or higher** equates to a C in college and could still qualify you for college credit. Anything less than a 3 won't qualify you.

But the College Board likes the word "qualified" so here's how it puts it:

    5 — Extremely well qualified

    4 — Well qualified

    3 — Qualified

    2 — Possibly qualified

    1 — No recommendation

Scoping out the scoring process

You might be wondering how they come up with the 1 to 5 score. If you weren't wondering, well, we were, and we figured while we're at it that we might as well make the information available to you.

The multiple-choice section is scored by a computer. The answer sheets are scanned and the computer adds the number of correct responses and subtracts a fraction for each wrong answer. You don't get penalized for answers left blank, but you subtract a fraction from your score for each wrong answer. Thus, make only educated guesses (see the section, "Getting through the questions . . . with as many right answers as possible" for more on making educated guesses.)

The free-response section is scored by real, live people during the first half of June. Really, they're living, breathing people. Basically, major special college professors and veteran super-duper AP teachers come together in the summer and have an AP scoring party. They all gather, distribute their pocket protectors, and get down to the fun of reading your responses.

The scores from the multiple-choice section and free-response section are combined to give you a composite score.

Chapter 2

Preparing Yourself for the Exam

. . . .

In This Chapter

▶ Packing the tools you need

▶ Following a study schedule

▶ Building your confidence

▶ Relieving stress

. . . .

Preparing yourself for the AP exam involves several things, and they are all equally important. First you should familiarize yourself with AP test (in case you haven't done so already, check out Chapter 1). In this chapter, we take you a step further by explaining what you need to take with you to the test as well as how to study and mentally prepare for the test.

Taking Your Tools to the Test

Although you're not packing to go to a remote island in Fiji (albeit, a great visualization skill to use right before the test! See "Practicing visualization" at the end of this chapter), we do want you to put the same amount of thought into what you need to bring and not bring to the test site on test day. Pay special attention to the lists in the following sections. Being prepared in advance will minimize your stress level and assure that you have what you need.

Packing what you need

It's not advisable to ask the test proctor to wait 20 minutes while you run home to get your forgotten calculator. You definitely need to be sure you have the permitted tools for the test on test day. The night before the test try to pack the following list of items:

✔ **Pencils with erasers (#2):** used for the multiple-choice test

✔ **Black pens (with nonfelt tip):** For the free-response questions

✔ **Your Social Security number:** This number identifies you

✔ **Your school code:** Your school counselor or AP chemistry teacher can give you your school code. The AP folks will give you one if you are home-schooled

✔ **Photo I.D.:** Just in case they suspect that you paid your neighbor to take the test for you

✔ **A watch:** You will need to pay close attention to the time

✔ **A scientific calculator:** For use on the free-response section

✔ **Snacks and drinks:** Quiet snacks, like soft granola bars, are a better choice than loud snacks like cookies. First, you won't disrupt others, and second, you won't come crashing down after a sugar high. Water is a better choice than sugar-filled soda for the same reason.

✔ **Appropriate clothing:** You don't know whether there will be an arctic freeze or global warming at the test site, so be prepared for both. Layering your clothes works for either instance.

✔ **Tissues:** No one wants to be distracted by sleeve wipes or sniffing. Bring something to wipe your nose!

Knowing what not to bring

The following list needs just as much attention as the list of what you can bring (see the section above, "Packing what you need"). The below items are some things you definitely *don't* want to bring. You run the risk of being thrown out of the test site for bringing some of these items, so read closely:

✔ **Scratch paper:** The AP will let you use portions of the exam booklets to take notes.

✔ **Notes, books, dictionaries, highlighters, or cheat sheets:** Leave it all at home, folks!

✔ **Electronic devices like MP3 players, cell phones, beepers, and watches that beep:** No one wants to get interrupted by your Beyoncé ringtone.

✔ **Your smart, nerdy friend to take the test for you:** For some reason they don't like that substitution.

✔ **Energy drinks:** Unless you want to come crashing down and have a brain melt, leave them in the fridge.

Needing Special Attention

Normal is so last year. If you require special accommodations to take the AP exam, you are not alone. Not everyone takes the test under the same conditions. You may have special circumstances that require that the test be administered to you in a different way. Here is a brief list of special instances that might warrant an adjustment to the AP chemistry exam:

✔ **Learning disabilities:** If you have a diagnosed learning disability, you may be able to get special accommodations. You may have extended time, large print, a reader, and frequent breaks, among other things, but you must specifically request learning disability accommodations on your application form. The accommodations you request will depend on your specific, diagnosed learning disability. You should make sure that your school has a SSD (Services for Students with Disabilities) Coordinator's Form on file with the College Board. You must fill out this form and send it to the College Board. Allow seven weeks for the pros at College Board to review your request.

To get special testing

• You must have been formally diagnosed with a learning disability by a professional.

• You must have a current, individualized educational plan at school.

• In most cases, the evaluation and diagnostic testing should have taken place within five years of the request for accommodations.

- You must also describe the comprehensive testing and techniques used to arrive at the diagnosis, including test results with subtest scores.

- Your best bet is to log on to the College Board Web site (`www.collegeboard.com/ssd/student/index.html`) to see all up-to-the-minute requirements for accommodations.

✔ **Physical disabilities:** If you have a physical disability, you may be allowed to take a test in a special format — in Braille, in large print, or on an audio cassette or CD. Follow the same instructions detailed above to request accommodations, and if you have further questions contact the College Board directly for more information.

✔ **Religious obligations:** If your religion prohibits you from taking a test on a specific day, you may test on an alternate date. Again, the College Board folks can guide you in the right direction for alternate dates.

✔ **Military duty:** If you're an active military person you don't need to complete the normal registration form. Instead, ask your Educational Services Officer about the testing through DANTES (Defense Activity for Non Traditional Educational Support).

Using Long-Term Strategies for Training and Survival

When you decide to take the AP chemistry exam you're basically committed to getting hitched. Think of the test day as your wedding day and the months leading up to it your engagement. You can consider the day after the test your divorce date, if you will, but for all intents and purposes, you are pretty darn committed to nurturing your AP perpetual partner until then.

You need to plan about ten months in advance for this test. You cannot study for this test like you might a regular test in an ordinary class. In other words, you can't cram — it just plain doesn't work for the AP tests. The AP people are a smart group — they know how to ask questions in such a way as to eliminate (or not "qualify") the students who didn't plan in advance, and simply stayed up all night with a few cans of Red Bull as support. About now, you may be wondering what you should be doing for those ten months. Well, don't despair: We've outlined the entire ten months for you in the following sections. Consider us your AP wedding planners!

Early planning

The early-planning phase takes you from September to October. In the following list, we've given you some solid ideas on what to do during each month as you plan for your upcoming exam:

✔ **September:** As soon as you are able, enroll in an AP chemistry class. If your school doesn't offer one, speak with your counselor and see if there is a community college chemistry course you could take. If an AP course is not available to you, get your hands on a college level chemistry textbook, but keep in mind that nothing comes close to a passionate teacher teaching you the concepts face-to-face.

✔ **November:** Because you've been studying chemistry for a few months, you should be ready to take a diagnostic test. You can use one of the sample tests in this book or you can go to `www.collegeboard.com` and find some there. Taking such tests gives you an early indicator of your strengths and weaknesses, which can help guide your studying in the months to come. Your first diagnostic score can also help to identify what concepts you need to pay more attention to in this book.

✔ **December:** Begin reading this book from the beginning. You will find that being in the classroom for the last four months has provided you a wealth of information. You should already be familiar with many of the concepts covered in the first third of this book.

Midyear planning

You lay the groundwork for taking the exam in the earlier stages of planning to discover your strengths and weaknesses. Although you've been honing your knowledge and skills the last four months, you still need to continue to prepare from January to May:

✔ **January through March:** Continue reviewing this book and take another practice test in early March. You still have two months before your wedding day, so your score on this test will help you pinpoint further concepts from this book that you might still need to review or re-review.

✔ **April and May:** Take another practice test in early May. Make this your last practice test. Whatever you do, *do not* take another practice test the week of the real test. Like a long-distance runner who has been preparing all year for the big race, she doesn't do the race right before the race!

Up-to-the-minute planning

Last-minute planning does not equal cramming! For ten months, you should've been preparing for your exam, and the last few days and minutes count as well. Make the best use of your final time before the test by following this schedule:

✔ **The night before the test:** Relax the night before the test. You might be tempted to pull an all-nighter, but our experience has proven that if you don't know the material the night before the test, you aren't going to learn it the night before the test. Pamper your brain instead by eating a good dinner, watching a movie, reading a good book. If you want to glance at this book one last time, that's cool, but just don't turn it into a cram-fest.

✔ **The morning of the test:** Listen to your mama and eat breakfast, please. We know you may be nervous and your appetite might be slim, but your brain needs the energy. If you're too nervous, or Heaven forbid, you've gotten up late, consider bringing some healthy breakfast items for the car ride. It can actually help keep your mind less stressed.

✔ **Test day:** Take the test.

✔ **June:** Wait patiently for your score . . . very patiently. Did we mention to wait *patiently?*

Feeling Confident: It's All in Your Head

You have been studying and preparing for this test for a long time. You know more than you think you do. You might not remember every single detail of every single concept, but you know at least something about something. If you come to the test feeling completely insecure, please remind yourself that you did the work, you studied, and you will be fine.

Anticipating the outcome

Like we have been saying throughout this chapter, the AP exam tests core concepts and material. There are no surprises. After finishing this book, you will have been exposed to all the material you need to excel on this test. Anticipate the best possible outcome.

Remembering: This is a test . . . This is only a test

Listen, your score on the AP test is not going to change your life. It's not a do-or-die type of experience. Your life's success is not tied into the outcome of this test. Treat this test as you would treat any other test you have taken (you've taken hundreds, right?). And don't worry, you don't know the AP folks. They are not going to snicker as you walk by and say under their breath, "There goes the one that didn't know what Dalton's Law of Partial Pressure was." Although the AP guys don't know you, we realize that your peers do. And it's inevitable that "score gossip" will creep into your conversations with your friends. Comparing scores is not advisable, as it creates stress and underlying feelings about your chemistry capabilities. Remembering that this is only a test and not a life-defining moment will help you stay balanced and keep it all in perspective.

Surviving During the Test with These Four Stress-Busters

It's completely normal to feel some fear the day of the test. Everyone does. Research has shown us that a little anxiety is actually good for the brain. A touch of anxiety helps the brain stay ultra-focused and attentive. A little anxiety tells the brain, "Hey, brain, if I'm feeling this way, whatever I'm doing must be important, so I better stay alert." However, excess anxiety and panic actually have the opposite effect. High levels of fear make the brain go into flight-or-fight mode. Your brain tells your body to conserve your higher-order thinking skills in an effort to send extra physical strength to the body in case of emergency. So, if you're feeling too much anxiety during the test, here are some stress-busters that will calm those nerves.

Counting to four

Breathing is grossly underrated. Take a deep breath until your belly expands, hold it in four counts, and then expel the air for four counts. Be sure not to exhale like you're in the middle of an aerobic workout so as not to disturb the other test-takers. Oh, and some breath mints wouldn't hurt either.

Try not to take shallow breaths, which can cause you to become even more anxious because your body is deprived of oxygen. Hyperventilating or sucking on your inhaler during the exam is grossly overrated.

Stretching

Several stretches can help relieve stress and make your body feel more relaxed and comfortable. Do the practices in the list discretely so you don't disrupt others around you — in other words, refrain from doing the downward-dog position from your yoga class:

✔ Rotate your head around to stretch out and relax your neck muscles. (We suggest keeping your eyes closed so the proctor doesn't think you are trying to cheat.)

✔ Hunch and roll your shoulders to help relax your back and spine. You'll be sitting for quite some time, so maintaining good posture is crucial.

✔ Shake out your hands like you have writer's cramp. Imagine that all your tension and stress is going out through your fingertips.

✔ Extend and push out your legs like you're pushing something away with your heels.

✔ With your legs stretched out in front of you, point your toes back toward your knees and hold that position for a count of three.

Practicing visualization

Right before the exam or during a short break, practice creative visualization. Close your eyes and imagine yourself in the test room cheerfully looking at questions that you know the answer to, filling in the answers, finishing early, and double-checking your work. Picture yourself leaving the exam room full of energy, and then getting your score and rejoicing. Think of how proud of you your parents are (if thinking about them stresses you out leave this one alone). Imagine not having to take early chemistry in college. The goal is to associate the AP exam with feelings of joy.

Don't do this visualization exercise during the test. You'll just waste time and lose concentration.

Stop, drop, and, change your mind

Any time you feel yourself starting to panic or think negative thoughts, make a conscious decision to

1. Say to yourself, "Stop, self! Don't dwell on anything negative."

2. Drop that negative thought like a hot Bunsen burner.

3. Then switch over to a positive train of thought.

For example, suppose you catch yourself thinking, "Why didn't I pay more attention to chemical equilibrium?" Change that script to, "I've got most of this right, maybe I'll get this right, too. No sense worrying now, overall I'm rocking my world!"

Part II
Building Matter from the Ground Up: Atoms and Bonding

The 5th Wave   By Rich Tennant

DOCTOR HORBUS OBSERVES THE CHEMICAL REACTION OF SULFURIC ACID IN A DRIBBLE BEACON TO PROF. DUNSON'S NEW LAB COAT.

In this part . . .

Taking something apart is a great way to learn how to put it together. In this part, we unsnap the building blocks of matter so you can understand how matter is reassembled. The first step is to get acquainted with the building blocks themselves: atoms. Because all atoms are themselves made of three types of even tinier building blocks, we dissect the atom. Then, starting with just three wee types of particles, we begin to show you how the lovely complexity of the universe is built. Particles assemble into elements, and elements bond into compounds. As they bond, elements follow certain rules. Understanding these rules, these few and simple rules, takes you a long way to understanding the rest of chemistry.

Chapter 3

Looking Under the Atomic Hood: Atomic Structure

For hundreds of years, scientists have operated under the idea that all matter is made up of smaller building blocks called *atoms*. So small, in fact, that until the invention of the electron microscope, the only way to find out anything about these tiny, mysterious particles was to design a very clever experiment. These experiments allowed chemists to get around the fact that they couldn't exactly corner a single atom in a back alley somewhere and study it alone — they had to study the properties of whole gangs of atoms and try to guess what individual ones might be like. Through some remarkable cleverness and some incredibly lucky shots in the dark, chemists now understand a great deal about the atom, and guess what? You need to, too! Because we don't want to leave you hanging, in this chapter we introduce you to everything you need to know about atoms, their properties, and what they are made of.

Picking Apart Atoms and the Periodic Table

The following sections begin by giving you the basics on how atoms are structured so that you can not only get to know their components, but also, you can easily figure out how atoms help to classify elements. Even better, we give you the ultimate road map for the periodic table, showing you how the periodic table can give you some valuable info that helps you predict certain properties of elements.

Getting a gander at what makes up an atom

For now, picture an atom as a microscopic building block. Atoms come in all shapes and sizes, and you can build larger structures out of them. However, like building blocks, atoms are extremely hard to break. In fact, there is so much energy stored inside an atom that breaking one in half is the process that drives a nuclear explosion.

All substances consist of the 120 unique varieties of atoms, each of which is made up of a combination of three types of *subatomic particles:*

✔ **Protons:** Protons have equal and opposite charges to electrons and have very nearly the same mass as neutrons.

✔ **Electrons:** Electrons have equal and opposite charges to protons, and electrons are much lighter than protons and neutrons.

✔ **Neutrons:** Neutrons are neutral and have the same mass as protons.

We summarize the must-know information about the three subatomic particles in Table 3-1.

Table 3-1		The Subatomic Particles
Particle	*Mass*	*Charge*
Proton	1amu	+1
Electron	$\frac{1}{1836}$ amu	−1
Neutron	1amu	0

Atoms always have an equal number of protons and electrons, which makes them overall electrically neutral. Many atoms, however, actually prefer to have an unbalanced number of protons and electrons, which leaves them with an overall charge. We discuss these charged atoms, called *ions,* even further in the section "Exercising Electrons: Ions and Electron Configuration," later in this chapter.

The atom can still safely be called the smallest possible unit of an element because after you break an atom of an element into its subatomic particles, it loses the basic properties that make that element unique.

So what does all of this mean for the structure of an atom? What does an atom actually look like? It took scientists a very long time to figure it out through clever experimentation and tricky math, and over time a succession of models grew closer and closer to an accurate description:

✔ **The Thompson model,** also called the "Plum Pudding" model, pictured discrete, negatively charged electrons evenly distributed through a positively charged medium that composed the rest of the atom. The electrons were like plums in a positive pudding.

✔ **The Rutherford model** modified the Thompson model by making clear that most of the volume of the atom is empty space, with a large amount of charge concentrated at the center of the atom.

✔ **The Bohr model** built on the Rutherford model by describing the compact, central charge as a nucleus composed of distinct proton and neutron particles. The positive charge of the nucleus derived from the protons. Bohr envisioned electrons as discrete particles that orbited the nucleus along distinct paths, like planets in orbit around the sun.

✔ **The Quantum Mechanical model** modified the Bohr model, pointing out that electrons do not orbit the nucleus like planets around the sun. Instead, they occupy their orbitals in a cloudlike manner; one can only describe their location in terms of probability, with some "dense" regions having a very high probability of having an electron and other regions having lower probability.

Classifying elements with mass and atomic numbers

Two very important numbers associated with an atom are the *atomic number* and the *mass number*. Chemists tend to memorize these numbers like sports fans memorize baseball stats, but clever chemistry students do not need to resort to memorization when they have the all-important periodic table at their disposal. Here are the basics about atomic numbers and mass numbers:

✔ **The atomic number is the number of protons in the nucleus of an atom.** Atomic numbers identify elements, because the number of protons is what gives an element its unique identity. Changing the number of protons (and therefore the atomic number) changes the identity of the element. Atomic numbers are listed as a subscript (on the bottom) to the left side of an element's chemical symbol.

Notice that the atoms of the periodic table are lined up according to their atomic number. Atomic numbers increase by one each time you move to the right in the periodic table, continuing their march, row by row, from the top of the table to the bottom.

✔ **The mass number is the sum of the protons and neutrons in the nucleus of an atom.** Subtracting the atomic number from the mass number gives you the number of neutrons in the nucleus of an atom. Different atoms of the same element can have different numbers of neutrons, and we call those different atoms *isotopes.* Mass numbers are listed as a superscript (on the top) to the left side of an element's chemical symbol.

The mass number gives the mass of an individual atom in atomic mass units, amu, where 1 amu = 1.66×10^{-27} kg.

Why, you may wonder, don't we care about electrons? Remember that an electron has only $\frac{1}{1836}$ of the mass of a proton or neutron. So, though electrons are incredibly important from the standpoint of charge, they make little contribution to mass. Atomic masses in amu are never whole numbers for a variety of reasons, including the fact that most elements exist in nature as a combination of different isotopes. The percent of each isotope relative to the whole is called its *natural abundance.* The nonwhole number atomic masses you see listed on the periodic table are the average of all isotopes of a given element, in which each isotope is weighted by its natural abundance within the average.

TIP

In order to specify the atomic and mass numbers of an element, chemists will typically write that element in the form

$$^A_Z X$$

where Z is the atomic number, A is the mass number, and X is the chemical symbol for that element.

Grouping elements within the periodic table

The whole purpose of the periodic table, aside from providing interior decoration for chemistry classrooms, is to organize elements to help predict and explain their properties. Notice in Figure 3-1 that the atoms of the periodic table are lined up according to their atomic numbers, as if their teacher has just taken a roll call to make sure they are all in their proper places. Here are some characterisitics of how the periodic table is organized:

✔ Atomic numbers increase by one each time you move to the right in the periodic table and when they run out of space in one row, they move down one and continue increasing to the right again.

✔ The rows are called *periods*.

✔ The columns are called *groups*.

The elements within any group have very similar properties. The properties of the elements emerge mostly from their different numbers of protons and electrons, and from the arrangement of their electrons. As you move across any period, you pass over a series of elements whose properties change in a predictable way. Here are the groups you can find on the periodic table:

✔ **Group IA (alkali metals):** Starting on the far left of the periodic table is group IA (notice the group label atop each column), also known as the alkali metals. These elements, the most metallic of all elements, are very reactive, and are never found naturally in a pure state, but always in a bonded state.

✔ **Group IIA (alkaline earth metals):** Just like the alkali metals, the alkaline earth metals are highly reactive, and not found free in nature.

✔ **Group B . . . mostly (transition metals):** The large central block of the periodic table is occupied by the transition metals. Transition metals have properties that vary from extremely metallic, at the left side, to far less metallic, on the right side. The rightmost boundary of the transition metals is shaped like a staircase, shown in bold in the figure, and the staircase separates the elements further:

   • **Metals:** Elements to the left of the staircase (except hydrogen) are metals. Metals tend to be solid, shiny, malleable (easily shaped), and ductile (easily drawn out into wire). Metals also conduct electricity and heat and easily give up electrons.

   • **Nonmetals:** Elements to the right of the staircase are nonmetals, which have properties opposite to those of metals: They also tend to be solids, but are generally dull, hard, and do not conduct electricity.

   • **Metalloids** or **semi-metals:** Elements bordering the staircase are called metalloids or semi-metals because they have properties between metals and nonmetals: They are somewhat ductile and a little shiny and can conduct electricity, but not as well as a metal.

✔ **Group VIIIA (noble gases):** The most extreme nonmetals are the noble gases which are inert, or extremely unreactive.

✔ **Group VIIA (halogens):** One column to the left of the noble gases, is another key family of nonmetals, the halogens. In nature, the reactive halogens tend to bond with metals to form salts, like sodium chloride, NaCl.

Be aware that some periodic tables use a different system for labeling the groups, in which each column is simply numbered from 1 to 18, left to right.

Rummaging through reactivity in the periodic table

The properties of elements change as a function of the numbers of protons and electrons in the element. Increasing numbers of protons increases the positive charge of the nucleus, which contributes to *electronegativity,* the pull exerted by the nucleus of one atom on electrons in a bond with a second atom. Increasing numbers of electrons changes the reactivity of the element in predictable ways, based on how those electrons fill up successive energy levels.

The outermost electrons (in the highest energy level, which is discussed later) are called *valence electrons.* Valence electrons are the ones most important in determining whether an element is reactive or unreactive and are the electrons involved in bonding. Because much of chemistry is about the making and breaking of bonds, valence electrons are the most important particles for chemistry. The details of how electrons fill up the various energy levels are covered later in this chapter, in the section "Exercising Electrons: Ions and Electron Configuration." For now, understand that atoms are most stable when their valence shells (or clouds) are completely filled with electrons.

Chemistry happens as a result of atoms attempting to fill the valence shells. Alkali metals are especially reactive because they need only give up one electron to have a completely filled valence shell. Halogens are especially reactive because they need only acquire one electron to have a completely filled valence shell. The elements within a group tend to have the same number of valence electrons, and for that reason tend to have similar chemical properties. Elements in the A groups have the same number of valence electrons as the roman numeral of their group. For example, magnesium in Group IIA has two valence electrons.

Rounding up the atomic radius in the periodic table

In addition to reactivity, another property that varies across the table is atomic radius, or the geometric size (not the mass) of the atoms. As you move down a column or to the right along a row on the periodic table, elements have both more protons and more electrons. As you move down the table the added electrons occupy discretely higher energy levels. So, atomic radius tends to decrease as you move to the right because the increasing positive charge of the nuclei pulls inwardly on the electrons of that energy level. As you move down the table, atomic radius tends to increase because, even though you are adding positively charged protons to the nucleus, you are now adding electrons to higher and higher energy levels, which are farther and farther from the nucleus and therefore feel a much lower attraction to the nucleus.

PERIODIC TABLE OF THE ELEMENTS

1 IA								
1 **H** Hydrogen 1.00797	2 IIA							
3 **Li** Lithium 6.939	**4** **Be** Beryllium 9.0122							
11 **Na** Sodium 22.9898	**12** **Mg** Magnesium 24.312	3 IIIB	4 IVB	5 VB	6 VIB	7 VIIB	8 VIIIB	9 VIIIB
19 **K** Potassium 39.102	**20** **Ca** Calcium 40.08	**21** **Sc** Scandium 44.956	**22** **Ti** Titanium 47.90	**23** **V** Vanadium 50.942	**24** **Cr** Chromium 51.996	**25** **Mn** Manganese 54.9380	**26** **Fe** Iron 55.847	**27** **Co** Cobalt 58.9332
37 **Rb** Rubidium 85.47	**38** **Sr** Strontium 87.62	**39** **Y** Yttrium 88.905	**40** **Zr** Zirconium 91.22	**41** **Nb** Niobium 92.906	**42** **Mo** Molybdenum 95.94	**43** **Tc** Technetium (99)	**44** **Ru** Ruthenium 101.07	**45** **Rh** Rhodium 102.905
55 **Cs** Cesium 132.905	**56** **Ba** Barium 137.34	**57** **La** Lanthanum 138.91	**72** **Hf** Hafnium 179.49	**73** **Ta** Tantalum 180.948	**74** **W** Tungsten 183.85	**75** **Re** Rhenium 186.2	**76** **Os** Osmium 190.2	**77** **Ir** Iridium 192.2
87 **Fr** Francium (223)	**88** **Ra** Radium (226)	**89** **Ac** Actinium (227)	**104** **Rf** Rutherfordium (261)	**105** **Db** Dubnium (262)	**106** **Sg** Seaborgium (266)	**107** **Bh** Bohrium (264)	**108** **Hs** Hassium (269)	**109** **Mt** Meitnerium (268)

Lanthanide Series

58 **Ce** Cerium 140.12	**59** **Pr** Praseodymium 140.907	**60** **Nd** Neodymium 144.24	**61** **Pm** Promethium (145)	**62** **Sm** Samarium 150.35	**63** **Eu** Europium 151.96
90 **Th** Thorium 232.038	**91** **Pa** Protactinium (231)	**92** **U** Uranium 238.03	**93** **Np** Neptunium (237)	**94** **Pu** Plutonium (242)	**95** **Am** Americium (243)

Actinide Series

Figure 3-1:
The periodic
table.

			13 IIIA	14 IVA	15 VA	16 VIA	17 VIIA	18 0
								2 He Helium 4.0026
			5 B Boron 10.811	6 C Carbon 12.01115	7 N Nitrogen 14.0067	8 O Oxygen 15.9994	9 F Fluorine 18.9984	10 Ne Neon 20.183
10 VIIIB	11 IB	12 IIB	13 Al Aluminum 26.9815	14 Si Silicon 28.086	15 P Phosphorus 30.9738	16 S Sulfur 32.064	17 Cl Chlorine 35.453	18 Ar Argon 39.948
28 Ni Nickel 58.71	29 Cu Copper 63.546	30 Zn Zinc 65.37	31 Ga Gallium 69.72	32 Ge Germanium 72.59	33 As Arsenic 74.9216	34 Se Selenium 78.96	35 Br Bromine 79.904	36 Kr Krypton 83.80
46 Pd Palladium 106.4	47 Ag Silver 107.868	48 Cd Cadmium 112.40	49 In Indium 114.82	50 Sn Tin 118.69	51 Sb Antimony 121.75	52 Te Tellurium 127.60	53 I Iodine 126.9044	54 Xe Xenon 131.30
78 Pt Platinum 195.09	79 Au Gold 196.967	80 Hg Mercury 200.59	81 Tl Thallium 204.37	82 Pb Lead 207.19	83 Bi Bismuth 208.980	84 Po Polonium (210)	85 At Astatine (210)	86 Rn Radon (222)
110 Uun Ununnilium (269)	111 Uuu Unununium (272)	112 Uub Ununbium (277)	113 Uut §	114 Uuq Ununquadium (285)	115 Uup §	116 Uuh Ununhexium (289)	117 Uus §	118 Uuo Ununoctium (293)

64 Gd Gadolinium 157.25	65 Tb Terbium 158.924	66 Dy Dysprosium 162.50	67 Ho Holmium 164.930	68 Er Erbium 167.26	69 Tm Thulium 168.934	70 Yb Ytterbium 173.04	71 Lu Lutetium 174.97
96 Cm Curium (247)	97 Bk Berkelium (247)	98 Cf Californium (251)	99 Es Einsteinium (254)	100 Fm Fermium (257)	101 Md Mendelevium (258)	102 No Nobelium (259)	103 Lr Lawrencium (260)

§ Note: Elements 113, 115, and 117 are not known at this time, but are included in the table to show their expected positions.

Exercising Electrons: Ions and Electron Configuration

The outermost shell of an element, which is the highest energy level, is the *valence shell*. Elements are so insistent about having filled valence shells that they'll actively gain or lose valence electrons to do so. Atoms that gain or lose electrons in this way are called *ions*. When atoms become ions, they lose the one-to-one balance between their protons and electrons, and therefore acquire an overall charge.

✔ Atoms that lose electrons (like metals) acquire positive charge, becoming **cations**, such as Na^+ or Mg^{2+}. You can remember this with the simple phrase "cats have paws (pos +)," so cations are positively charged.

✔ Atoms that gain electrons (like nonmetals) acquire negative charge, becoming **anions**, such as Cl^- or O^{2-}.

In the following sections we show you how to figure out which elements gain or lose electrons as well as how to diagram and write out electron configurations.

Knowing which elements gain or lose electrons

You can discover what kind of ion many elements will form simply by looking at their position on the periodic table. With the exception of row one (hydrogen and helium), all elements are most stable with a full shell of eight valence electrons, known as an *octet*. Atoms tend to take the shortest path to a complete octet, whether that means ditching a few electrons to achieve a full octet at a lower energy level, or grabbing extra electrons to complete the octet at their current energy level. In general, metals tend to lose electrons, and nonmetals tend to gain electrons. However, a partner atom is usually required for these processes to happen.

You can predict just how many electrons an atom will gain or lose to become an ion (but things do get unpredictable in the transition metal region and then get more predictable with the nonmetals). Elements tend to lose or gain as many electrons as necessary to have valence shells resembling the elements in Group VIIIA, the noble gases:

✔ Group IA (alkali metal) elements lose one electron.

✔ Group IIA (alkaline earth metal) elements lose two electrons.

✔ Group VIIA (halogen) elements gain one electron.

✔ Group VIA elements tend to gain two electrons.

✔ Group VA elements tend to gain three electrons.

Figuring out electron configuration

An awful lot of detail goes into determining just how many electrons an atom has and just which energy levels those electrons occupy. Several different schemes exist for annotating all of this important information, but the *electron configuration* is a type of shorthand that captures much of the pertinent information. Each electron gets a symbol code that indicates the type of energy and shape (called an electron orbital) that it will have in the atom.

Here's how electron configuration works:

✔ Each numbered row of the periodic table corresponds to a different *principal energy level,* with higher numbers indicating higher energy. Remember higher energy means farther from the nucleus.

✔ Within each energy level, electrons can occupy different *subshells* (or clouds). Different types of subshells have slightly different energy levels.

✔ Each subshell has a certain number of *orbitals.*

   • Each orbital can hold up to two electrons, but electrons won't double up within an orbital unless no other open orbitals exists within the same subshell.

   • Electrons fill up orbitals from the lowest energies to the highest, to try to keep everything in the lowest energy state, like everything else in chemistry.

The four types of subshells or clouds, named *s, p, d,* and *f,* fill with electrons in the following ways:

✔ Row 1 consists of a single 1*s* subshell. A single electron in this subshell corresponds to the electron configuration of hydrogen, written as $1s^1$. The superscript written after the symbol for the subshell indicates how many electrons occupy that subshell. Filling the subshell with two electrons, $1s^2$, corresponds to the electron configuration of helium. Each higher principal energy level contains its own *s* subshell (2*s,* 3*s,* and so on), and these subshells are the first to fill within those levels.

✔ In addition to *s* subshells, principal energy levels 2 and higher contain *p* subshells. There are three *p* orbitals at each level, accommodating a maximum of six electrons. Because the three *p* orbitals (also known as p_x, p_y, and p_z) have equal energy in an isolated atom they are each filled with a single electron before any receives a second electron. The elements in rows 2 and 3 on the periodic table contain only *s* and *p* subshells. The *p* subshells of each energy level are filled only after the *s* subshell is filled.

✔ Rows 4 and higher on the periodic table include *d* orbitals, of which there are five at each principal energy level, accommodating a maximum of 10 electrons. *d* subshell electrons are a major feature of the transition metals. These are also equal in energy unless the atom is involved in bonding.

✔ Rows 5 and higher include *f* orbitals, numbering seven at each level, accommodating a maximum of 14 electrons. *f* subshell electrons are a hallmark of the *lanthanides* and the *actinides,* the two rows of elements detached from the rest of the periodic table, and set aside at the bottom.

After you get to row 4, the exact order in which you fill the energy levels can get a bit confusing. To keep things straight, it's useful to refer to the Aufbau filling diagram, shown in Figure 3-2. Start at the bottom of the diagram and work your way up from the lowest arrows to the highest. For example, always start by filling 1*s*, then fill 2*s*, then 2*p*, then 3*s*, then 4*s*, then 3*d*, and so on.

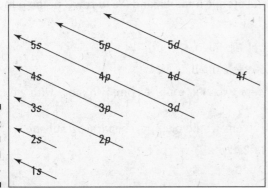

Figure 3-2: The Aufbau filling diagram.

Noting exceptions to the Aufbau filling diagram

Sadly, there are a few exceptions to the tidy picture presented by the Aufbau filling diagram. Copper, chromium, and palladium are notable examples. These exceptional electron configurations arise from situations where electrons get transferred from their proper, Aufbau-filled subshells to create half-filled or fully filled sets of *d* subshells; these half- and fully filled states are slightly more stable than the states produced by strict Aufbau-based filling.

Two conditions typically lead to exceptional electron configurations:

✔ Successive orbital energies must lie close together, as is the case with 3*d* and 4*s* orbitals, for example.

✔ Shifting electrons between these energetically similar orbitals must result in a half-filled or fully filled set of identical orbitals, an energetically happy state of affairs.

Here are a few examples:

✔ Strictly by the rules, chromium should have the following electron configuration: $[Ar]3d^44s^2$.

✔ Because shifting a single electron from 4*s* to the energetically similar 3*d* level half-fills the 3*d* set, the actual configuration of chromium is $[Ar]3d^54s^1$.

✔ For similar reasons, the configuration of copper is not the expected $[Ar]3d^94s^2$, but instead is $[Ar]3d^{10}4s^1$.

Writing out electron configuration

To come up with a written electron configuration you

1. **Determine how many electrons the atom in question actually has.**

2. **You assign those electrons to subshells, one electron at a time, from the lowest energy subshells to the highest.**

   In a given type of subshell (like a 2*p* or 3*d* subshell, for example) you only place two electrons within the same subshell when there is no other choice. For example, only at oxygen ($1s^22s^22p^4$) would electrons begin to double up in the 2*p* subshells.

3. **Write out the configuration based on the assignment of electrons.**

   From lowest subshell to highest, you write the subshell and add the number of electrons assigned to it in superscript, like oxygen, $1s^22s^22p^4$.

For example, suppose you want to find the electron configuration of carbon:

1. **Determine the amount of electrons.** Carbon has six electrons, the same as the number of its protons as described by its atomic number, and it's in row 2 of the periodic table.

2. **Assign electrons to subshells.**

   First, the *s* subshell of level 1 is filled.

   Then, the *s* subshell of level 2 is filled.

   These subshells each accept two electrons, leaving two more with which to fill the *p* subshells of level 2.

   Each of the remaining electrons would occupy a separate *p* subshell.

3. **Write out the configuration: $1s^22s^22p^2$.**

Ions have different electron configurations than their parent atoms because ions are created by gaining or losing electrons. Atoms tend to gain or lose electrons so they can achieve full valence octets, like those of the noble gases. Guess what? Many of the resulting ions have precisely the same electron configurations as those noble gases. So, by forming the Br^- anion, bromine achieves the same electron configuration as the noble gas, krypton. This is called being *isoelectronic*. You can use this concept to condense electron configurations, which can get a bit long to write. For example, $[Ne]3s^23p^3$ is the condensed configuration for phosphorous. The expanded electron configuration for phosphorous is $1s^22s^22p^63s^23p^3$. The condensed form simply means that the atom's electron configuration is just like that of neon (Ne), with additional electrons filled into subshells $3s$ and $3p$ as annotated. Warning: Don't try to get too creative and write something like $[Be]2p^4$ for oxygen. *Only noble gases* can be used for the base element in the condensed form! On the AP test be careful — if the problem asks for the complete formula, the condensed version does not receive credit.

Locating electrons with quantum numbers

Chemists use a set of four numbers called collectively the *quantum numbers* to describe the state of any particular electron within an electron configuration of an atom. No two electrons in the same atom can share the same set of quantum numbers. In other words, there is no such thing as identical electron twins. They *must* differ from one another in at least one of their quantum numbers. Note that chemists invented this scheme to describe electrons in an atom. There are not "slots" to fill in the atom.

- The **principal quantum number,** which indicates the principal energy level of an electron, is denoted with the variable n and can be any integer number greater than or equal to 1. Electrons that share the same principal quantum number are said to be in the same *shell*.

- The **azimuthal quantum number,** l, which denotes the subshell which an electron occupies, has only four values: 0, 1, 2, or 3 corresponding to s, p, d, and f subshells respectively. Any particular n value, or *shell*, has subshells ranging from $l = 0$ to $l = n-1$. The third shell, for example, has subshells with l values 0, 1, and 2. In other words, the third shell has s, p, and d subshells.

- The **magnetic quantum number,** m_l, builds off of the azimuthal quantum number and describes the three-dimensional geometry of each subshell. The magnetic quantum number describes the number of *orbitals* in each subshell, and each orbital can be occupied by at most two electrons. The magnetic quantum number has values ranging from $-l$ to l for each subshell. In other words, a p subshell (which has an azimuthal quantum number of $l = 1$) has m_l values of -1, 0, and 1, while a d subshell with an azimuthal quantum number of $l = 2$ has m_l values of -2, -1, 0, 1, and 2. This rule means that a p subshell has three orbitals in which electrons may exist, while a d subshell has five.

- **The spin quantum number,** s, has only two values: $-\frac{1}{2}$ and $\frac{1}{2}$. Each orbital shares the same n, l, and m_l numbers by definition and, because no two electrons can share the same set of quantum numbers, the spin quantum number must be where the two electrons in an orbital differ. Two electrons can be in one orbital as long as one has a spin of $-\frac{1}{2}$ and the other has a spin of $\frac{1}{2}$. Counting up the m_l and s quantum numbers leads you directly to the number of total electrons that can fit in each subshell. If an s orbital has an l value of 0, for example, this means that it has a single m_l value of 0. Because two electrons (one with each type of spin) can fit in this $m_l = 0$ lobe, an s orbital will always fit two and only two electrons no matter what its principle quantum number is. Similarly, a p subshell has three m_l values (-1, 0, and 1) and each of those can have two electrons of opposing spin so a p subshell can have up to six electrons.

Isolating Info on Isotopes

Isotopes are atoms of the same element, but they have different atomic masses, which must translate to different numbers of neutrons. If that's all there is to them, then why do chemists get so titillated by isotopes? A neutron is, after all, a *neutral* particle, so you wouldn't think adding one would change much about an atom. Indeed, much of the time, adding a neutron does not alter the properties of an element significantly, but merely makes it slightly heavier.

Occasionally, however, adding another sinister little neutron can push an atom over to the dark side of radioactivity. If an atom has just the right number of neutrons, it becomes unstable, or radioactive, which means that it will break down into another element at a predicable rate. These radioactive isotopes are called radioisotopes for short. The more unstable the added neutron makes the atom, the faster it tends to decay into another.

Although the word "radioactivity" conjures up images of three-headed frogs and fish with ten eyes, the truth is that it doesn't deserve its bad rep. Many radioactive elements are harmless. Some radioactive isotopes are very useful, such as those used in medical imaging.

Regardless of how you feel about radioactivity, we give you some useful calculations related to radioisotopes and istopes that you definitely need to check out in the sections that follow.

Calculating the remainder of a radioisotope

Science and medicine are stuffed with useful, friendly applications for radioisotopes. Many of these applications center on the predictable decay rates of various radioisotopes. These predictable rates are characterized by *half lives*. The half life of a radioisotope is simply the amount of time it takes for exactly half of a sample of that isotope to decay into daughter nuclei. For example, if a scientist knows that a sample originally contained 42mg of a certain radioisotope and measures 21mg of that isotope in the sample four days later, then the half life of that radioisotope is four days. The half lives of radioisotopes range from seconds to billions of years.

Radioactive dating is the process scientists use to date samples based on the amount of radioisotope remaining. The most famous form of radioactive dating is carbon-14 dating, which has been used to date human remains and other organic artifacts. However, radioisotopes have also been used to date the Earth, the solar system, and even the universe.

Table 3-2 lists some of the more useful radioisotopes, along with their half lives and decay modes, which are described in detail later in the chapter.

Table 3-2	Common Radioisotopes, Half Lives, and Decay Modes	
Radioisotope	*Half Life*	*Decay Mode*
Carbon-14	5.73×10^3 years	beta
Iodine-131	8.0 days	beta, gamma
Potassium-40	1.25×10^9 years	beta, gamma
Radon-222	3.8 days	alpha
Thorium-234	24.1 days	beta, gamma
Uranium-238	4.46×10^9 years	alpha

Comparing carbon masses

Consider the element carbon, for example, which has three naturally occurring isotopes. carbon-12 (or carbon with six protons and six neutrons), written as:

$$^{12}_{6}\text{C}$$

is boring old run-of-the-mill carbon, and it accounts for 99 percent of all of the carbon out there. carbon-13 (or carbon with six protons and *seven* neutrons), written as:

$$^{13}_{6}\text{C}$$

is a slightly more exotic, though equally dull, isotope, which makes up most of the remaining 1 percent of carbon atoms. Taking on an extra neutron makes carbon-13 slightly heavier than carbon-12, but does little else to change it. However, even this minor change has some very real scientific consequences. Scientists compare the ratio of carbon-12 to carbon-13 in meteorites to help them determine where that meteorite came from. These ratios have been especially useful for identifying Martian meteorites. Earth is significantly more massive than Mars, and therefore, has a stronger magnetic field, allowing it to hold onto its atmosphere. Mars, on the other hand, is too small to hold onto the upper part of its atmosphere. carbon-12 is lighter than carbon-13, so it floats up to the upper atmosphere of Mars, where the solar wind comes along like the big bad wolf and blows it away. This leaves Mars with a higher percentage of carbon-13 than you would find on Earth. So, minerals on Mars that take carbon from the atmosphere and turn it into rock end up with a little more carbon-13 and a little less carbon-12 than similar minerals on earth. When a meteor from Mars lands on Earth, scientists can verify its origin by testing the ratio of carbon-12 to carbon-13.

carbon-14, the most exotic and interesting isotope of carbon, has been very important in the process of radioactive dating. Carbon is one of the building blocks of organic matter, including the human body. Only one out of every trillion or so carbon atoms is the carbon-14 (or carbon with six protons and *eight* neutrons) radioisotope, which looks like

$$^{14}_{6}\text{C}$$

You have many, many trillions of trillions of trillions of carbon atoms in your body, which means that you contain trillions of atoms of radioactive carbon! Now before you go checking the mirror to see if you have sprouted a third eye, rest assured that this radioactive carbon will not harm you in any way. In fact, it is what allows scientists to determine the age of fossils.

As an example, take Matilda the Mammoth, who met her untimely end 4,500 years ago. While Matilda was alive, the carbon in her body was constantly being replenished, so she was always made of about 99 percent carbon-12, 1 percent carbon-13, and 0.0000000001 percent carbon-14. However, when poor Matilda kicked the bucket, the biological processes that were replacing the carbon in her body came to an abrupt end. With their supply of carbon-14 cut off, Matilda's bones slowly lost their carbon-14 as it broke down through radioactive decay into nitrogen. Paleontologists digging up Matilda 4,500 years later will run straight to their friendly neighborhood chemist, Dr. Isotopian, and ask him to tell them how much carbon-14 is left in Matilda. Because carbon-14 breaks down at a very predictable rate, Dr. Isotopian is able to guess to within a few hundred years exactly how long ago Matilda kicked the bucket.

To calculate the remaining amount of a radioisotope, use the following formula where A_0 is the amount of the radioisotope that existed originally, t is the amount of time the sample has had to decay and T is the half life:

$$A = A_0 \times (0.5)^{1/T}$$

Calculating the average mass of an element

That number beneath each element in the periodic table is calculated using percentages such as the carbon abundances mentioned earlier, which allow scientists to give an *average* atomic mass for the isotope. Certain elements, such as chlorine, have several commonly occurring isotopes, so their average atomic masses are rarely close to whole integers. Many elements, however, such as carbon, have one very commonly occurring isotope and several rare isotopes, resulting in an average atomic mass that is very close to the mass of the most common isotope.

To calculate an average atomic mass you

1. **Make a list of the masses of each isotope, noting the percentage, in decimal form, at which each isotope is found in nature, a quantity called the *relative abundance* of the isotope.**

2. **Multiply the mass and the relative abundance together.**

3. **Add up the results of each calculation in step 2 to get the average atomic mass.**

On the AP chemistry exam, you will be expected to know how to calculate this weighted average or to determine at which isotope is the most common based on that weighted average.

The most commonly occurring isotope *often* has a mass number closest to the average atomic mass. Be careful when the average atomic mass differs from the nearest whole number by more than about 0.1, however. Such numbers can lead you to false conclusions regarding the most abundant isotope. Chlorine, for example, has an average atomic mass of 35.5amu. This may lead you to conclude that the isotopes

$$^{35}_{17}\text{Cl}$$

$$^{36}_{17}\text{Cl}$$

both have roughly 50 percent abundances because the average atomic mass is almost exactly halfway between 35 and 36. In reality, however

$$^{35}_{17}\text{Cl}$$

exists with roughly 75 percent abundance, while

$$^{37}_{17}\text{Cl}$$

exists with roughly 25 percent abundance.

$$^{36}_{17}\text{Cl}$$

is about as common as a Chemistry teacher without chalk on the butt of his pants.

Running into Radioactivity

Atomic numbers are like name tags, identifying elements by telling you the number of protons in the nucleus of that element. Adding or removing a proton to the nucleus of an atom changes its atomic number (and thus its identity), and atoms are very fond of their identities, so adding or removing protons from an atom is usually impossible.

However, you *can* change atomic numbers:

✔ Certain large elements can be split in half, and certain small elements can be smashed together in processes called *nuclear fission* and *nuclear fusion,* respectively. This splitting or joining of atoms releases a tremendous amount of energy and is not something that you should try at home, even if you do have a nuclear reactor in your basement.

✔ The only other way for an atom to change its atomic number (and therefore its identity) is to "decay" into something else through a radioactive process, and this happens about as often as chemists remember to brush their hair in the mornings. There are three key ways in which an element can decay, and these are described following.

Many elements in the periodic table have unstable versions called *radioisotopes.* These radioisotopes decay into other, generally more stable elements at periodic intervals in a process called *radioactive decay.* There are many radioisotopes in existence; however not all radioisotopes are created equal. Radioisotopes break down through three separate decay processes. Yep, you guessed it: We're gonna tell you all about 'em.

Alpha decay

The first decay process, called *alpha decay,* involves the emission of an alpha particle by the nucleus of an unstable atom. Emitting an alpha particle, which is nothing more exotic than the nucleus of a helium atom (two protons and two neutrons), causes the atomic number of the daughter nucleus to decrease by two and the mass number to decrease by four. The following pattern shows the decay of the parent nucleus X into a daughter nucleus Y and an alpha particle.

$$_B^A X \rightarrow \, _{B-2}^{A-4} Y + \, _2^4 He$$

Beta decay

The second type of decay, called *beta decay,* comes in three forms:

- **Beta plus decay:** A proton in the nucleus is converted into a neutron, a positron (e^+), and a tiny weakly interacting particle called a neutrino (υ), resulting in the atomic number decreasing by one. The mass number, however, *does not change,* even though the actual mass does change very slightly. A proton becoming a neutron, after all, should have absolutely no effect on the mass number because both protons and neutrons are *nucleons,* or particles contained in the nucleus, and converting one to the other does not change the overall number of particles in the nucleus.

  Beta plus decay follows the form of the equation below, with the atomic number being decreased by one and the mass number remaining the same.

  $$_B^A X \rightarrow \, _{B-1}^A Y + e^+ + \upsilon$$

- **Beta minus decay:** A neutron is converted into a proton, emitting in this case an electron and an *anti*neutrino ($\bar{\upsilon}$). Again, the mass number remains the same throughout the decay because the number of nucleons remains the same; however the atomic number *increases* by one. In case you haven't guessed, this is the reverse of beta plus decay.

  Beta minus decay follows the form of the equation below, with the atomic number being increased by one and the mass number remaining the same.

  $$_B^A X \rightarrow \, _{B+1}^A Y + e^- + \bar{\upsilon}$$

- **Electron capture:** The final form of beta decay occurs when an inner electron is captured by an atomic proton, converting it into a neutron and emitting a neutrino in the process.

  Electron capture follows the form of the equation below, with the atomic number being decreased by one and the mass number remaining the same.

  $$_B^A X + e^- \rightarrow \, _{B-1}^A Y + \upsilon$$

All three of these processes involve the emission or capture of an electron or a positron and result in a change in the atomic number of the daughter atom.

Gamma decay

The last form of decay, termed *gamma decay,* involves the emission of a high-energy form of light, called a gamma ray, by an excited nucleus. Although it does not change the atomic number or the mass number of the daughter nucleus, gamma decay often accompanies alpha or beta decay and it is a means for the nucleus of a daughter atom to reach its lowest possible energy state. The general form of gamma decay is shown below, where

$$_B^A X *$$

represents the excited state of the parent nucleus, and the greek letter gamma (γ) represents the gamma ray.

$$_B^A X * \rightarrow _B^A Y + \gamma$$

The forces that hold atomic nuclei together are extremely powerful, and the energy required to join two light nuclei together or split one heavy nucleus in two is tremendous. The only place where nuclear reactions happen in nature is in the very center of stars like our sun, where extreme temperatures cause atoms of hydrogen, helium, and other light elements to smash together and join into one. This extremely energetic process, called *nuclear fusion,* is what ultimately causes the sun to shine and provides the outward pressure required to support the sun against gravitational collapse.

Nuclear fission, on the other hand, occurs only rarely in nature. At a site in central Africa called Oklo a natural nuclear reactor has been found to have occurred many, many, years ago. Humans first harnessed the tremendous power released when splitting an atom through fission during the United States's Manhattan Project, which led to the development of the first atomic bomb. Fission has since been used for more benign purposes in nuclear power plants, which produce energy much more efficiently and cleanly than traditional fossil fuel-burning power plants.

Getting Electrons Excited

Special circumstances can drive electrons to configurations other than their ideal *ground state* (or lowest energy) configurations. You will need to know the causes and consequences of electrons being driven from their ground states and the equations that will allow you to qualitatively analyze those transitions, all of which are laid out for you in this chapter.

Excited state

Electrons can be driven to higher energy levels than they would otherwise occupy by absorbing energy in the form of heat, light, or electric current. An electron can only absorb an amount of energy exactly equivalent to the difference in energy between the state it is already occupying and another nearby energy state. Electrons in these higher-than-expected energy levels are said to be in an *excited state.* Excited electrons quickly release amounts of energy equivalent to the difference between their excited state and a lower energy state by emitting a *photon,* or particle of light, with an equivalent energy. The *wavelength* (which in the realm of visible light is equivalent to color) of this photon characterizes the energy difference between the excited and lower energy state. It is only by measuring these that chemists know about the energy structure of atoms. There isn't any way to learn about the inside of an atom by just looking at a ground state, stable one.

An electron that should be in the 3s subshell, for example, may be kicked up into the 4f subshell, where it wants to remain just about as badly as you want to remain on the couch during your Great Aunt Maybell's slideshow of her summer at the lake. In order to get back to the 3s subshell where it belongs, it must emit an amount of energy exactly equal to the energy difference between the 4f and the 3s subshells. This energy difference corresponds to a very specific wavelength, or color, of light because of the relationship between the energy and wavelength of a photon of light.

Converting energies to wavelengths

Max Planck developed the equation

$$E = \frac{hc}{\lambda}$$

to show the relationship between the wavelength and the energy of a photon.

However, this equation can also be written as

$$E = h\upsilon$$

This alternative equation is based on another convenient relationship between the properties of light, specifically the relationship between wavelength and frequency:

$$c = \lambda\upsilon$$

Here's an explanation of the variables in those two equations:

Energy (E) is expressed in Joules (J).

h is a constant called Planck's constant equal to 6.63×10^{-34} m^2 × kg/s.

Frequency (υ) is expressed in reciprocal seconds (1/s or s^{-1}) also known as Hertz (1 Hz = 1 s^{-1}).

c is the speed of light, specifically 3.0×10^8 m/s.

Wavelength (λ) is expressed in meters.

Niels Bohr, in order to complete his model of the atom, used Planck's ideas to explain why the orbits of electrons in the hydrogen atom are *quantized,* or occur only at certain distances from the nucleus. Planck had proposed that the energy of light particles (*photons*) came in *quanta,* or discrete values. Bohr found that the energy levels of the electron orbits were also quantized and followed the formula

$$E_n = \frac{-2.178 \times 10^{-18}}{n^2} \text{ joules}$$

where n is the principal quantum number. You will be expected to recognize and use this formula for predicting energy levels in the hydrogen atom on the AP exam, but make sure that you note that it applies to the *hydrogen atom alone* and will give you an incorrect answer if you apply it to any other element.

Emission spectra

Although it sounds like the next blockbluster sci-fi movie, Planck's formula $E = h\upsilon$ also helps explain the phenomenon of atomic *emission spectra,* which are characterized by bright lines of light of very specific colors. Chemists had realized long before Bohr and Planck came along that if you used a prism or spectrograph to disperse the light emitted when a pure solid element is burned or a pure gaseous element has an electronic current passed through it, then every individual element would emit a characteristic sequence of bright-colored lines called its *spectrum.* No two excited elements emitted identical spectra, so spectra became a definitive method for identifying elements.

An element's characteristic spectrum contains lines of specific colors because every element has its own unique set of allowed energies and therefore its own unique set of energy transitions. Because excited electrons must emit photons in order to reach the ground state and photon energies correspond directly to wavelengths, or colors, the appearance of every atomic spectrum is unique in the color and brightness of each line. The brighter a spectral line is, the more likely that transition is to occur. Most elements have hundreds of allowed transitions and emit light from infrared to X-ray wavelengths. The specific wavelength of light emitted when an electron drops to a lower energy level can be found by taking the energy difference between the excited and ground states and translating it into a wavelength using Planck's formula and substituting this energy change (Δ) for E as in

$$\Delta E = \frac{hc}{\lambda}$$

Chapter 4

Answering Questions about Atomic Structure

The concepts reviewed in Chapter 3 are the most basic in all of chemistry, and they are also some of the most important and testable concepts on the AP chemistry exam. A strong understanding of the most fundamental aspects of modern chemistry is essential to your success on the exam, and the concepts presented in Chapter 3 provide the foundation for the rest of introductory chemistry. Questions that appear regularly in both multiple-choice and free-response questions involve

✔ Periodic table trends

✔ Nuclear chemistry

✔ The structure of the atom and nucleus

Many AP questions that do not ask directly about atomic structure nonetheless assume that you have a solid understanding of its concepts.

In this chapter, we highlight the most important points of Chapter 3 for you, and then allow you to try your hand at AP chemistry–style test questions about atomic structure. Keep your eye out for these question types and others that build off of them in the practice tests (Chapters 29 and 30) at the end of the book as well.

Reinforcing the Foundation for Atomic Structure

For the AP exam, you should be familiar with the concepts of atomic structure as well as how to use and manipulate any equations and/or constants related to atomic structure. In this section we review these important concepts covering atomic structure so that you're prepared for some AP-style exam questions we give you later in this chapter.

Make sure that you thoroughly understand all of the basic components of the atom and how scientists arrived at their understanding of the modern atom. Burn the following major points into your brain before the exam:

- Atomic structure came to be understood gradually through the work of Thompson, Rutherford and Bohr:

  - Thompson's model, called the *plum pudding* model of the atom, consisted of blobs of negative charge in a soup of positive charge.

  - Rutherford shot bullets (alpha particles) at tissue paper (gold foil) and had to dodge them every once in a while as they came bouncing back at him. He thereby discovered that the majority of the mass and one of the types of charge (the positively charged protons) of the atom were concentrated at its very center in the nucleus.

  - Bohr said that the negative charges in an atom (the electrons) orbited the positively charged nucleus like the planets orbit the sun.

  - Later models see the electrons in atoms as cloudlike and not as particles that orbit like planets around the sun.

- Proton number is what determines the identity of an atom and is equivalent to its atomic number.

- If you add the number of neutrons to the number of protons in an atom, you will get its mass number, equivalent to the number of massive particles in its nucleus (*nucleons,* or protons and neutrons).

- All neutral atoms have equal numbers of positive and negative charges.

- Atoms with more electrons than protons have excess negative charge and are called *anions.*

- Atoms with too few electrons and excess positive charge are called *cations.*

- Atoms can decay via three processes:

  - *Alpha decay* involves the emission of a helium nucleus, resulting in a daughter atom with a decrease of two in the atomic number and four in the mass number.

  - *Beta decay* happens through three separate methods, all of which result in an atomic number change of one.

  - *Gamma decay* is simply the emission of a high-energy particle of light by an excited atom and does not result in any changes in the important numbers defining an atom.

- *Nuclear fusion* involves the joining of two light elements into one heavier element.

- *Nuclear fission* involves the splitting of an enormous atom into two smaller atomic pieces.

- *Electron configurations* show the locations of the electrons in an atom, organized by shells and subshells:

  - Electrons always occupy these shells and subshells from lowest energy to highest, often according to the Aufbau diagram.

  - Atoms of elements in the copper and chromium families, however, are prone to "steal" an electron from their neighboring *s* orbitals in order to achieve a half full or completely full *d* orbital, which is the more stable state. These elements have *exceptional electron configurations,* which differ slightly from those predicted by the Aufbau diagram.

  - Unruly electrons can jump to higher energy levels by absorbing energy, which they must then emit in order to reach the *ground state.* They release this energy in the form of particles of light called *photons,* which have very specific *wavelengths.*

Testing Your Knowledge

Now that you've absorbed everything you need to know about atoms, try answering some questions. Remember that you can look back at the formulas, variables, and constants, listed on the Cheat Sheet at the front of this book, but try not to look back at the rest of the text until after you check your answers.

The best way that you can practice is by going through the questions provided in this chapter in one sitting and then checking them at the end to identify which areas you still need to review.

Questions 1 through 4 refer to atoms of the following elements.

(A) Berylium

(B) Boron

(C) Carbon

(D) Oxygen

(E) Fluorine

1. In the ground state, has two electrons in one (and only one) of the *p* orbitals.

2. Has the largest atomic radius.

3. Has the largest value of the first ionization energy.

4. Has the smallest second ionization energy.

5. The half-life of ^{3}H is about 12 years. How much of a 4mg sample will remain after 36 years?

(A) 0.25mg

(B) 0.5mg

(C) 1mg

(D) 2mg

(E) 4mg

6. All of the following statements about the oxygen family of elements are true **except**:

(A) The electron configuration of the valence shell of the atom is ns^2np^4.

(B) It contains all nonmetals.

(C) The atomic radii decrease with decreasing atomic number.

(D) Electronegativity increases with decreasing atomic number.

7. Which of the following is the correct electron configuration for molybdenum?

(A) $1s^22s^22p^63s^23p^64s^23d^{10}4p^65s^24d^4$

(B) [Kr] $5s^24d^4$

(C) $1s^22s^22p^63s^23p^64s^23d^{10}4p^65s^14d^5$

(D) [Kr] $5s^14d^4$

8. Which of the following sequence of decays might lead to the creation of

$$^{234}_{91}Pa$$

from

$$^{238}_{92}U$$

(A) Alpha then gamma decay

(B) Alpha then beta decay

(C) Alpha decay only

(D) Beta decay only

9(a). There are three common isotopes of naturally occuring magnesium as indicated in the table below.

Isotope	Mass (amu)	Percent Abundance
Mg–24	24.0	79.0%
Mg–25	25.0	10.0%
Mg–26	26.0	11.0%

Using the information above, calculate the average atomic mass of magnesium.

9(b). A major line in the emission spectrum of magnesium corresponds to a wavelength of 518.3nm. Calculate the energy in Joules of the transition resulting in the emission of that spectral line.

The table below shows the first three ionization energies for two mystery elements.

	First Ionization Energy (kJ mol^{-1})	Second Ionization Energy (kJ mol^{-1})	Third Ionization Energy (kJ mol^{-1})
Element 1	520	7300	11820
Element 2	900	1760	14850

10(a). Elements 1 and 2 are both in the second period. What are their identities? Explain your reasoning.

10(b). Write the full electron configuration for element 1 after it has been ionized once. What other element shares the same electron configuration in its neutral state?

Checking Your Work

You've done your best. Now check your work. Make sure to read the explanations thoroughly for any questions you got wrong — or for any you got right by guessing.

1. (D). There are three *p* subshell orbitals, and an atom that has been forced to double up on electrons in one and only one *p* orbital must have four electrons in its *p* subshell. The electron configuration of oxygen is $1s^22s^22p^4$, so it fits the bill.

2. (A). Atomic radius decreases as you move to the right across a period, therefore the element in the list with the lowest atomic number (in other words, the farthest to the left), will have the largest atomic radius.

3. (E). The element with the largest value of the first ionization energy should be the one that is least likely to give up an electron. Halogens are the most likely candidates for high first ionization energies because they are the closest to achieving noble gas configurations through gaining electrons. Fluorine is the only halogen in the group.

4. (A). The element with the lowest second ionization energy should be one that is eager to give up its second electron. This corresponds to elements in the alkaline earth metal column, which need to be ionized twice in order to achieve noble gas configurations.

5. (B). Thirty-six years is equivalent to three half lives for ^{3}H. This means that the amount of radioactive material in the sample will have decreased by half three times. This leaves 2mg at 12 years, 1mg at 24 years, and 0.5mg at 36 years.

6. (B). The first statement is true because elements in the oxygen family always have full s subshells and four electrons in their p subshells by virtue of sharing the same horizontal location in the periodic table. B is not true. Several of the oxygen family's elements fall below the "staircase" marking the division between nonmetals and metals on the periodic table, indicating that they are metals or semimetals. C is a deceptively tricky statement because, while the atomic radius of elements in the same *family* do indeed decrease with decreasing atomic number, elements in the same period have precisely the opposite relationship, so you must be careful to keep them straight. D is also true because electronegativity increases to the right in a period and toward the top of a family.

7. (C). One might expect molybdenum to have the electron configuration shown in A and B (which are equivalent). However, molybdenum, being of the chromium family, falls under the umbrella of exceptional electron configurations. Its 4d orbital will abscond with an electron, which would normally belong to the 5s orbital in order to make it more stable. This leaves you with the electron configuration $1s^2 2s^2 2p^6 3s^2 3p^6 4s^2 3d^{10} 4p^6 5s^1 4d^5$, which can also be written as $[Kr]5s^1 4d^5$.

8. (B). The net result in the decay of

$$^{238}_{92}U$$

to

$$^{234}_{91}Pa$$

is a decrease of four in mass number and a decrease of one in atomic number. Because neither beta nor gamma decay result in a decrease in mass number, an alpha decay must occur, which would leave you with

$$^{234}_{90}Th$$

To then increase the atomic number by one from the initial alpha decay,

$$^{234}_{90}Th$$

must undergo a beta decay.

9(a). 24.3amu. Multiply each mass by its percent abundance in decimal form, giving you

$$(24.0 \times 0.79) + (25.0 \times .010) + (26.0 \times 0.11) = 24.32amu$$

9(b). 3.84×10^{-19}J. This problem is a direct application of the equation $E = h\nu$, and $c = \lambda\nu$. Plugging the speed of light as well as the given wavelength into the second equation and solving for frequency gives you

$$\nu = \frac{c}{\lambda} = \frac{3.0 \times 10^8 ms^{-1}}{518.3 \times 10^{-9} m} = 5.79 \times 10^{14} Hz$$

Take this frequency value and plug it in turn into the first equation to give you the energy value of the spectral line in question:

$$E = h \times \nu = 6.63 \times 10^{-34} Js \times 5.79 \times 10^{19} Hz = 3.84 \times 10^{-14} J$$

10(a). Element 1 is lithium and Element 2 is berylium. It requires roughly 14 times more energy to remove the second electron from element 1 than it did to remove the first, implying that the first electron was given up relatively easily and that the atom is very reluctant to give up another. This is consistent with the alkali metals, which only need to lose one electron to achieve a noble gas configuration. Once any element achieves noble gas configuration, it is reluctant to lose it by giving up another electron, explaining the large jump to the second ionization energy. Because you are told that these elements are in the second period, element 1 must be lithium. As for Element 2, it requires only twice as much energy to remove the second electron as was required to remove the first, but then it requires eight times more energy to remove the third. This implies that the first and second electrons are considerably easier to remove than the third, which is consistent with an alkaline earth metal. In this case, that metal must be berylium, which is the only alkaline earth metal in the second period.

10(b). $1s^2$, helium. The general electron configuration of lithium is $1s^2 2s^1$. When lithium is ionized, however, it loses an electron, achieving the electron configuration of the noble gas helium, whose electron configuration is simply $1s^2$.

Chapter 5

Working and Playing Together: Bonding

· ·

In This Chapter

▶ Taking or embracing electrons

▶ Naming compounds

▶ Getting a glimpse of ionic bonds

▶ Cozying up to covalent bonds

▶ Arming yourself with info on polarity and electonegativity

· ·

Chemistry textbooks would be much thinner if it weren't for bonding. As appealing as thinner textbooks may be, they'd come at the cost of other appealing things, such as cars, chocolate (Heaven help us!), and being alive. Bonds of different types between different elements and in different arrangements underlie the stunning diversity and complexity of matter. Although that complexity may sound daunting, we make taking a look at bonding a breeze by breaking down just how bonding builds things up.

Bringing Atoms Together: Bonding Basics

Sometimes atoms hang out together. Sometimes they form close friendships, sometimes only loose alliances. Close mutual attractions between atoms, ones strong enough to hold them persistently together, are called *bonds*. Isn't that sweet?

Why do atoms buddy up and bond with one another? In short, pairs of atoms seek stability. In particular, atoms prefer to dwell in states with more stable electron configurations. (Chapter 3 describes what makes for a stable electron configuration in case you need to check it out . . . and you do for the exam.) Atoms just tend to be most stable when their valence shells are completely filled with electrons so each atom does its best to fill its *valence shell,* its outermost shell.

Just like the different types of bonds you form in life — bonds between buddies, sisters and brothers, your children, your coworkers (okay, maybe that's a stretch) — atoms form different types of bonds as well, and we'll give you the skinny on those types of bonds in the following sections.

Attracting the opposite: Ionic bonds

Ionic bonds form between oppositely charged ions. *Ions* themselves form when an atom either gains or loses valence electrons in pursuit of filling its valence shell. Positively charged ions are electrostatically attracted to negatively charged ions, so the two nestle together to form an ionic bond. In a sense, atoms transfer whole electrons to create ions and form these bonds (more on cations and anions in Chapter 3). Sometimes, an ion consists of more than one kind of atom. These multiatom ions are called *polyatomic ions*, which have won many awards for their extreme inconvenience to chemistry students. Table 5-1 lists the most common polyatomic ions, grouped by ionic charge. They'll pop up frequently. They'll annoy you until you simply buckle down and memorize them. You can find out more about ionic bonds later in the chapter in the section, "Eyeing Ionic Bonds."

Table 5-1	Common Polyatomic Ions		
−1 Charge	*−2 Charge*	*−3 Charge*	*+1 Charge*
Dihydrogen phosphate ($H_2PO_4^-$)	Hydrogen phosphate (HPO_4^{2-})	Phosphite (PO_3^{3-})	Ammonium (NH_4^+)
Acetate ($C_2H_3O_2^-$)	Oxalate ($C_2O_4^{2-}$)	Phosphate (PO_4^{3-})	
Hydrogen sulfite (HSO_3^-)	Sulfite (SO_3^{2-})		
Hydrogen sulfate (HSO_4^-)	Sulfate (SO_4^{2-})		
Hydrogen carbonate (HCO_3^-)	Carbonate (CO_3^{2-})		
Nitrite (NO_2^-)	Chromate (CrO_4^{2-})		
Nitrate (NO_3^-)	Dichromate ($Cr_2O_7^{2-}$)		
Cyanide (CN^-)	Silicate (SiO_3^{2-})		
Hydroxide (OH^-)			
Permanganate (MnO_4^-)			
Hypochlorite (ClO^-)			
Chlorite (ClO_2^-)			
Chlorate (ClO_3^-)			
Perchlorate (ClO_4^-)			

Sharing electrons: Covalent bonds

Covalent bonds form between atoms that share electrons, rather than transfer them. These shared electrons are in orbits that surround both bonded atoms. Sharing may or may not be equal, depending on which atom "wants" the electrons more. The chemical term for electron greed is *electronegativity*. Differences in electronegativity between atoms largely determine what kind of bond forms between them. As described more completely in the section, "Getting a grip on electronegativity," differences in electronegativity determine whether a bond is ionic or covalent.

Keeping it solid with metals: Metallic bonding

Okay, we're not referring to being groupies with a heavy metal band. You can do that after the AP exam. Instead, *metallic bonding* occurs within metallic solids (big surprise, we know). Metallic bonding helps to explain the unique properties of metals:

✔ The *electron-sea model* of metallic bonding describes metallic solids as lattices of metal cations immersed within a fluid "sea" of mobile electrons. Although the electrons are constrained to the cation lattice, they can move freely within it. This model provides a satisfying intuitive explanation for metals' electrical conductivity because the electrons within the lattice aren't constrained to orbitals around individual atoms, but are in delocalized orbitals.

✔ For example, a sample of solid copper holds itself together with metallic bonds. The electrons in this sample don't clearly "belong" to any particular copper atom, but can move through the sample. So, if you draw out the sample into a wire, you can use the mobile electrons to conduct electric current.

So what do you get when you bond? In life, you get families and friends. In chemistry, you get compounds. Chemical *compounds* are the consequence of bonding. A compound consists of bonded atoms, and has properties that are different from a simple mixture of the identical, nonbonded atoms. In other words, the structure of a compound confers properties to that compound. The relationship between structures and properties is a major theme of chemistry. That theme revolves around the concept of the chemical bond. In the next section we begin to deal with the basics of compounds, especially how to recognize and interpret their names and formulas.

Naming Names: Nomenclature, Formulas, and Percent Composition

Okay, so our parents always told us not to call people names, but in chemistry, there are exceptions (and we don't mean your chemistry teacher). There are more known chemical compounds than there are residents of New York City so something must be done to keep these compounds straight. In the early days of chemistry, far fewer compounds were known, so chemists assigned those compounds *common names*, like "ammonia." Calling a compound by its common name is essentially like calling it "Reggie." Eventually, it became clear that common names, while charming, made a mess of things because these names told you nothing about the compound. So, modern chemists prefer *systematic names*, ones assigned by certain rules, which tell you a great deal about a compound. The process of using rules to assign names to compounds is called chemical *nomenclature*.

In the next sections you'll learn to move easily back and forth between the chemical formula for a compound (H_2O) and its systematic name (dihydrogen oxide). For common compounds, there is always the possibility of a charmingly messy common name (water). Finally, you'll learn to crunch the numbers of percent composition, one of the common ways chemists begin to define an unknown compound.

Knitting names to formulas: Nomenclature

Different naming rules apply to inorganic compounds (ones not based on multi-carbon structures) versus organic compounds (ones that *are* based on multi-carbon structures). The system presented in this chapter applies to inorganic compounds; organic compound nomenclature is described in Chapter 25.

The main idea behind systematic naming is that the poor, addled chemist, drowning in compounds, can move easily between the name of a compound and that compound's *formula*.

Formulas can be one of two types:

✔ **Empirical:** An empirical formula lists the kinds of atoms in a compound and gives the ratio of atoms to each other — not necessarily their actual numbers.

✔ **Molecular:** A molecular formula also lists the kinds of atoms in a compound, but gives the actual number of each kind of atom within one molecule of that compound. The compounds nitrogen monoxide and dinitrogen tetroxide make clear the difference between empirical and molecular formulas; just check out Table 5-2.

Table 5-2	Checking Out the Differences between Empirical and Molecular Formulas	
Compound	Empirical Formula	Molecular Formula
Octane	C_4H_9	C_8H_{18}
Nitrogen monoxide	NO	NO
Dinitrogen tetroxide	NO_2	N_2O_4

When you write out the molecular formula, if the number of atoms within a molecule of the compound is one (1), you don't write the numeral one (1), as it's implied. Nitrogen monoxide only has one atom within each molecule of that compound so you write its molecular formula simply as NO.

A great number of chemical compounds are binary compounds, ones that are built of only two kinds of elements. Binary compounds can be either ionic (held together by ionic bonds) or molecular (held together by covalent bonds). Ionic compounds have only empirical formulas because they are not composed of distinct molecules. Molecular compounds have both empirical and molecular formulas.

Naming binary compounds

Many chemical substances are binary compounds, consisting of only two kinds of elements. The generic formula for a binary compound is X_AY_B, where X and Y are the different elements, and A and B are the relative (and sometimes actual) amounts of each element within one representative unit of the compound. A representative unit might be a molecule or it might be a *formula unit*, the smallest repeating unit of an ionic compound.

Figure 5-1 provides a visual summary of the method for assigning systematic names to inorganic compounds, but we also give you the steps for naming a binary compound below.

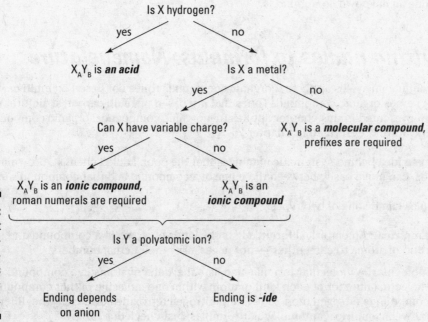

Figure 5-1: Flowchart for assigning systematic names to inorganic compounds.

To name a binary compound, ask yourself these questions:

1. **Is X hydrogen?** Compounds that contain hydrogen cations (H⁺) often fall into a class of compounds called *acids*. Sadly, many common acids are still saddled with common names. Table 5-3 summarizes many of these names. Although these are common names (and are therefore inconvenient), some patterns exist that help make sense of things:

   • Acids composed with monoatomic (single-atom) anions use the prefix *hydro-* and the suffix *-ic*.

   • Acids composed with polyatomic anions with names ending in *-ate* use the suffix *-ic*.

   • Acids composed with polyatomic anions with names ending in *-ite* use the suffix *-ous*.

   • If a polyatomic anion includes the prefix *per-* or *hypo-*, that prefix transfers to the acid name.

   • Be careful. If a compound like HCl is in a gaseous state, it is *not* called hydrochloric acid.

2. **Is X a nonmetal or a metal?** If X is a nonmetal, you're dealing with a molecular compound. As such, you'll need to use prefixes to describe the number of each type of element, X and Y. Table 5-4 summarizes the prefixes you'll most commonly encounter. If X is a metal, then you're dealing with an ionic compound — proceed to step 3.

   The name of a molecular compound usually lists the element furthest to the left within the periodic table as the first. If both elements occur in the same column, the lower one is usually listed first. The second element in the name of the compound receives the suffix *-ide*.

   Prefixes are always given on both elements unless there is only one of the first element; in that case, the "mono" prefix is dropped.

   Nitrogen monoxide (NO) and dinitrogen tetroxide (N_2O_4) are examples of molecular compounds named according to these rules.

3. **Is X a metal that can form cations of different charges?** Although alkali metals and alkaline earth metals reliably form only one kind of cation, metals in the center of the periodic table (especially the transition metals) can sometimes form different charges of cations. If a compound contains one of these metals, you must specify the charge of the cation within the compound name. To do so, use Roman numerals within parentheses after the name of the metal; the Roman numeral corresponds to the size of the positive charge. For example, the element iron, Fe, frequently occurs as either a +2 cation or a +3 cation. The Roman numeral system describes Fe^{2+} as iron (II) and Fe^{3+} as iron (III).

   Unlike the case with molecular compounds, ionic compound names do not include prefixes to describe the number of each kind of atom within the ionic compound. Why not? Because ionic compounds always occur in the combination of ions that results in zero overall charge. Because ionic compounds have zero overall charge, you can use the charges of the individual ions to determine the formula of the ionic compound — and vice versa. Binary ionic compounds add the suffix *-ide* to the name of the second element.

   Here are some examples of binary ionic compound formulas and names:

   • Na^+ and Cl^- combine to form NaCl, known as sodium chloride.

   • Fe^{2+} and O^{2-} combine to form FeO, known as iron (II) oxide.

   • Fe^{3+} and O^{2-} combine to form Fe_2O_3, known as iron (III) oxide.

4. **Is Y a polyatomic ion?** If Y is a polyatomic ion, you have permission to be momentarily annoyed. After the moment has passed, either recall the name of the polyatomic ion from memory (as you'll have to do on the AP Chemistry exam), or refer to a handy table like Table 5-1.

For example, if Y is SO_4^{2-}, then you're dealing with the polyatomic ion called *sulfate*. If your compound is Na_2SO_4, then the name is *sodium sulfate*.

Table 5-3	Some Common Acids
Name	*Formula*
Carbonic acid	H_2CO_3
Hydrochloric acid	HCl
Nitric acid	HNO_3
Phosphoric acid	H_3PO_4
Sulfuric acid	H_2SO_4
Perchloric acid	$HClO_4$
Choric acid	$HClO_3$
Chlorous acid	$HClO_2$
Hypochlorous acid	HClO

Table 5-4	Prefixes for Binary Molecular Compounds		
Prefix	*Number of Atoms*	*Prefix*	*Number of Atoms*
Mono	1	Hexa	6
Di	2	Hepta	7
Tri	3	Octa	8
Tetra	4	Nona	9
Penta	5	Deca	10

Calculating percent composition

Formulas and names emphasize the numbers of different kinds of *atom*s within a compound. *Percent composition* emphasizes the *mass* of different kinds of atoms within a compound. Naming by percent composition simply means that you make a list of each kind of atom in a compound accompanied by the percent of the compound's total mass contributed by that kind of atom.

To calculate percent composition you

1. **Determine the mass for each mole of the compound.** Do this by counting the number of each kind of element within the compound, finding the elements' atomic masses from the periodic table, and then adding up the individual masses of all the elements.

2. **Determine the mass for each mole of each atom.** Again, find the elements' atomic masses on the periodic table.

3. **Multiply the mass of the atom by however many of that atom the compound contains.**

4. **Divide the mass contributed by an individual element by the total mass of the compound.**

5. **Multiply the result of step 4 by 100%.**

6. **Repeat steps 2 through 5 for each element type in the compound.**

7. **Write out the percent composition by listing each element type alongside that element's percent mass.**

Consider the compound water, H_2O.

- Each mole (6.022×10^{23}) of water molecules has mass 18.02g.

- Each mole of water molecules contains one mole of oxygen atoms and two moles of hydrogen atoms.

- Each mole of oxygen atoms has mass 16.00g.

- Each mole of hydrogen atoms has mass 1.008g.

So, you calculate the percent composition of water

Oxygen: (16.00g mol^{-1} / 18.02g mol^{-1}) × 100.0% = 88.79%

Hydrogen: [(2 × 1.008g mol^{-1}) / 18.02g mol^{-1}] × 100.0% = 11.21%

More briefly, the percent composition is O 88.79%, H 11.21%.

So, by sheer number of atoms, water is mostly hydrogen. But by mass — in other words, by percent composition — water is overwhelmingly oxygen.

Eyeing Ionic Bonds

Ionic bonds form between cations (positively charged ions) and anions (negatively charged ions). The strength of an ionic bond derives from electrostatic attraction between ions of opposite charge. Ionic compounds form extended, three-dimensional lattices, such as the one shown in Figure 5-2. The exact geometrical arrangement of ions in an ionic lattice results from an interplay of factors, all conspiring to maximize the favorable (as in attractive) interactions between the ions.

Figure 5-2:
A lattice of Na$^+$ and Cl$^-$ ions within the ionic compound sodium chloride, NaCl.

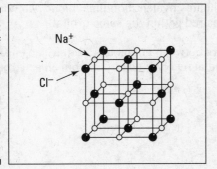

Na$^+$

Cl$^-$

The NaCl lattice shown in Figure 5-2 consists of repeating, two-atom units of Na$^+$ and Cl$^-$. You can imagine that each sodium atom has donated an electron to a neighboring chlorine atom. Thinking about ionic bonding in this way makes clear why ionic compounds typically form between a metal and a nonmetal. Metals tend to lose valence electrons easily. Nonmetals tend to gain extra valence electrons avidly. Ions scratch each other's backs.

The strength of ionic bonds within an ionic compound is expressed by *lattice energy*. Lattice energy represents the amount of energy required to completely separate the component ions of one mole of an ionic compound into gaseous ions. Larger positive lattice energies correspond to stronger ionic bonds.

Electrostatic attraction increases not only with the magnitude of opposing charges, but also as the distance between those charges decreases. In other words, greater quantity of charge attracts more strongly than less quantity of charge, and closer charges attract more strongly than distant charges. The overall charge of an ion effectively acts as a point charge at the atom's center. The centers of bigger-sized atoms can't nestle as closely together as those of smaller-sized atoms. As a result, ionic bonds tend to be stronger between ions with greater magnitude of charge and between ions of smaller size. So, the strongest ionic bonds are between two small, highly charged ions.

Considering Covalent Bonds

Unlike ionic bonds, where atoms either lose or gain electrons, covalent bonds are a a kinder, gentler bond. Covalent bonds form when atoms share valence electrons. Atoms do this kind of thing because it helps them to fill their valence shells. Covalent bonds tend to form between atoms that do not completely give up electrons. In other words, covalent bonds tend to form between nonmetals.

Sharing electrons . . . or not

The attractive force of a covalent bond arises from the attraction of the shared electrons to the positively charged nuclei of the bonded atoms. Within bonds, electrons don't act as truly distinct particles, but are distributed into "clouds" of varying density. The shared electrons of a covalent bond distribute with higher density in the region directly between nuclei, as shown in Figure 5-3.

Each single covalent bond houses two shared electrons. In a standard covalent bond, each bonded atom contributes one electron. So, each atom gains one electron (that of its bonding partner) in the bargain.

Sometimes one atom donates both electrons to a covalent bond, with the other atom contributing no electrons. This kind of bond is called a *coordinate covalent bond*. Atoms with *lone pairs* of electrons often engage in coordinate covalent bonding. A *lone pair* consists of two electrons that are not used in bonding paired within the same orbital.

Even though covalent bonding usually occurs between nonmetals, metals can engage in coordinate covalent bonding. Usually, the metal receives electrons from an electron donor called a *ligand*.

Figure 5-3:
Distribution of electron density within the electron cloud of a covalent bond.

+   +

Getting to know structural formulas

Atoms can share more than a single pair of electrons. When atoms share two pairs of electrons, they form a double bond, and when they share three pairs of electrons they form a triple bond. A solid line drawn between element symbols serves as shorthand in structural formulas to indicate that atoms are covalently bonded. So, the covalently bonded atoms of water, carbon dioxide, and dinitrogen can be indicated as shown in Table 5-5.

Table 5-5	Examples of Structural Formulas	
Compound	*Molecular Formula*	*Structural Formula*
Water	H_2O	H-O-H
Carbon dioxide	CO_2	O=C=O
Dinitrogen	N_2	N≡N

Rules for determining the number of covalent bonds between atoms and for estimating the geometric arrangement of covalently bonded atoms are described in Chapter 7.

Measuring the strength of covalent bonds

Bond enthalpy describes the strength of covalent bonds. The bond enthalpy (ΔH) is an estimate of the amount of energy required to break the bond. In the case of diatomic molecules with single covalent bonds (such as Cl_2), the bond enthalpy is a very good estimate. In the case of polyatomic molecules (such as CH_4, which contains four distinct C-H bonds), the bond enthalpy is an *average bond enthalpy*, an estimate averaged over the four bonds of the molecule.

Larger bond enthalpies correspond to stronger covalent bonds. Typically, atoms held together with more bonds and/or with stronger bonds approach each other more closely than do atoms held together with fewer and/or with weaker bonds.

Bond enthalpies are easily measured, so chemists frequently use them to help determine the strength of bonds and to estimate bond distances. On the AP exam, you might be asked to make the same kinds of estimates from a set of bond enthalpy data.

Separating Charge: Polarity and Electronegativity

Polarity refers to an uneven distribution of charge. In chemical bonds, polarity arises from a difference in *electronegativity* between bonded atoms. More electronegative atoms draw greater electron density toward themselves. You might think that atoms with more protons (more positively charged nuclei) are always more electronegative, but this isn't the case. Why? The electronegativity of an atom derives from the positive charge of its nucleus *and* from the extent to which that positive charge is offset or "shielded" by the successive layers of electron density that surround that nucleus. Because atoms with many protons also tend to have a greater number of electron shells, the large positive charge of these atoms' nuclei is partially offset by the negatively charged electron shells.

Getting a grip on electronegativity

Within a given row of the periodic table, electronegativity tends to increase from left to right. Within a given column of the table, electronegativity tends to increase from bottom to top. These trends are only overall patterns because electronegativity is influenced by more subtle factors. The electronegativities of the elements are shown in Figure 5-4.

Figure 5-4: Electronegativities of the elements.

1 H 2.1																	
3 Li 1.0	4 Be 1.5											5 B 2.0	6 C 2.5	7 N 3.0	8 O 3.5	9 F 4.0	
11 Na 0.9	12 Mg 1.2											13 Al 1.5	14 Si 1.8	15 P 2.1	16 S 2.5	17 Cl 3.0	
19 K 0.8	20 Ca 1.0	21 Sc 1.3	22 Ti 1.5	23 V 1.6	24 Cr 1.6	25 Mn 1.5	26 Fe 1.8	27 Co 1.9	28 Ni 1.9	29 Cu 1.9	30 Zn 1.6	31 Ga 1.6	32 Ge 1.8	33 As 2.0	34 Se 2.4	35 Br 2.8	
37 Rb 0.8	38 Sr 1.0	39 Y 1.2	40 Zr 1.4	41 Nb 1.6	42 Mo 1.8	43 Tc 1.9	44 Ru 2.2	45 Rh 2.2	46 Pd 2.2	47 Ag 1.9	48 Cd 1.7	49 In 1.7	50 Sn 1.8	51 Sb 1.9	52 Te 2.1	53 I 2.5	
55 Cs 0.7	56 Ba 0.9	57 La 1.1	72 Hf 1.3	73 Ta 1.5	74 W 1.7	75 Re 1.9	76 Os 2.2	77 Ir 2.2	78 Pt 2.2	79 Au 2.4	80 Hg 1.9	81 Tl 1.8	82 Pb 1.9	83 Bi 1.9	84 Po 2.0	85 At 2.2	
87 Fr 0.7	88 Ra 0.9	89 Ac 1.1															

Decreasing →

↑ Increasing

Electronegativities of the Elements

The greater the difference in electronegativity between bonded atoms, the more polar is the bond between those atoms. Covalent bonds that are very polar more closely resemble ionic bonds than do covalent bonds that are less polar. In fact, no real physical distinction exists between ionic and covalent bonds — ionic bonds are simply so polar that it becomes useful to imagine that one atom has emerged entirely victorious from the tug-of-war between competing nuclei for electron density.

Although different sources use slightly different numbers to make the split between polar and nonpolar, usually, differences in electronegativity are interpreted with the following categories:

- ✔ A difference in electronegativity between bonded atoms of less than about 0.3 leads to a description of the bond as "nonpolar."

- ✔ A difference in electronegativity ranging from 0.3 to about 1.7 leads to a description of "polar."

- ✔ A difference in electronegativity greater than about 1.7 leads to the description "ionic."

Digging into dipoles

Within a polar covalent bond, electrons are distributed unevenly between the two atoms. The more electronegative atom is surrounded by greater electron density and assumes a *partial negative charge*. The less electronegative atom draws correspondingly less electron density and assumes a *partial positive charge*. Partial negative and partial positive charges are indicated by the symbols $\delta-$ and $\delta+$, respectively. Separation of charge along the line connecting two bonded atoms (the *bond axis*) creates a *bond dipole*. Bond dipoles are often indicated in one of two different ways, as shown in Figure 5-5:

Figure 5-5: Two different depictions of a bond dipole in the HCl molecule.

$$\delta+ \quad \delta- \qquad \xrightarrow{\hspace{1.5cm}}$$
$$\text{H} — \text{Cl} \qquad \text{H} — \text{Cl}$$

The size of a bond dipole is measured quantitatively by the *dipole moment*, μ. A dipole moment measures the polarity of a bond by taking into account two key factors:

- ✔ How much charge is separated along the bond axis

- ✔ How far apart the charge is separated

Polar bonds have a larger dipole moment than nonpolar bonds. Dipole moments are vector quantities, which means that they have both size and direction. Within a molecule, different bonds may point in different directions, and these differences can be important.

Individual bonds' dipoles sum over all the bonds of a molecule, resulting in a *molecular dipole*. In addition to the *permanent dipoles* created by polar bonds, *instantaneous dipoles* can flicker into and out of existence within nonpolar bonds and molecules. Both kinds of dipoles play important roles in the interactions between molecules. Permanent dipoles lead to *dipole-dipole interactions* and to *hydrogen bonds*. Instantaneous dipoles lead to attractive *London forces*.

In addition to ion-ion interactions, these are the forces that must be overcome in order to turn a liquid into a gas or a solid into a liquid. Different compounds have different boiling points and different melting points because they engage in different collections of interactions. Here is a summary of the intermolecular forces that contribute to boiling and melting points, and examples of compounds in which each kind of force dominates.

Force	Compound	Melting point	Boiling point
Ion-ion	NaCl	1074 K	1738 K
Hydrogen bonding	H_2O	273 K	373 K
Dipole-dipole	H_2CO	156 K	254 K
London forces	CH_4	91 K	112 K

Chapter 6

Answering Questions on Bonding

. . . .

In This Chapter

▶ Reviewing important points on bonding

▶ Breaking yourself in with some practice questions on bonding

▶ Finding out what you did right . . . and not so right

. . . .

Chapter 5 introduced you to chemical bonds, the attractive forces that bring atoms together to form compounds. Atoms come together to form different kinds of bonds and a multitude of different compounds. Yet, for all the complication, the basic rules of the game are pretty straightforward. Review them here and test your mastery of the material with the practice questions.

Brushing Up on Bonding Forget-Me-Nots

We've included a helpful summary of the points from Chapter 5 that you need to remember. So, do some bonding of your own with the following list:

✔ *Compounds* are collections of atoms held together by bonds.

   • *Ionic bonds* arise from electrostatic attraction between ions of opposite charge (between anions and cations).

   • *Covalent bonds* arise when atoms share electrons.

   • *Metallic bonding* involves a sea of electrons that moves freely through a lattice of metal cations.

✔ Ionic bonds tend to form between a metal and a nonmetal.

✔ Covalent bonds tend to form between nonmetals.

✔ Inorganic compounds are named in a systematic way (see the figure in Chapter 5). Naming and recognizing compounds will be much easier if you are familiar with the common polyatomic ions (see the table in Chapter 5).

✔ Solid ionic compounds consist of a highly ordered lattice of ions. The strength of an ionic bond within a lattice is expressed by *lattice energy.* Large, positive lattice energies correspond to strong ionic bonds.

✔ Other factors being equal, ionic bonds tend to be stronger between ions with more charge, and between smaller ions.

✔ Electrons within covalent bonds distribute with greater density in the region between the two atomic nuclei.

✔ Usually, each atom of a covalent bond contributes one electron per bond. When one atom contributes both electrons, we say the bond is a *coordinate covalent bond.*

✔ The strength of covalent bonds is described by *bond enthalpy*. Large, positive bond enthalpies correspond to strong covalent bonds.

✔ *Polar bonds* are marked by unequal sharing of electrons. *Electronegativity* is the tendency of an atom to draw electrons to itself.

- • Bonds between atoms with extreme differences in electronegativity are ionic.

- • Bonds between atoms with significant differences in electronegativity are polar covalent.

- • Bonds between atoms with insignificant differences in electronegativity are nonpolar covalent.

✔ The polarity of a bond is described by the *bond dipole*, $\mu = Qd$, where Q is the magnitude of the separated charge and *d* is the distance of separation.

✔ Bond dipoles sum within molecules to produce *molecular dipoles*. Dipole-dipole and ion-dipole interactions are major contributors to the properties of different compounds, especially boiling points and melting points.

Testing Your Knowledge

Try your hand at these practice questions on bonding. Do yourself a favor by not looking back in the text to get the questions correct on the first try. Do your best, and then check your answers afterward so that you can discover what you need to review.

1. Which of the following are compounds that might reasonably form from combining iron and oxygen?

    (I) Fe_2O_3

    (II) Fe_3O_2

    (III) FeO

    (A) I only

    (B) II only

    (C) III only

    (D) I and II only

    (E) I and III only

2. Which of the following is least likely to be a stable compound?

    (A) CF_4

    (B) S_2O

    (C) PCl_3

    (D) SiO_3

    (E) NO

3. Which of the following compounds is least likely to form?

    (A) $Na_2Cr_7O_2$

    (B) $KC_2H_3O_2$

    (C) Li_2CN

    (D) $Rb_2C_2O_4$

    (E) HNO_2

4. Which of the following is correctly named?

   (A) CsCl; cesium (I) chloride

   (B) Fe_2O_3; iron (II) oxide

   (C) CBr_4; carbon quatrobromide

   (D) NO_2; dioxygen mononitride

   (E) MnO_2; manganese (IV) oxide

5. Which empirical formula best fits the following percent composition? 50.00% carbon, 5.59% oxygen, 44.41% hydrogen

   (A) $C_5H_6O_4$

   (B) C_3HO_2

   (C) $C_3H_4O_2$

   (D) $C_4H_3O_2$

   (E) $C_3H_5O_2$

Questions 6 through 10 refer to the following types of bonds:

   (A) Ionic bonds

   (B) Polar covalent bonds

   (C) Nonpolar covalent bonds

   (D) Coordinate covalent bonds

   (E) Metallic bonds

6. Form in compounds like N_2 and O_2

7. Tend to support electrical conductivity

8. Tend to form strong, brittle compounds

9. Tend to form from an atom containing a lone pair of electrons

10. Form in compounds like CO and H_2S

Checking Your Work

You've done your best. Now check your work. Make sure to read the explanations thoroughly for any questions you got wrong — or for any you got right by guessing.

1. (E). Both Fe_2O_3 (iron (III) oxide) and FeO (iron (II) oxide) can form, the former from the Fe^{3+} cation and the latter from the Fe^{2+} cation. Fe_3O_2 is a bad choice because the overall charge of this compound is not neutral; even using Fe^{2+}, there is an excess of positive charge because each oxygen can contribute only –2 charge.

2. (D). As shown, SiO_3 is least likely to be stable. Although the SiO_3^{2-} (silicate) anion exists, the compound shown in the question is neutral. In order to fill its valence shell and remain neutral, silicon requires four covalent bonds.

3. (C). Li_2CN is least likely to occur because the "ion math" doesn't work. Lithium forms cations with +1 charge (Li^+) and the cyanide ion has –1 charge (CN^-). So, a more likely version of this compound is LiCN.

4. (E). If a manganese ion with +4 charge is used (Mn^{4+}), the compound can form; that ion is specified by the Roman numeral IV. As described in the answer for question 1, Fe_2O_3 is correct as long as the Fe^{3+} ion is used. But the Roman numeral II in the name specifies the Fe^{2+} ion. Answer A is incorrect because the alkali metal cesium forms only ions with +1 charge, so no Roman numeral is used in its name. The correct names for the compounds in choices C and D are carbon tetrabromide and nitrogen dioxide, respectively.

5. (C). Percent composition refers to percent weight. So, a 100.g sample of the mystery compound would contain 50.00g carbon, 44.41g oxygen, and 5.59g hydrogen. To convert these masses to relative numbers of each type of atom, divide each mass by the molar mass of that atom:

50.00g carbon / (12.01g mol^{-1}) = 4.163 mol carbon

5.59g hydrogen / (1.01g mol^{-1}) = 5.53 mol hydrogen

44.41g oxygen / (16.00g mol^{-1}) = 2.776 mol oxygen

Next, take the smallest mol value of the set (2.776 mol in this case) and divide all the mol values by that number. You get: 1.5 mol carbon, 1 mol oxygen, and 2.0 mol hydrogen. You can't build compounds with fractions of an atom, so multiply all results by the smallest number that will produce a set of whole numbers. In this case, multiplying all values by 2 does the trick, yielding: 3 mol carbon, 2 mol oxygen, and 4 mol hydrogen. Finally, assemble the atom symbols and the whole numbers into an empirical formula: $C_3H_4O_2$.

6. (C). Covalent bonds between identical atoms are nonpolar because each atom has identical electronegativity.

7. (E). The sea of mobile electrons that is the defining feature of metallic bonding supports electrical conductivity; the mobile electrons can move in response to applied voltage.

8. (A). Ionic bonds are strong, but the extreme order of ionic bonding lends itself to fracturing.

9. (D). Coordinate covalent bonds form when one atom contributes both electrons to the bond; lone pairs are perfect for this task, because they include two electrons not otherwise involved in bonding.

10. (B). Polar covalent bonds form between the atoms in compounds like these because the atoms have significantly different electronegativities — but not so different that ionic bonds form instead.

Chapter 7

Doing Atomic Geometry: Molecular Shapes

Molecular shapes help determine how molecules interact with each other and within themselves. For example, molecules that stack nicely on one another are more likely to form solids. And two molecules that can fit together so their reactive bits lie closer together in space are more likely to react with one another. In this chapter, we describe how to get a sense of the shape of a molecule, starting with only its molecular formula. Then, we explore a few ways that molecular shape can impact molecular properties.

Drawing Dots and Lines with Lewis

Valence electrons are critical for bonding. If that sentence confuses you, then you need to amble through Chapter 5 before moving further into this chapter. Because valence electrons are the ones important for bonding, chemists use *Lewis electron dot structures:* symbols that represent valence electrons as dots surrounding an atom's chemical symbol.

You should be able to draw and interpret electron dot structures for atoms so that you understand how the dots relate to the valence shell electrons possessed by an atom. By keeping track of how filled or unfilled an atom's valence shell is, you can predict whether it is likely to bond with other atoms in an attempt to fill its valence shell.

Drawing electron dot structures

Figure 7-1 shows the electron dot structures for elements in the periodic table's first two rows. Electrons in an atom's valence shell are represented by dots surrounding the symbol for the element. Completely filled shells are surrounded by eight (8) dots (or two dots, in the lone case of helium). Valence shells progressively fill moving from left to right in the table. This pattern repeats itself with each successive row of the periodic table, as each row corresponds to adding a new outermost valence shell. Because atoms are most stable (think "happiest") when their valence shells are full, they tend to seek full-shell states in one of two ways. Atoms may gain or lose electrons (forming anions or cations, respectively) to end up with filled shells. Or, atoms may covalently bond with each other so that each atom can lay claim to the electrons within the bond.

To draw the electron dot structure of any element

1. **Write the element's name.**

2. **Count the number of electrons in that element's valence shell.**

3. **Draw that number of dots around the chemical symbol for the element.** As a general rule, space the dots evenly around the element's symbol.

Figure 7-1:
Electron dot structures for elements in the first two rows of the periodic table.

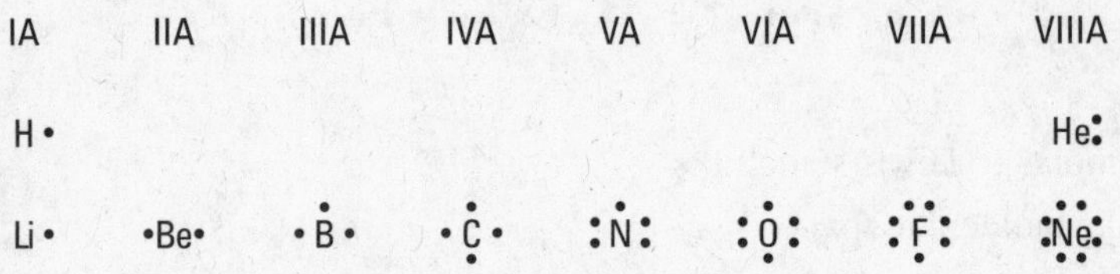

IA	IIA	IIIA	IVA	VA	VIA	VIIA	VIIIA
H •							He ••
Li •	•Be•	•B•	•C•	:N:	:O:	:F:	:Ne:

Constructing Lewis structures

Just as electron dot structures show the number of valence electrons that surround individual atoms, Lewis structures use dots and lines to show the distribution of electrons around all the atoms within a compound. Lewis structures are great tools for figuring out which molecules are reasonable (filling up all the atomic valence shells) and which molecules aren't so reasonable (leaving some valence shells unhappily unfilled).

If you know the molecule's formula you can figure out the correct Lewis structure for that molecule. The following example gives you the steps you need to work out a Lewis structure. This example uses formaldehyde, CH_2O. You can follow along with Figure 7-2:

1. **Add up all the valence electrons for all the atoms in the molecule.**

   These are the electrons you can use to build the structure. Account for any extra or missing electrons in the case of ions. For example, if you know your molecule has +2 charge, remember to subtract two from the total number of valence electrons. In the case of formaldehyde, C has four valence electrons, each H has one valence electron, and O has six valence electrons. The total number of valence electrons is 12.

2. **Pick a "central" atom to serve as the anchor of your Lewis structure.**

   The central atom is usually one that can form the most bonds, which is often the atom with the most empty valence orbital slots to fill. In larger molecules, some trial-and-error may be involved in this step, but in smaller molecules, some choices are obviously better than others. For example, carbon is a better choice than hydrogen to be the central atom because carbon tends to form four bonds, whereas hydrogen tends to form only one bond. In the case of formaldehyde, carbon is the obvious first choice because it can form four bonds, while oxygen can form only two, and each hydrogen can form only one.

3. **Connect the other, "outer" atoms to your central atom using single bonds only.**

   Each single bond counts for two electrons. In the case of formaldehyde, attach the single oxygen and each of the two hydrogen atoms to the central carbon atom.

4. **Fill the valence shells of your outer atoms. Then put any remaining electrons on the central atom.**

In our example, carbon and oxygen should each have eight electrons in their valence shells; each hydrogen atom should have two. However, by the time we fill the valence shells of our outer atoms (oxygen and the two hydrogens), we have used up our allotment of 12 electrons.

5. **Check whether the central atom now has a full valence shell.**

   If the central atom has a full valence shell, then your Lewis structure is drawn properly — it's formally correct even though it may not correspond to a real structure. If the central atom still has an incompletely filled valence shell, then use electron dots (nonbonding electrons) from outer atoms to create double and/or triple bonds to the central atom until the central atom's valence shell is filled. Remember, each added bond requires two electrons. In the case of our formaldehyde molecule, we must create a double bond between carbon and one of the outer atoms. Oxygen is the only choice for a double-bond partner, because each hydrogen can accommodate only two electrons in its shell. So, we use two of the electrons assigned to oxygen to create a second bond with carbon.

1. $C(4 e^-) + H(1 e^-) + H(1 e^-) + O(6 e^-) = 12e^-$

2. Carbon is the central atom; it can form more bonds (4) than O, H.

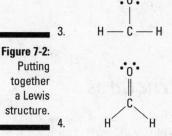

Figure 7-2: Putting together a Lewis structure.

3.

4.

Getting bonds straight with line structures

Atoms involved in covalent bonds share electrons such that each atom ends up with a completely filled valence shell, as described more fully in Chapter 5. The simplest and best studied covalent bond is the one formed between two hydrogen atoms (dihydrogen). Separately, each atom has only one electron with which to fill its $1s$ orbital. By forming a covalent bond, each atom lays claim to two electrons within the molecule of dihydrogen. Covalent bonds can be represented in different ways, as shown for dihydrogen in Figure 7-3.

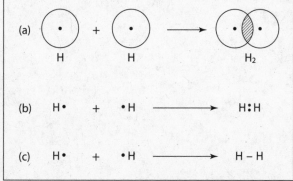

Figure 7-3: Three representations of the formation of a covalent bond in dihydrogen.

Lewis structures for compounds more complicated than dihydrogen can get pretty busy-looking. Most Lewis structures get rid of the dizzying effect of swarms of electron dots by representing shared electron pairs — that is, covalent bonds — as lines. Each two-dot electron pair that is shared between two atoms is rewritten as a nice, clean line connecting the atom symbols. Only nonbonding electrons (lone pairs) are left as dots. Whether a Lewis structure for a compound uses electron dot pairs or lines to represent bonds, those dots or lines refer to preceisely the same entities: covalent bonds.

Atoms can share more than a single pair of electrons. When atoms share two pairs of electrons, they are said to form a *double bond,* and when they share three pairs of electrons they are said to form a *triple bond.* Examples of double and triple bonds are shown with Lewis structures using both electron dots and lines in Figure 7-4.

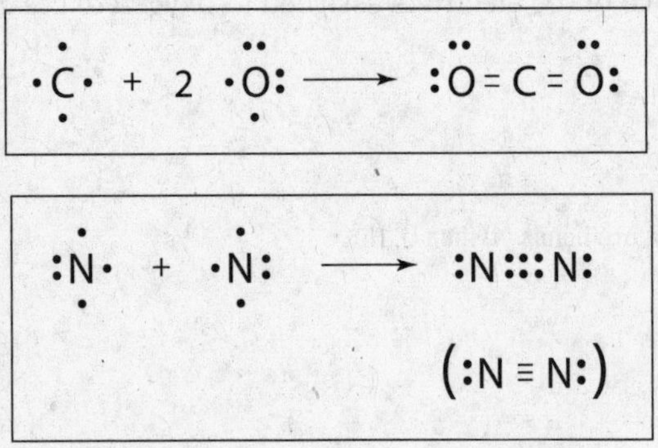

Figure 7-4:
The formation of double bonds in carbon dioxide and triple bonds in dinitrogen.

Rummaging through resonance structures

Sometimes a given set of atoms can covalently bond with each other in multiple ways to form a compound. In these cases, you can draw several possible valid Lewis structures for the compound. This situation leads to something called *resonance.* Each of the possible Lewis structures is called a *resonance structure.* The actual structure of the compound is a *resonance hybrid,* a sort of average of all the resonance structures. Resonance is particularly common in molecules where much bonding occurs in a side-by-side manner with p-orbital electrons from adjacent atoms. These kinds of bonds are called π bonds (see "Overlapping Orbitals to Form Valence Bonds " for more on that type of bond).

When resonance structures involve π bonds, electrons in the bonds can become "delocalized," forming an extended orbital that bonds more than two atoms. The benzene molecule, shown in Figure 7-5, is a classic example of resonance and electron delocalization.

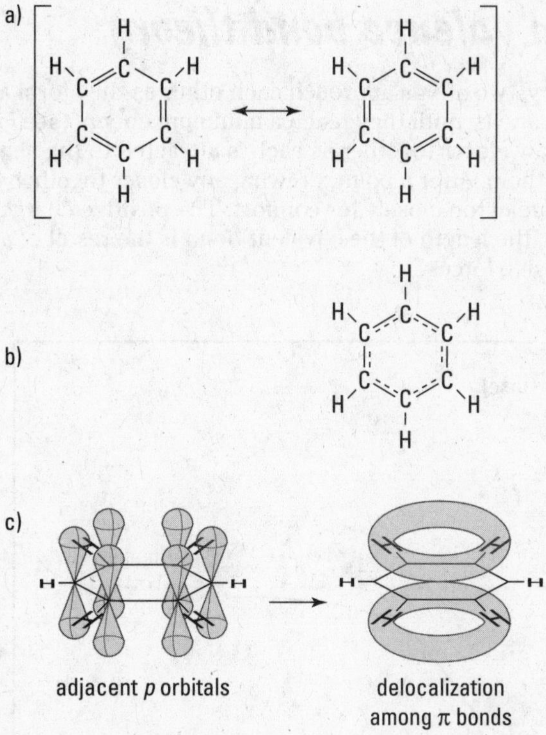

Figure 7-5:
The reso-
nance
structures
(a) reso-
nance
hybrid
(b) and
electron
delocaliza-
tion of
benzene(c).

adjacent *p* orbitals

delocalization
among π bonds

Overlapping Orbitals to Form Valence Bonds

When we say that covalent bonds "share" valence electrons (if you're asking yourself, "What in the world are covalent bonds?" see Chapter 5 for more info), what do we really mean? We mean that shared electron pairs now flit about within overlapping atomic orbitals (if your head is orbiting right now trying to figure out what atomic orbitals are, check out Chapter 3).

From a molecular formula you can figure out a Lewis structure. From the Lewis structure and by using valence bond theory and VSEPR theory you can often get a pretty good idea of the shape of a molecule, which is a major factor in determining other properties, like polarity, phase behavior, and the tendency to interact or react with other molecules. Going from a Lewis structure to a molecular shape requires a little know-how, however. In the next few sections, we give you the skills you need.

We realize the word "theory" sounds stuffy and perhaps intimidating, but, hey, we didn't pick it. However, we take the stuffiness out of the valence bond theory as well as the VSEPR theory in the following sections so that you can predict the shape of molecules.

Getting a grasp on valence bond theory

According to valence bond theory, two atoms approach each other as they form a covalent bond, overlapping their electron shells, until they reach a minimum energy (see Figure 7-6). The positively charged nuclei draw closer together as each is attracted to the negative charge of the electrons between them. After a point, drawing any closer together would crowd their positively charged nuclei too closely for comfort. The positive charges begin to repel at such short distances. So, the length of the covalent bond is the result of a balancing act between attractive and repulsive forces.

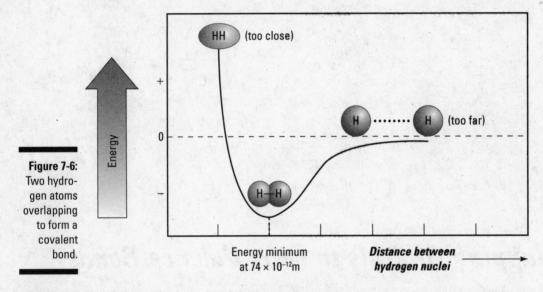

Figure 7-6: Two hydrogen atoms overlapping to form a covalent bond.

Different kinds of bonds result from the different ways the orbitals can overlap in space. The kind of symmetry the resulting bond has with the *bond axis,* (an imaginary line that connects the centers of the bonded atoms) determines what kind of bond is formed:

- ✔ **Sigma bond:** When orbitals overlap in a way that is *completely* symmetrical with the bond axis, a σ bond (sigma bond) is formed. Sigma bonds form when *s* or *p* orbitals overlap in a head-on manner. Single bonds are usually sigma bonds.

- ✔ **Pi bond:** When orbitals overlap in a way that is symmetrical with the bond axis *in only one plane,* a π bond (pi bond) is formed. Pi bonds form when adjacent *p* orbitals overlap above and below the bond axis.

Sigma bonds are stronger than pi bonds because the electrons within sigma bonds lie entirely between the two atomic nuclei, simultaneously attracted to both. A double bond is one sigma bond and one pi bond, and a triple bond is one sigma bond and two pi bonds.

Figure 7-7 shows you an example of a sigma bond forming from two *s* orbitals and the formation of a pi bond from two adjacent *p* orbitals.

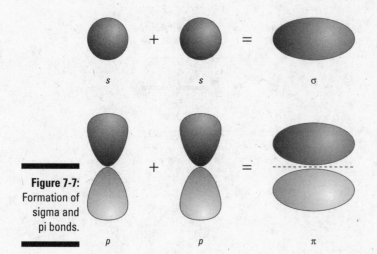

Figure 7-7:
Formation of
sigma and
pi bonds.

Shaping up with the VSEPR theory

Sigma and pi bonds form by overlapping the valence orbitals of atoms (see "Getting a grasp on valence bond theory" for more on sigma and pi bonds), so the overall shape of a molecule depends largely on the geometric arrangement of valence orbitals around each atom. That's where VSEPR theory comes into play. Now don't worry: We'll start with the hard part: VSEPR stands for *valence shell electron pair repulsion.* Okay, now it gets easier. VSEPR is simply a model that helps predict and explain why molecules have the shapes they do.

Only use VSEPR theory on the "*p*-block" elements. These elements are the ones in Groups IIIA, IVA, VA, VIA, VIIA, and VIIIA (except helium) on the right side of the periodic table. For the most part, this restriction means that you use VSEPR theory to predict geometry around nonmetal atoms.

In the following sections, we break down the basics of VSEPR theory so you can understand and predict the shape of molecules.

Predicting shapes with VSEPR theory

Simply by combining the principle that valence electron pairs (lone or bonds) want to get as far apart as they can (thus "repulsion" in the acronym VSEPR) with the fact that lone pairs repel more strongly than do bonding pairs, you can predict an impressive array of molecular shapes. The total number of valence electron pairs determines the overall geometry around a central atom. The distribution of these pairs between bonding and nonbonding orbitals adjusts that geometry. So VSEPR theory can predict several shapes that appear over and over in real-life molecules. These shapes, shown in Figure 7-8, resemble those observed in real-life molecules.

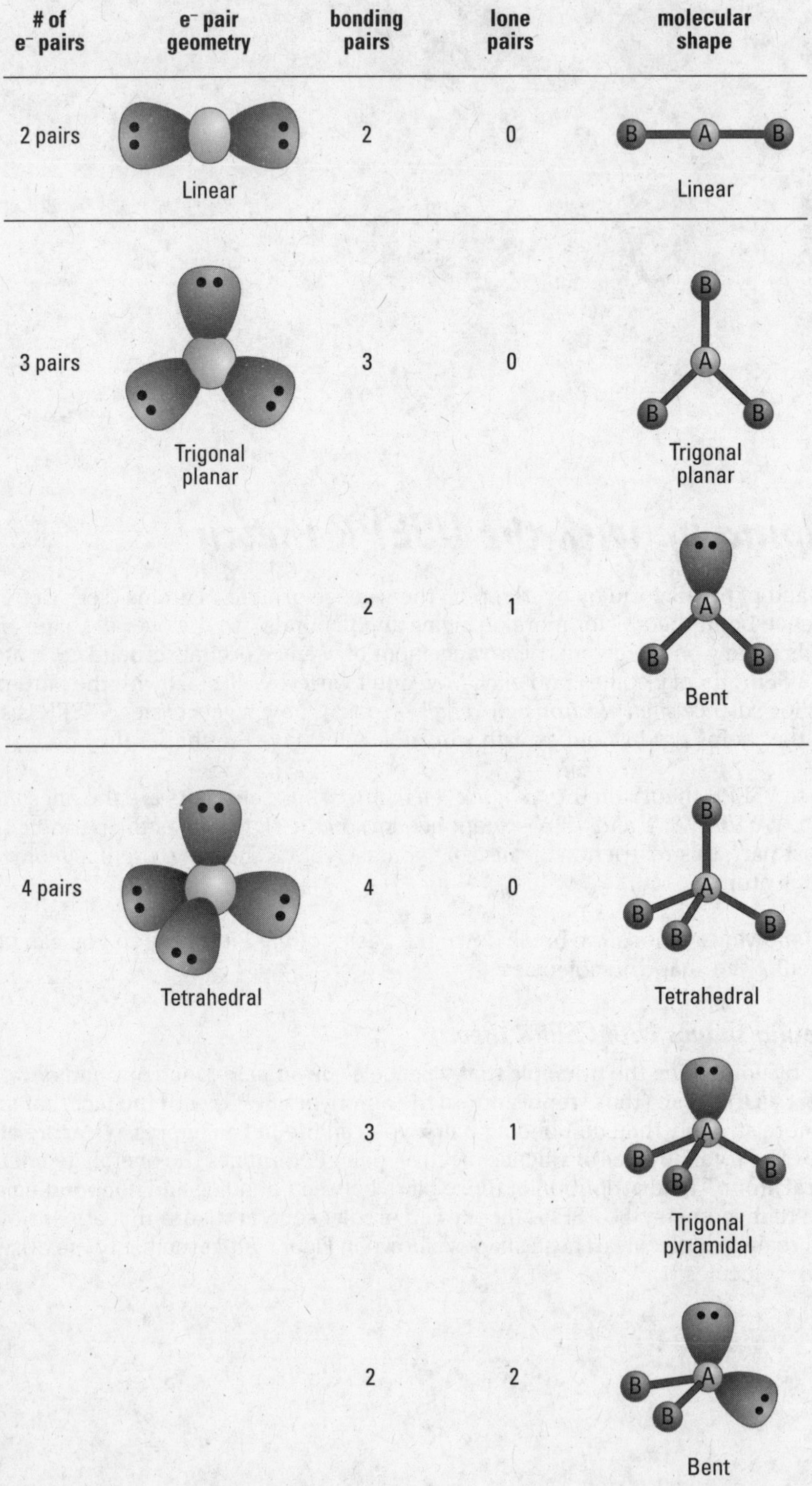

# of e⁻ pairs	e⁻ pair geometry	bonding pairs	lone pairs	molecular shape
2 pairs	Linear	2	0	Linear
3 pairs	Trigonal planar	3	0	Trigonal planar
		2	1	Bent
4 pairs	Tetrahedral	4	0	Tetrahedral
		3	1	Trigonal pyramidal
		2	2	Bent

Figure 7-8: Molecular shapes predicted by VSEPR theory.

Making molecular shapes with hybridization

VSEPR theory is pretty good at making predictions about what shapes will emerge from a set of mostly equivalent valence shell electron pairs. But wait — what about the difference between valence electrons in *s* orbitals versus those in *p* orbitals? These electrons don't seem equivalent. That's when hybridization jumps in. *Hybridization* refers to the mixing of atomic orbitals into new, hybrid orbitals. Valence electron pairs occupy equivalent hybrid orbitals.

Check out the electron configuration of carbon in Figure 7-9. Carbon contains a filled $1s$ orbital, but this is an inner-shell orbital, so it doesn't impact the geometry of bonding. However, the valence shell of carbon contains one filled $2s$ orbital, two half-filled $2p$ orbitals, and one empty *p* orbital. Not the picture of equality, so the valence orbitals must form new shapes or hybridize.

Different combinations of orbitals produce different hybrids. One *s* orbital mixes with two *p* orbitals to create three identical sp^2 hybrids. One *s* orbital mixes with one *p* orbital to create two identical *sp* hybrids. Centers with sp^3, sp^2, and *sp* hybridization tend to possess tetrahedral, trigonal planar, and linear shapes, respectively.

Figure 7-9:
The electron configuration of carbon.

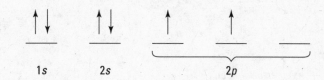

Doing the math to predict molecular shapes

The shapes of real molecules emerge from the geometry of valence orbitals — the orbitals that bond to other atoms. Here's how to predict this geometry:

1. **Count the number of lone pairs and bonding partners an atom actually has within a molecule.** You can do this by looking at the Lewis structure.

   In formaldehyde ($H_2C = O$), for example, carbon bonds with two hydrogen atoms and double bonds with one oxygen atom. So, carbon effectively has three valence orbitals.

2. **Next, inspect the electron configuration, looking for the mixture of orbital types (like *s* and *p*) that the valence electrons occupy.**

   Carbon has four valence electrons in $2s^2 2p^2$ configuration. Two valence electrons occupy an *s* orbital, and one electron occupies each of two identical *p* orbitals. The *s* orbital isn't equivalent to the *p* orbitals. So, we mix the *s* orbital with the two *p* orbitals to create three identical sp^2 hybrid orbitals, as shown in Figure 7-11.

Note that the total number of orbitals doesn't change; in the example, formaldehyde has three valence orbitals before mixing and still has three valence orbitals after mixing. VSEPR theory predicts that electrons in three identical orbitals mutually repel to create a trigonal planar geometry like the one shown in Figure 7-8, which is the shape of the formaldehyde molecule.

Elements in periods 3 and below of the periodic table can engage in *hypervalency*, meaning that they can possess more than eight electrons in their valence shell. Hypervalency complicates things a bit, but the basic principles of VSEPR theory still apply — electron pairs distribute themselves as far apart as possible around the hypervaent atom. Depending on the distribution of bonding and nonbonding electron pairs, hypervalent atoms participate in "expanded octet" geometries like octahedral, square planar and T-shaped, as shown in Figure 7-10.

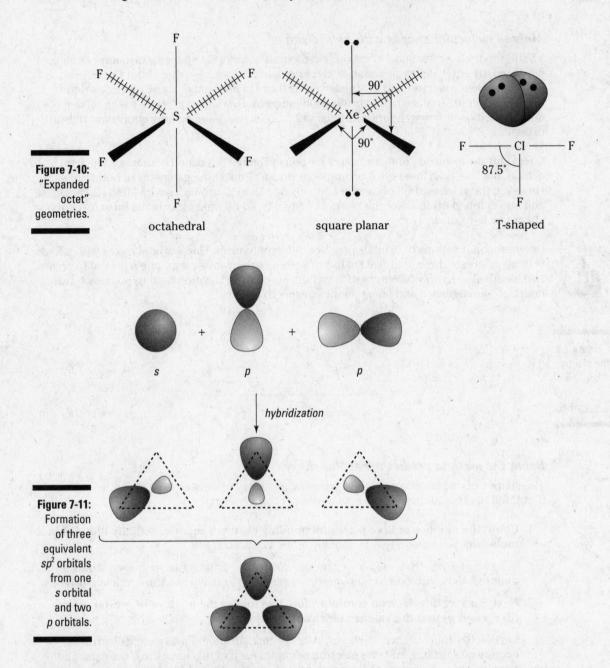

Figure 7-10: "Expanded octet" geometries.

octahedral

square planar

T-shaped

Figure 7-11: Formation of three equivalent sp^2 orbitals from one s orbital and two p orbitals.

hybridization

Polarity and Isomers

You can often get a pretty good idea of the shape of a molecule (if you can't, see the section earlier in this chapter, "Overlapping Orbitals to Form Valence Bonds"). Knowing the shape of a molecule helps you to understand the molecule's other properties, like polarity, phase behavior, and the tendency to interact or react with other molecules. These are exactly the kinds of things that chemists and the rest of us actually *care* about, whether we know it or not. For example, life on Earth depends largely upon the unique polarity and shape of the water molecule. The tasty combination of oil and vinegar in a raspberry vinaigrette depends on the phase behavior of the oil and vinegar components.

Polarity

The shape of a molecule and its polarity have a fairly simple relationship; the polarity of a molecule emerges from the polarities of the bonds within that molecule. As described in Chapter 5, differences in the electronegativity of atoms create polarity in bonds (which causes those bonds to be polar bonds) between those atoms. Each polar bond possesses a *bond dipole,* which is represented by this symbol: μ. Depending of the shape of the molecule, individual bond dipoles may or may not lead to an overall molecular dipole, a polarity in the molecule as a whole.

You can calculate a bond dipole by multiplying the amount of charge separated along that bond (Q) by the distance of the separation, *d*. Because the SI units of charge and distance are Coulombs (C) and meters (m), this calculation yields units of Coulomb-meters. A Coulomb-meter is a ridiculously big unit compared to the size of bond dipole, so bond dipoles are typically reported in *debyes* (D), where $1 \text{ D} = 3.336 \times 10^{-30}$ C·m.

You won't be asked to calculate a bond dipole on the AP exam, but knowing how bond dipoles are calculated helps you to understand how they add up within a molecule.

You won't be asked to calculate a bond dipole on the AP exam, but knowing how bond dipoles are calculated helps you to understand how they add up within a molecule.

The precise way in which individual bond dipoles contribute to the overall *molecular dipole* depends on the shape of the molecule. Dipoles add as *vectors*. Basically, this means

- ✓ If two equivalent bond dipoles point in opposite directions, they cancel each other out.

- ✓ If the two bond dipoles point in the same direction, they add.

- ✓ If two bond dipoles point at a diagonal to one another, the horizontal and vertical components of the dipoles add (or cancel) separately.

The molecular dipoles of carbon dioxide and water demonstrate vector addition of bond dipoles nicely, as shown in Figure 7-12:

- ✓ Each molecule consists of a central atom bonded to two identical partners.

  - • In the case of carbon dioxide, carbon double-bonds to two oxygens to produce a linear molecule (thanks to *sp* hybridization!). The two C–O bond dipoles of carbon dioxide cancel each other out, so the molecular dipole is zero.

  - • In the case of water, a single oxygen atom single-bonds to two hydrogens that leaves the oxygen with two lone pairs.

- ✓ The bond pairs and the lone pairs (which are sp^3 hybridized) lead to a nearly tetrahedral orbital geometry and result in a bent shape of the molecule as you trace from hydrogen to oxygen to hydrogen.

- ✓ The two C-O bond dipoles of carbon dioxide cancel each other out, so the molecular dipole is zero.

- ✓ Only the horizontal components of the two H-O bond dipoles of water cancel each other out.

- ✓ The vertical components of those bond dipoles add, making the molecular dipole of water the sum of those components.

Because carbon dioxide has no molecular dipole, it interacts very weakly with itself and therefore is a gas at room temperature. Because water has a significant molecular dipole, it acts as a good solvent for other polar compounds.

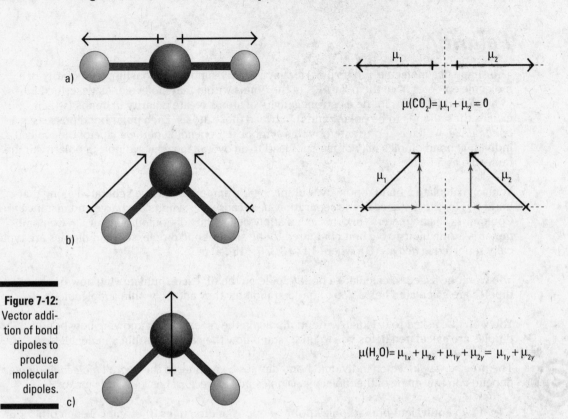

Figure 7-12:
Vector addition of bond dipoles to produce molecular dipoles.

$\mu(CO_2) = \mu_1 + \mu_2 = 0$

$\mu(H_2O) = \mu_{1x} + \mu_{2x} + \mu_{1y} + \mu_{2y} = \mu_{1y} + \mu_{2y}$

Isomers

Molecular shapes can contribute to chemical diversity (an abundance of compounds with differing properties) through the creation of *isomers*. Isomers are compounds made up of the same atoms but put together in different arrangements. Isomers with the same molecular formula may freeze or boil at different temperatures, may undergo different kinds of reactions, and may have important differences in situations where exact shape is critically important. Distinguishing between isomers is especially important in organic chemistry and biochemistry (see Chapter 25 for more detail on these) because these branches of chemistry involve many large molecules built with many individual bonds. With many bonds, there are many opportunities for isomerism. It turns out that big organic molecules can be very selective, reacting specifically with one isomer but not with others.

The following list gives you an overview of the different kinds of isomers, and how each is related to the others:

✔ **Structural isomers** have the same set of atoms, but the atoms are connected differently.

✔ **Stereoisomers** have the same bonds, but different three-dimensional arrangements of atoms. Among stereoisomers, you have

 • **Geometric isomers:** Geometric isomers can occur when atoms are restricted from rotating freely about their bonds, such as in the case of a double bond or when bulky groups of atoms bump into each other. In these cases, atoms or groups of atoms can get trapped on one side or the other of the bond, creating geometrically distinct versions of the molecule.

 • **Optical isomers:** Optical isomers (also called *enantiomers*) are mirror images of one another, but ones you can't superimpose — in the same way you can't match your left hand with your right hand when both palms face in the same direction. Molecules that have this handedness are called *chiral*.

Chapter 8

Answering Questions on Molecular Shapes

Chapter 7 was packed with ideas, starting with the way valence electrons distribute themselves around individual atoms, building up through the way atoms connect themselves into molecules with particular shapes and finishing with the way molecular shapes help determine the properties of different compounds. In other words, Chapter 7 was a lot to take in one bite. Here is where you chew. In the following sections we review the highlights of Chapter 7, give you practice problems to help sear important concepts into your brain, and give you full explanations for the answers to help you swallow what you've chewed.

Shaping Up the High Points

Before you attempt any questions in this chapter, you should review the following sections of important points from Chapter 7.

Electron dot structures and Lewis structures

Electron dot structures and *Lewis structures* use dots and/or lines to represent valence electrons and/or lines to represent valence electron pairs.

Basic steps to follow when drawing a Lewis structure are

1. **Add up all the valence electrons for all the atoms in the molecule.**

2. **Pick a "central" atom to serve as the anchor of your Lewis structure.**

3. **Connect the other, "outer" atoms to your central atom using single bonds only.**

4. **Fill the valence shells of your outer atoms. Then put any remaining electrons on the central atom.**

5. **Check whether the central atom now has a full valence shell.**

Compounds that can engage in multiple, valid Lewis structures participate in *resonance*, based on an old idea that the molecules might actually be in equilibrium between the multiple structures. Although we now know that not to be the case, the term has stuck. Individual *resonance structures* contribute to an overall, averaged structure called a *resonance hybrid*. Adjacent *p*-orbital electrons are commonly involved in resonance, in which they participate within delocalized π bonds.

Valence bond theory

Valence bond theory describes covalent bonds in terms of overlap between atomic orbitals:

- ✔ Orbitals that overlap in a manner completely symmetric to the bond axis form *sigma bonds*.

- ✔ Orbitals that overlap in a manner that is symmetric to the bond axis in only one plane form *pi bonds*.

VSEPR theory

VSEPR theory explains molecular shapes in terms of mutual repulsion between electron pairs (both bonding and nonbonding pairs) around a central atom:

- ✔ Lone (nonbonding) pairs repel more strongly than bonding pairs.

- ✔ Valence electrons in different kinds orbitals (like *s* and *p* orbitals) *hybridize* into mixed, equivalent orbitals. The total number of hybridized orbitals is the same as the original number of unmixed orbitals.

- ✔ One *s* orbital mixes with one *p* orbital to create two identical *sp* hybrids.

  sp hybridized centers will have linear geometry.

- ✔ One *s* orbital mixes with two *p* orbitals to create three identical sp^2 hybrids.

  sp^2 hybridized centers will have trigonal planar geometry.

- ✔ One *s* orbital mixes with three *p* orbitals to create four identical sp^3 hybrids.

  sp^3 hybridized centers will have tetrahedral geometry.

Bond dipoles

Individual *bond dipoles* arise from differences in electronegativity in bonded atoms. Except for homonuclear diatomic molecules (like O_2, H_2, and so on), all atom-atom bonds will have a at least a small bond dipole. Within a molecule, bond dipoles may or may not contribute to an overall *molecular dipole* depending on geometry:

- ✔ Bond dipoles add as vectors within molecules, building on each other or canceling each other out depending on geometry.

- ✔ Dipole moment (μ) = amount of separated charge (Q) × distance separated (*d*); the equation looks like this: $\mu = Qd$

Isomers

Isomers are compounds made up of the same atoms, but are put together in the following different arrangements:

- *Structural isomers* have different bonds.

- *Stereoisomers* have the same bonds, but different three-dimensional arrangements of atoms.

  - *Geometric isomers* can occur when atoms are restricted from rotating freely about their bonds.

  - *Optical isomers* (also called *enantiomers*) are nonsuperimposable mirror images of one another and are therefore *chiral*.

Testing Your Knowledge

Now that you know your sp^2's from your sp^3's, and have your VSEPR in shape, give these practice questions a try.

1. Which of the following correctly lists the molecules in order of increasing polarity?

   (A) H_2O, H_2S, HF, H_2

   (B) H_2, HF, H_2S, H_2O

   (C) H_2, H_2S, HF, H_2O

   (D) H_2, H_2S, H_2O, HF

   (E) H_2S, H_2O, HF, H_2

2. Which of the following molecules has bond angles closest to 109.5°?

   (A) BF_3

   (B) SiS_2

   (C) CCl_4

   (D) H_2O

   (E) CO_2

3. Identify the incorrect set of resonance structures.

A.

B.

C.

D.

E.

4. Group IVA elements tend to hybridize in which of the following ways?

   (A) sp

   (B) sp^2

   (C) sp^3

   (D) s^2p^2

   (E) sp^2d

5. Within formaldehyde, CH_2O, carbon possesses which of the following traits?

   I. sp^2 hybridization

   II. Three equivalent sigma bonds

   III. Trigonal planar bonding geometry

   (A) I only

   (B) II only

   (C) III only

   (D) I and III only

   (E) I, II and III

Questions 6 through 10 refer to the following list of geometries:

(A) Linear

(B) Bent

(C) Trigonal planar

(D) Tetrahedral

(E) Trigonal bipyramidal

6. Characteristic of four electron pairs, two bonding and two nonbonding

7. Typical of *sp* hybridization

8. Accounts for the nonpolarity of SiF_4

9. Nitrate anion

10. PCl_5

Questions 11 through 14 refer to the following molecule, and the accompanying set of choices:

(A) I only

(B) II only

(C) III only

(D) I and II only

(D) II and III only

11. Exhibit(s) sp^2 hybridization

12. Exhibit(s) sp^3 hybridization

13. Has a lone pair

14. Has a pi bond

Checking Your Work

Don't play the angles. Clear up any lingering confusion by checking your answers, making sure you understand the explanations for any questions you got wrong — or the ones that you got right by guessing.

1. (D). H_2 is the least polar because its sole covalent bond occurs between atoms that are identical, and which therefore have identical electronegativity. H_2S and H_2O are close, but H_2O is slightly more polar because oxygen is slightly more electronegative than sulfur. The "bent" geometry of both H_2S and H_2O is key to the presence of a molecular dipole in each

case — a linear shape would lead to canceling of the bond dipoles, resulting in nonpolar molecules. The sole covalent (nearly ionic) bond in HF is the most polar of them all.

2. (C). Bond angles of 109.5° signify a perfect tetrahedron. Many molecules approach this ideal, but perfect tetrahedral geometry typically requires that the central atom (carbon, in this case) possess four identical bonding partners (chlorines, in this case).

3. (E). The last set of resonance structures, for CO_3^{2-} anion, are incorrect. Each of the structures includes an oxygen atom with an excess of valence electrons. The oxygen atom double-bonded to carbon in each structure should have only two lone pairs, not three.

4. (C). Group IVA atoms tend to engage in four covalent bonds. Because of hybridization, these bonds involve four identical sp^3 orbitals.

5. (D). In formaldehyde, the central carbon indeed exhibits sp^2 hybridization. This hybridization leads to a trigonal planar (though not perfectly so) molecular shape. However, the three sigma bonds are not identical, for two reasons. First, C-O bonds differ in polarity from C-H bonds. Second, the C-O bond axis includes a pi bond; double-bonding between carbon and oxygen brings those two atoms closer together than does a single bond.

6. (B). In water, for example, the two lone pairs of oxygen and the two O-H bonds lead to just such a bent shape.

7. (A). In an sp hybridized atom, there are only two bond axes. The electrons in these axes mutually repel to achieve the maximum separation afforded by linear geometry.

8. (D). Because of its perfect tetrahedral geometry, silicon tetrafluoride is nonpolar, despite the fact that each silicon-fluorine bond is polar.

9. (C). Nitrate displays resonance between three sp^2 hybridized resonance structures. The hybrid structure therefore has a trigonal planar shape consistent with sp^2 hybridization.

10. (E). This compound may have thrown you. The central phosphorous of PCl_5 is "hypervalent," meaning that it engages in more than the three covalent bonds you'd expect from its position on the periodic table. But even if the hypervalency freaked you out, take heart: the basic principles of VSEPR still apply. Whatever the reasons for pentavalent phosphorous, the fact remains that the electrons within the five bond axes mutually repel, seeking the shape with maximum separation. That shape is trigonal bipyramidal.

11. (C). Carbon III engages in a double bond (with carbon) and in two single bonds (with hydrogen and nitrogen). This arrangement leads to three bonding axes and sp^2 hybridization.

12. (D). Carbon I engages in four separate single bonds, one with another carbon and three with hydrogens. Four separate single bonds around a central carbon are a hallmark of sp^3 hybridization. Nitrogen II may at first appear to be sp^2 hybridized, because it engages in only three bonds — but the nitrogen also contains a lone pair, and is thus sp^3 hybridized.

13. (B). Carbons I and III keep their four valence electrons occupied within various combinations of covalent bonds. Because it has one more valence electron than carbon, Nitrogen II engages in only three covalent bonds, reserving two electrons as a nonbonding "lone pair."

14. (C). Double bonds, like the carbon-carbon double bond in which Carbon III participates, include both sigma and pi bonds.

Part III
Colliding Particles to Produce States

The 5th Wave   By Rich Tennant

"...finally, those researchers working after hours should limit their investigation to the behavior of protons and electrons, and hereafter refrain from putting eggs in the particle accelerator."

In this part . . .

Chemistry likes little things, but not to the exclusion of big things. For example, chemists take note of the facts that the mountains are made of solid rock, the oceans brim with liquid brine and the sky is filled with gaseous air. These facts beg explanation. It turns out that the explanations are in terms of little things. In this part, we explain how energy interacts with particles to help them settle on an identity of state—solid, liquid, gas, and a few exotic others. Once this big decision has been made, there are other decisions to be made about things like pressure, temperature and volume. All these details are interrelated, and in this part we show you how.

Chapter 9

Putting a Finger to the Wind: Gases

• •

In This Chapter

▶ Keeping up with kinetic theory

▶ Going through the Gas Laws

▶ Digging in Dalton's Law of Partial Pressures

▶ Checking out diffusion and Graham's Law

▶ Encountering ideal and nonideal gases

• •

At first pass, gases may seem to be the most mysterious of the states of matter. Nebulous and wispy, gases easily slip through our fingers. For all their diffuse fluidity, however, gases are actually the best understood of the states. The key to understanding gases is that they all tend to behave in the same ways — physically, if not chemically. For example, gases expand to fill the entire volume of any container in which you put them. Also, gases are easily compressed to fill smaller volumes.

Before taking the AP chemistry exam, you will need to become familiar with a body of equations called the ideal gas laws and should be able to choose the equation appropriate to a particular situation and manipulate it. You will also need to have a solid understanding of the underlying theory that describes the behavior of gases, including kinetic molecular theory and the differences between real and ideal gases. This chapter will familiarize you with all of these concepts, after that you should tackle Chapter 10, which includes questions on gases similar to those which will appear on the actual exam.

Moving and Bouncing: Kinetic Molecular Theory of Gases

Imagine two billiard balls, each glued to either end of a spring. How many different kinds of motion could this contraption undergo? You could twist along the axis of the spring. You could bend the spring or stretch it. You could twirl the whole thing around, or you could throw it through the air. Diatomic molecules have just such a structure and can undergo these same kinds of motions when you supply them with energy. As collections of molecules undergo changes in energy, those collections move through the states of matter — solid, liquid, and gas. Transitions between phases of matter will be described in detail in Chapter 11, while this chapter will lay out the basics of kinetic theory.

The basics of kinetic theory

Kinetic theory, the body of ideas that explains the internal workings of gases and the phenomena caused by them, first made a name for itself when scientists attempted to explain and predict the properties of gases. Kinetic theory also explains the behavior of solids and liquids, and Chapter 11 will cover how it applies to each of those phases. Kinetic theory as applied to gases is particularly concerned with how the properties of a gas change with varying temperature and pressure. The underlying idea behind kinetic theory is the concept of kinetic energy, or the energy of motion. The particles of matter within a gas (atoms or molecules) undergo vigorous motion as a result of the kinetic energy within them.

Gas particles have a lot of kinetic energy and constantly zip about, colliding with one another or with other objects. This is a complicated picture, but scientists simplify things by imagining an ideal gas where

- ✔ Particles move *randomly*.
- ✔ The only type of motion is *translation* (moving from place to place, as opposed to twisting, vibrating, and spinning).
- ✔ When particles collide, the collisions are *elastic* (perfectly bouncy, with no loss of energy).
- ✔ Gas particles *neither attract nor repel* one another.

Gases that actually behave in this simplified way are called *ideal*.

The model of ideal gases explains why gas pressure increases with increased temperature. By heating a gas, you add kinetic energy to the particles. As a result, the particles collide with greater force upon other objects, so those objects experience greater pressure. In other words, as temperature increases, the average kinetic energy of gas particles increases and pressure increases proportionally as well.

Statistics in kinetic theory

The study of all matter, and gases in particular, is a study of statistics. Gas properties are measured in averages rather than absolutes. Note that above we spoke of temperature as being proportional to the average kinetic energy of the gas particles, not the kinetic energy of an individual gas particle. You could spend your whole life searching for a collection of gas molecules in which each individual particle is moving at the same speed as the next. Because gas particles collide and transfer energy between one another at random, neighboring gas particles generally experience different numbers of collisions and end up with different kinetic energies. If you were to measure the velocities of a whole slew of gas molecules, you would find that although they will have a distribution of varying velocities, they will be clustered around an average value. Remember that a liter of gas at STP contains 6.02×10^{23} gas particles, which is a lot to average!

Examine the two velocity distributions in Figure 9-1. Both show a similar bell-shaped curve (called a Maxwell-Boltzman distribution). Note that the probability of finding a gas particle at a certain velocity drops off precipitously as that velocity gets farther from the average velocity peak. Although both of the distributions in Figure 9-1 are Gaussians, there is one key difference between them: The figure on the left represents the velocity distribution of gas particles at a low temperature, and the figure on the right shows a similar distribution at a higher temperature.

Note the features of the low-temperature distribution:

✔ The peak of the velocity distribution is at a lower velocity.

✔ The velocity distribution shows a pronounced peak with a relatively narrow range of velocities surrounding it.

Note the features of the high-temperature distribution:

✔ The peak of the velocity distribution is at a higher velocity.

✔ The velocity distribution shows a wider, flattened curve with a large range of velocities.

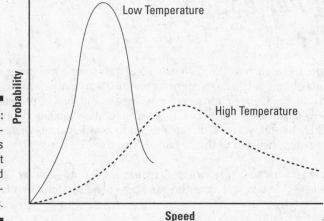

Figure 9-1:
Velocity distributions for gases at high and low temperatures.

Particle velocities translate into average kinetic energies through the equation:

$$KE = \tfrac{1}{2}\,mv^2$$

This is the kinetic energy of an individual gas molecule, where m is the mass of the molecule and v is the velocity. To calculate an average kinetic energy, we therefore need an expression for the average velocity.

$$u = \sqrt{\frac{3RT}{M}}$$

In this expression u is the average velocity of a collection of particles, and is alternately written as u_{rms} (the root-mean-square velocity), R is the ideal gas constant 8.314 J K^{-1} mol^{-1}, T is temperature, and M is the molar mass. Although we will not derive this expression here, take a minute to look at it and assure yourself that it makes sense. The expression has temperature in the numerator of the velocity calculation, which means that as temperature increases, average velocity should as well. This is consistent with kinetic theory. Molar mass, however, appears in the denominator of the equation, which means that as molar mass increases, average velocity decreases. This too makes sense; kinetic energy depends on both mass and velocity, so heavier particles move more slowly than lighter particles at the same kinetic energy. You will also often see this expression written as the following, which is simply a variation of the equation above and will yield the same result:

$$u = \sqrt{\frac{3kT}{m}}$$

Here, Boltzmann's constant k (1.38×10^{-23} J K^{-1}) replaces the ideal gas constant R, and m, the mass of an individual gas particle, replaces molar mass M.

The final equation that you may consider useful when calculating the kinetic energy of a system of molecules is the following expression for the kinetic energy *per mole* of a gas, which can be derived by plugging the first expression you were given for the average velocity of gas particles into the general equation for kinetic energy and solving for kinetic energy per mole, using this:

$$KE = \frac{3}{2}RT$$

Remember to always use $R = 8.314 \text{ J K}^{-1} \text{ mol}^{-1}$ for these kinetic energy problems. If you use the wrong units for the gas constant, you will lose points on the AP exam.

Aspiring to Gassy Perfection: Ideal Gas Behavior

Like all ideals (the ideal job, the ideal mate, and so on), ideal gases are entirely fictional. All gas particles occupy some volume. All gas particles have some degree of interparticle attraction or repulsion. No collision of gas particles is perfectly elastic. But lack of perfection is no reason to remain unemployed or lonely. Neither is it a reason to abandon kinetic molecular theory of ideal gases. In this chapter, you will be introduced to a wide variety of applications of kinetic theory, which come in the form of the so-called "gas laws."

Relationships between four factors: pressure, volume, temperature, and number of particles are the domain of the gas laws. We take a look at the gas laws in the sections that follow.

Boyle's Law

The first of these relationships to have been formulated into a law is that between pressure and volume. Robert Boyle, an Irish gentleman regarded by some as the first chemist (or "chymist," as his friends might have said), is typically given credit for noticing that gas pressure and volume have an inverse relationship:

Volume = Constant × (1/Pressure)

This statement is true when the other two factors, temperature and number of particles, are fixed. Another way to express the same idea is to say that although pressure and volume may change, they do so in such a way that their product remains constant. So, as a gas undergoes change in pressure (P) and volume (V) between two states, the following is true:

$$P_1 \times V_1 = P_2 \times V_2$$

The relationship makes good sense in light of kinetic molecular theory. At a given temperature and number of particles, more collisions will occur at smaller volumes. These increased collisions produce greater pressure. And vice versa. Boyle had some dubious ideas about alchemy, among other things, but he really struck gold with the pressure-volume relationship in gases.

Charles's Law

Lest the Irish have all the gassy fun, the French contributed a gas law of their own. History attributes this law to French chemist Jacques Charles. Charles discovered a direct, linear relationship between the volume and the temperature of a gas:

Volume = Constant × Temperature

This statement is true when the other two factors, pressure and number of particles, are fixed. Another way to express the same idea is to say that although temperature and volume may change, they do so in such a way that their ratio remains constant. So, as a gas undergoes change in temperature (T) and volume (V) between two states, the following is true:

$V_1 \div T_1 = V_2 \div T_2$

Not to be outdone by the French, another Irish scientist took Charles's observations and ran with them. William Thomson, eventually to be known as Lord Kelvin, took stock of all the data available in his mid-nineteenth century heyday and noticed a couple of things:

- First, plotting the volume of a gas versus its temperature always produced a straight line.

- Second, extending these various lines caused them all to converge at a single point, corresponding to a single temperature at zero volume. This temperature — though not directly accessible in experiments — was about –273 degrees Celsius. Kelvin took the opportunity to enshrine himself in the annals of scientific history by declaring that temperature as *absolute zero,* the lowest temperature possible.

This declaration had at least two immediate benefits. First, it happened to be correct. Second, it allowed Kelvin to create the Kelvin temperature scale, with absolute zero as the Official Zero. Using the Kelvin scale (where °C = K + 273), everything makes a whole lot more sense. For example, doubling the Kelvin temperature of a gas doubles the volume of that gas. When you work with any gas law involving temperatures, converting Celsius temperatures to Kelvin is crucial.

The Combined and Ideal Gas Laws

Boyle's and Charles's laws are convenient if you happen to find yourself in situations where only two factors change at a time. However, the universe is rarely so well behaved. What if pressure, temperature, and volume all change at the same time? Is aspirin and a nap the only solution? No. Enter the *Combined Gas Law:*

$$\frac{P_1 \times V_1}{T_1} = \frac{P_2 \times V_2}{T_2}$$

Of course, the real universe can fight back by changing another variable. In the real universe, for example, tires spring leaks. In such a situation, gas particles escape the confines of the tire. This escape decreases the number of particles, $n,$ within the tire. Cranky, tire-iron wielding motorists on the side of the road will attest that decreasing n decreases volume. This relationship is sometimes expressed as *Avogadro's Law:*

Volume = Constant × Number of particles

Combining Avogadro's Law with the Combined Gas Law produces the wonderfully comprehensive relationship:

$$\frac{P_1 \times V_1}{N_1 \times T_1} = \frac{P_2 \times V_2}{N_2 \times T_2}$$

The final word on ideal gas behavior summarizes all four variables (pressure, temperature, volume, and number of particles) in one easy-to-use equation called the *Ideal Gas Law:*

$$PV = nRT$$

Here, R is the gas constant, the one quantity of the equation that can't change. Of course, the exact identity of this constant depends on the units you are using for pressure, temperature, and volume. A very common form of the gas constant as used by chemists is $R = 0.08206 \text{L atm K}^{-1} \text{mol}^{-1}$ (which can also be expressed as $62.4 \text{ L torr mol}^{-1} \text{K}^{-1}$). Alternately, you may encounter $R = 8.314 \text{L kPa K}^{-1} \text{mol}^{-1}$.

Dalton's Law of Partial Pressures

Gases mix. They do so better than liquids and infinitely better than solids. So, what's the relationship between the total pressure of a gaseous mixture and the pressure contributions of the individual gases? Here is a satisfyingly simple answer: Each individual gas within the mixture contributes a partial pressure, and adding the partial pressures yields the total pressure. Fortunately, because the gases are ideal, the mass of the gas particles does not influence partial pressure. The number of molecules is the only factor that affects partial pressure. This relationship is summarized by *Dalton's Law of Partial Pressures,* for a mixture of a number of individual gases:

$$P_{\text{total}} = P_1 + P_2 + P_3 + \ldots + P_n$$

This relationship makes sense if you think about pressure in terms of kinetic molecular theory. Adding a gaseous sample into a volume that already contains other gases increases the number of particles in that volume. Because pressure depends on the number of particles colliding with the container walls, increasing the number of particles increases the pressure proportionally. Remember, this assumes that the individual gas molecules are behaving ideally.

You can also calculate the partial pressure of an individual gas A in a mixture of gases using the equation:

$$P_A = P_{\text{total}} \times X_A$$

Here X_A represents the mole fraction of the gas in question, or the number of moles of that gas divided by the total number of gas moles in the mixture.

Graham's Law and diffusion

"Wake up and smell the coffee." This command is usually issued in a scornful tone, but most people who have awakened to the smell of coffee remember the event fondly. The morning gift of coffee aroma is made possible by a phenomenon called *diffusion*. Diffusion is the movement of a substance from an area of higher concentration to an area of lower concentration. Diffusion occurs spontaneously, on its own. Diffusion leads to mixing, eventually producing a homogenous mixture in which the concentration of any gaseous component is equal throughout an entire volume. Of course, that state of complete diffusion is an equilibrium state; achieving equilibrium can take time.

Different gases diffuse at different rates, depending on their molar masses (see Chapter 7 for details on molar masses). The rates at which two gases diffuse can be compared using *Graham's Law.* Graham's Law also applies to *effusion,* the process in which gas molecules flow through a small hole in a container. Whether gases diffuse or effuse, they do so at a rate inversely proportional to the square root of their molar mass. In other words, more massive gas molecules diffuse and effuse more slowly than less massive gas molecules. So, for gases A and B

$$\frac{\text{Rate}_A}{\text{Rate}_B} = \frac{\sqrt{\text{molar mass}_B}}{\sqrt{\text{molar mass}_A}}$$

Getting Real: Nonideal Gas Behavior

Ideal gas equations are idealizations based on the oversimplified assumptions of ideal gas theory. Real gases, however, do not obey the ideal gas laws. At certain temperatures and pressures, they provide good approximations, but there are situations when a more sophisticated (and therefore more complicated) equation is needed.

Chemists use a rearranged version of the ideal gas law to determine whether or not they can use it.

$$\frac{PV}{RT} = n$$

For an ideal gas, *n* should equal one, so a gas for which the ideal gas law is a good approximation should have an *n* value close to one. At high pressures and at low temperatures, this becomes increasingly unlikely and a new equation is needed.

Assuming that a gas is ideal means assuming that its particles occupy no space and do not attract one another, but real gas particles defy both of those postulates: They do occupy space and they do attract one another. At high pressures, this fact becomes increasingly difficult to ignore because the space between gas particles becomes smaller and smaller and the assumption is only valid when the space between gas molecules is much greater than the size of an individual gas molecule. At high temperatures, the internal energy of a gas molecule is great enough to vastly outweigh the attraction to neighboring molecules, but at low temperatures, hence low kinetic energy, the effect of molecular attraction becomes increasingly significant. In these situations it is necessary to add in a correction for the volume of molecules and another for intermolecular attractions. These corrections yield a modified version of the ideal gas law called the van der Waals equation, which is given on the AP exam, so there is no need to memorize it!

$$\left(P + \frac{n^2 a}{V^2}\right)(V - nb) = nRT$$

This equation should make it clear why chemists prefer to use the ideal gas law to minimize necessary computations. Two new constants, called the van der Waals constants, appear in this equation. The constant *a* is a measurement of how strongly the gas molecules attract one another and has units of L^2atm mol^2. The constant *b* is a measurement of the volume occupied by a mole of gas molecules and has units of L mol. These constants vary as the identity of the gas in question varies. Larger and more complex molecules generally have larger *a* and *b* values because of their greater size and greater propensity to experience intermolecular forces, or interactions between molecules.

Chapter 10

Answering Questions on Gases

In This Chapter
▶ Remembering the important stuff
▶ Having a gas with some practice questions
▶ Going through answers with explanations

We blew through the gas laws and kinetic theory in Chapter 9, so review this condensed list of key concepts. After you think you've got these concepts under control, try tackling the AP Chemistry exam-style questions later in this chapter. Make sure to attempt the problems using only your cheat sheet as a reference in the same way you would use your list of formulas and constants on the real AP exam.

Doing a Quick Review

Here is a brief outline of the concepts presented in Chapter 9. Review it before attempting the practice questions and see Chapter 9 for the details of anything you're still a little rusty on:

✔ Kinetic theory states that as the temperature of a gas increases, so do the kinetic energies and velocities of its constituent particles.

✔ Ideal gases are those for which

- Particle motions are random.

- Particles undergo only translational motion and do not vibrate, spin, or twist.

- Collisions between particles are elastic, meaning that no kinetic energy is transformed.

- Neighboring gas particles do not exert forces (attraction or repulsion) on one another.

✔ The average kinetic energy of a gas particle is KE = ½mv^2, where KE is kinetic energy, m is mass, and v is velocity.

✔ Average particle velocity can be expressed as

$$u = \sqrt{\frac{3RT}{M}}$$

or as

$$u = \sqrt{\frac{3kT}{m}}$$

where u is the average velocity, R is the ideal gas constant, T is temperature, M is the molar mass, k is Boltzmann's constant, and m is the mass of an individual gas particle.

✔ The average kinetic energy per mole of a gas is KE = ⅜RT, where R is the ideal gas constant and T is the temperature.

✔ The Gas Laws

- **Boyle's Law**: $P_1 \times V_1 = P_2 \times V_2$, where P_1 and V_1 are the initial pressure and volume of the gas and P_2 and V_2 are the final

- **Charles' Law**: $V_1 \div T_1 = V_2 \div T_2$. Here again, V_1 and T_1 are the initial volume and temperature of the gas and V_2 and T_2 are the final.

- **The Combined Gas Law**:

$$\frac{P_1 \times V_1}{n_1 \times T_1} = \frac{P_2 \times V_2}{n_2 \times T_2}$$

where P_1, V_1, n_1 and T_1 are the initial pressure, volume, number of moles and temperature respectively and P_2, V_2, n_2 and T_2 are the final

- **The Ideal Gas Law**: $PV = nRT$, where P is pressure, V is volume, n is the number of moles, R is the ideal gas constant and T is temperature

- **Dalton's Law of Partial Pressures**: $P_{total} = P_1 + P_2 + P_3 + \ldots + P_n$, where P_1, P_2, P_3, etc. are the partial pressures of different components of the gas mixture

- **Graham's Law**:

$$\frac{\text{Rate}_A}{\text{Rate}_B} = \frac{\sqrt{\text{molar mass}_B}}{\sqrt{\text{molar mass}_A}}$$

✔ Chemists use the expression

$$\frac{PV}{RT} = n$$

to determine whether or not a real gas resembles an ideal gas closely enough to use the ideal gas laws as approximations, which correlates to n values near 1.

✔ For gases with n values far from 1, the more complicated expression

$$\left(P + \frac{n^2 a}{V^2} \right)(V - nb) = nRT$$

is used. Values of the parameters a and b must be known for the gas in question.

Testing Your Knowledge

The following practice problems can help you capture the wispy gas laws in your brain once and for all.

1. Which of the following accounts for the fact that gases generally do not behave ideally under high pressures?

(A) They begin to undergo rotational motion.

(B) Collisions between gas molecules become inelastic.

(C) Average molecular speeds increase.

(D) Intermolecular forces are greater when molecules are close together.

(E) Collisions with the walls of the container become more frequent.

2. Which of the following will increase the rms speed of a molecule of gas, assuming that all other variables are kept constant?

    I. Increasing pressure

    II. Increasing temperature

    III. Decreasing molar mass

    (A) I only

    (B) I and II

    (C) I and III

    (D) II and III

    (E) I, II, and III

3. If gas A effuses (goes through a small hole into a vacuum) at three times the rate of gas B, then

    (A) the molar mass of gas A is three times greater than the molar mass of gas B.

    (B) the molar mass of gas B is three times greater than the molar mass of gas A.

    (C) the molar mass of gas A is nine times greater than the molar mass of gas B.

    (D) the molar mass of gas B is nine times greater than the molar mass of gas A.

    (E) the molar masses of gas A and gas B are equal.

4. A sealed vessel contains 0.50 mol of neon gas, 0.20 mol hydrogen gas, and 0.3 mol oxygen gas. The total pressure of the the gas mixture is 8.0atm. The partial pressure of oxygen is

    (A) 0.24atm.

    (B) 0.3atm.

    (C) 2.4atm.

    (D) 3atm.

    (E) 4.8atm.

5. Molecules of an unknown gas diffuse at three times the rate of molecules of ammonia (NH_3) under the same conditions. What is the molar mass of the unknown gas?

6. A rigid 10.0L cylinder contains 2.20g H_2, 36.4g N_2 and 51.7g O_2.

    (a) What is the total pressure of the gas mixture at 30°C?

    (b) If the cylinder springs a leak, and the gases begin escaping, will the ratio of hydrogen remaining in the cylinder to the other two gases increase, decrease, or stay the same?

7. Samples of Br_2 and Cl_2 are placed in 1L containers at the conditions indicated in the diagram in Figure 10-1.

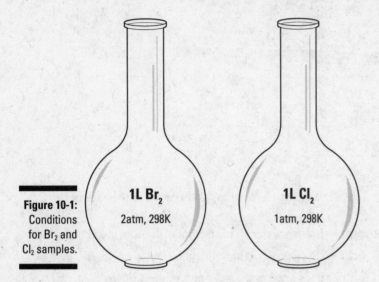

1L Br_2

2atm, 298K

1L Cl_2

1atm, 298K

(a) Which of the two gases has the greater kinetic energy per mole? Justify your answer.

(b) If the volume of the container holding the Cl_2 were decreased to 0.25L but the temperature remained the same, what would the change in pressure be? Assume that the gas behaves ideally.

8. Two flasks are connected by a stopcock as shown in Figure 10-2. The 10.0L flask contains CO_2 at a pressure of 5atm and the 5.0L flask contains CO at a pressure of 1.5atm. What will the total pressure of the system be after the stopcock is opened?

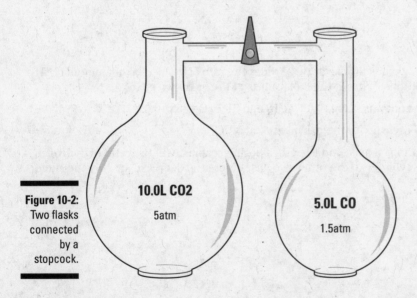

10.0L CO2

5atm

5.0L CO

1.5atm

Checking Your Work

You've done your best. Now check your work. Make sure to read the explanations thoroughly for any questions you got wrong — or for any you got right by guessing.

1. (D). High pressure gases do not behave ideally because their molecules are too close together to neglect intermolecular forces.

2. (C). Molecular speeds are proportional to temperature and inversely proportional to mass.

3. (D). Remember that Graham's Law states that gases effuse at a rate inversely proportional to the square root of their molar masses.

4. You are given enough information to find the mole fraction of each gas. Oxygen is 0.3 mol out of a total of 1.0mol of gas, so its mole fraction is 0.3. Multiply this by the total pressure to get a partial pressure of 2.4 for oxygen.

5. 153g mol. This is another Graham's Law problem. You know that the ratio of effusion rates is 3:1, so the left-hand side of Graham's Law will be 3. You can also easily calculate the molar mass of ammonia, which is 17g mol. Plug these values into the equation to get

$$\frac{3}{1} = \sqrt{\frac{x}{17}}$$

Square both sides to get rid of the square root and then multiply both sides by 17 to get 153g mol.

6. (a) 9.83atm. Begin by converting the masses you have been given to moles by dividing by their molar masses. Make sure to get in the habit of showing your work even on simple calculations such as these.

$$\frac{2.20\text{g H}_2}{1} \times \frac{1\text{mol H}_2}{2.00\text{g H}_2} = 1.10\text{mol H}_2$$

$$\frac{34.6\text{g N}_2}{1} \times \frac{1\text{mol N}_2}{28.0\text{g N}_2} = 1.23\text{mol N}_2$$

$$\frac{51.7\text{g O}_2}{1} \times \frac{1\text{mol O}_2}{32.0\text{g O}_2} = 1.62\text{mol O}_2$$

Next, add these values to yield the total number of moles of gas (3.95mol). Plug this and your known values for volume and temperature into the equation P = nRT÷V.

$$P = \frac{3.95\,mol \times 0.0821\frac{L \times atm}{mol \times K} \times 303K}{10.0L} = 9.83atm$$

(b) Decrease. Because hydrogen has the lowest molar mass, it will effuse the fastest under Graham's Law, so the ratio of hydrogen to the two other gases in the container, which escape more slowly, will decrease.

7. (a) They are the same. The average kinetic energy per mole of a sample of gas depends on temperature, and the two gases are at the same temperature.

(b) 0.25atm. This is a simple application of Charles's Law where $V_1 = 1.0L$, $V_2 = 0.25L$ and $P_1 = 1atm$. Plug these into the equation to yield

$$P_2 = \frac{V_2 \times P_1}{V_1} = \frac{0.25L \times 1atm}{1.0L} = 0.25atm$$

8. 3.8atm. Begin by finding the final pressure of each of the two gases when the entire volume is made available to them using Boyle's Law

P_2 of $CO_2 = P_1V_1 \div V_2 = 5.0atm \times 10.0L \div 15.0L = 3.3atm$

P_2 of $CO = P_1V_1 \div V_2 = 1.5atm \times 5.0L \div 15.0L = 0.5atm$

According to Dalton's Law of Partial Pressures, the final pressure of this gas mixture will simply be the sum of these two final pressures.

Chapter 11

Condensing Particles: Solids, Liquids, and Solutions

- -

In This Chapter

▶ Picturing the particles that make up liquids and solids

▶ Predicting what happens when you change temperature and pressure

▶ Seeing what happens when you mix particles to form solutions

- -

*W*hen asked, young children often report that solids, liquids, and gases are made up of different kinds of matter. This (mistaken) idea, not usually shared by older people, is understandable given the striking differences in the properties of these three states. But the AP chemistry exam leaves no room for charming misconceptions. In this chapter, we discuss how the properties of the condensed states — liquids and solids — emerge from the interactions between particles, and how these properties can change as temperature and pressure change. We also explore the kinds of interactions that occur between particles within homogenous mixtures called *solutions*. Like the properties of pure substances, the properties of solutions depend on the interactions of the particles that make them up.

Restricting Motion: Kinetic Molecular Theory of Liquids and Solids

Kinetic theory (as described in greater detail in Chapter 9) explains the properties of matter in terms of the energetic motions of its particles. Well, not *all* of the motions. Really, kinetic theory usually limits itself to a simplified version of matter, one in which infinitely small particles do nothing but zip around quickly and bump into one another or into the sides of a container. This description isn't ever literally true, but is sometimes a very good approximation of reality, especially for gases in large volumes.

Adding energy to a sample increases its temperature and increases the kinetic energy of the particles, which means that they move about more quickly and bump into things more vigorously. Removing energy from a sample (cooling it) has the opposite effect.

When atoms or molecules have less kinetic energy, or when that energy competes with other effects (like high pressure or strong attractive forces), the matter ceases to be in the diffuse, gaseous state and comes together into one of the *condensed states:*

✔ **Liquid:** The particles within a liquid are much closer together than those in a gas. As a result, applying pressure to a liquid does very little to change the volume. The particles still have an appreciable amount of kinetic energy associated with them, so they may undergo various kinds of twisting, stretching, and vibrating motions. In addition, the particles can slide past one another (translate) fairly easily, so liquids are fluid, though less fluid than gases. Fluid matter assumes the shape of anything that contains it, as shown in Figure 11-1. The particles of a liquid have *short-range order,* meaning that they tend to exhibit some degree of organization over short distances.

✔ **Solid:** The state of matter with the least amount of obvious motion is the solid. In a solid, the particles are packed together quite tightly and undergo almost no long-range translation. Therefore, solids are not fluid. Matter in the solid state still vibrates in place or undergoes other types of motion, depending on its temperature (in other words, on its kinetic energy). The particles of a solid have *long-range order,* meaning that they tend to exhibit organization over long distances. However, matter at a certain temperature must contain a specific amount of energy, regardless of its state. Temperature has no meaning without motion energy.

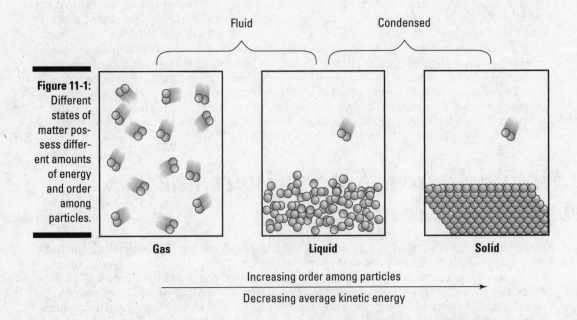

Figure 11-1: Different states of matter possess different amounts of energy and order among particles.

Fluid   Condensed

Gas   Liquid   Solid

Increasing order among particles
→
Decreasing average kinetic energy

The temperatures and pressures at which different types of matter switch between states depend on the unique properties of the atoms or molecules within that matter. But be careful! It is easy to get fooled by trying to compare different substances at different temperatures. Typically, particles that are very attracted to one another and have easily stackable shapes tend to be in condensed states (at a fixed temperature). Particles with no mutual attraction (or that have mutual repulsion) and with not-so-easily stackable shapes tend toward the gaseous state. Think of a football game between fiercely rival schools. When fans of either school sit in their own section of the stands, the crowd is orderly, sitting nicely in rows. Put rival fans into the same section of the stands, however, and they'll repel each other with great energy. But be sure you are comparing fans with the same amount of energy. Water as ice has less energy than as a liquid because it has a lower temperature. Technically if you could have a mole of water vapor, a mole of liquid water, and a mole of solid water (ice) all at the same temperature, the water molecules would all have the same kinetic energy!

Getting a Firm Grip on Solids

Solids all have less-apparent motion than their liquid or gaseous counterparts, but that doesn't mean all solids are alike. The forces between particles within solids as well as the degree of order in the packing of particles within solids vary greatly, giving each solid different properties. The sections that follow shed some light on both the forces that affect solids as well as the packing order that helps to determine a solid's properties.

Different types of solids and their properties

The properties of a solid depend heavily on the forces between the particles within it. The easiest property to compare is the melting point — that temperature at which the kinetic energy overcomes the strong forces of attraction holding the particles vibrating tightly in a solid.

✔ Several different forces determine different melting points of a solid:

- *Ionic solids* are held together by an array of very strong ionic bonds (see Chapter 5 for more about these bonds), and, therefore, tend to have high melting points — it takes a great deal of energy to pull apart the particles.

- *Molecular solids* consist of packed molecules that are less strongly attractive to each other, so molecular solids tend to have lower melting points.

- Some solids consist of many particles that are covalently bonded to one another in an extensive array. These *covalent solids* tend to be exceptionally strong due to the strength of their extensive covalent network. One example of a covalent solid is diamond. Covalent solids have *very* high melting points. Ever try to melt a diamond? (Chapter 5 has details about covalent bonds.)

✔ *Metallic solids* are made up of closely packed metal atoms. These atoms bond to one another more strongly than the particles of most molecular solids, but less strongly than the particles of covalent solids. Because metal atoms so easily give up valence electrons, the atoms within the lattice of a metallic solid seem to exist in a shared "sea" of mobile electrons. The positively charged metal nuclei are held together by their attraction to this negatively charged sea. Metallic solids can be be soft or relatively hard, are ductile and malleable, and are good conductors of heat and electricity. The orderly array of atoms in many metals can allow "sheets" of atoms to slide over one another easily, hence the ease with which metals can be made into wires (*ductility*) or beaten into thin foils (*malleability*).

Packing order in solids

The degree of order in the packing of particles within a solid can vary tremendously. How ordered the particles are within a solid determines how well defined its melting point is. If the particles are well ordered, then the whole sample tends to melt at the same temperature, but if different regions of the sample have different degrees of order, then those regions melt at different temperatures.

✔ Most solids are highly ordered, packing into neat, repeating patterns called *crystals*. The smallest packing unit, the one that repeats over and over to form the *crystalline solid*, is called the *unit cell*. Crystalline solids tend to have well-defined melting points.

The particles in crystalline solids tend to organize themselves into arrangements that make the most of the attractive forces between them. Usually, this means packing the particles as closely together as possible.

✔ *Amorphous solids* are those solids that lack an ordered packing structure. Glass and plastic are examples of amorphous solids. Amorphous solids tend to melt over a broad range of temperatures because some parts of the structures are more easily pulled apart than others.

When cooling a liquid through a phase transition into a solid, the rate of cooling can have a significant impact on the properties of the solid. The particles may need time to move into the extreme order with which they are packed together in crystalline solids. So, substances that are capable of forming crystalline solids may nevertheless freeze into amorphous solids if they are cooled rapidly. The particles may become trapped in disordered packing arrangements. Sometimes this adds considerable strength to a substance, so steels may be "hardened" by heating and sudden cooling.

A collection of different types of forces is very important in determining liquid-solid phase behavior. These forces are more important for the liquid-solid phase than in gases because liquids and solids are condensed states; the molecules within these states are in very close proximity.

In molecular solids, dipole-dipole forces, London dispersion forces, and hydrogen bonds play prominent roles (see Figure 11-2). At the same time, these forces are relatively weak compared to those that dominate in other kinds of solids. Because of the weakness of these forces, molecular solids are relatively soft and tend to have much lower melting points than other solids. In the list below, we describe how these forces work.

Forces at work in condensed states

The forces at work between the particles in a solid (or liquid) largely determine the properties of the substance. For the AP exam, you should definitely know each of the kinds of forces at work in solids and liquids, and be able to predict which forces are most important within a sample of a given compound. These forces include relatively weaker forces (dipole-dipole, London dispersion, and hydrogen bonding) and relatively stronger forces (ionic and covalent bonds).

Here are the intermolecular forces you should know:

✔ **Dipole-dipole forces** (see Figure 11-2) take place between molecules with permanent dipoles (separated regions of opposite charge). Oppositely charged parts of different molecules attract and regions with same type of charge repel. These forces tend to order the molecules.

✔ **London dispersion forces** (shown in Figure 11-2) take place when the positively charged nucleus of one atom attracts the electron cloud of another atom while the electron clouds of both atoms mutually repel one another. In other words, the two atoms induce dipoles in each other, and these *induced dipoles* (temporary dipoles created by the nearness of electron clouds) attract one another. It is more easy to redistribute the electrons of some molecules into an induced dipole than it is with others. In other words, some molecules are more *polarizable* (capable of having their electrons redistributed) than others. Polarizable molecules tend to take part more strongly in London dispersion forces.

✔ **Hydrogen bonds** are specific kinds of dipole-dipole attractions that take place between a hydrogen atom in a polar bond and a lone pair of electrons on an electronegative atom (see Chapter 5 for a refresher on electronegativity). Because it participates in a polar bond, the hydrogen has a partial positive charge, $\delta+$. Because it is electronegative, the atom that contributes the lone pair has a partial negative charge, $\delta-$. These partial charges attract. Hydrogen atoms that bond with fluorine, oxygen, and nitrogen are particularly prone to engage in hydrogen bonds. When these interactions take place *between molecules,* they significantly increase melting and freezing points. Water hydrogen bonds avidly to itself and to other molecules, as shown in Figure 11-2.

In addition to the relatively weak forces described above, ionic and covalent bonds (discussed in detail in Chapter 5) are strong forces that greatly affect the melting point of a compound:

✔ In ionic solids, **ionic bonds** (electrostatic interactions) provide a major source of attraction between particles. These types of solids tend to be hard but brittle (the ionic lattice can crack) and have very high melting points. *Lattice energy* is a measure of the strength of the interactions between ions in the lattice of an ionic solid. The larger the lattice energy, the stronger the ion-ion interactions.

✔ In covalent solids, particles are bound to each other within strong networks of **covalent bonds.** These solids are often exceptionally hard and have very high melting points.

Figure 11-2: Intermolecular interactions include (a) dipole-dipole interactions, (b) London dispersion forces, and (c) hydrogen bonds.

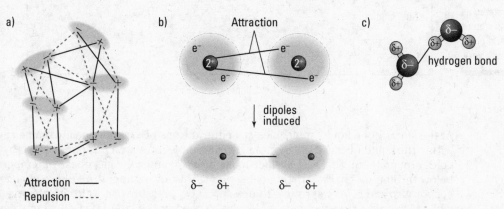

Moving Through States with Phase Diagrams

The previous sections described how microscopic interactions between particles can lead to large-scale differences in the properties of a sample, especially by causing the sample to be in different states (solid, liquid, or gas). This section describes some tools you can use to track the state of a sample as it moves through different regions of temperature, pressure, or added heat energy.

Phases and phase diagrams

Each state (solid, liquid, gas) is called a *phase.* When matter moves from one phase to another due to changes in temperature and/or pressure, that matter is said to undergo a *phase transition.* The way a particular substance moves through states as temperature and pressure vary is summarized by a *phase diagram.* Phase diagrams usually display pressure on the vertical axis and temperature on the horizontal axis. Lines drawn within the temperature-pressure field of the diagram represent the equilibrium boundaries between phases. A representative phase diagram is shown in Figure 11-3. Refer back to this figure as you read through the section.

Moving from liquid to gas is called *boiling,* and the temperature at which boiling occurs is called the *boiling point.* The normal boiling point is when this transition occurs at 1 atmosphere of pressure. Moving from solid to liquid is called *melting,* and the temperature at which melting occurs is called the *melting point.* The melting point temperature is the same as the *freezing point* temperature, but freezing implies matter moving from liquid to solid phase. The melting point a substance has at 1atm pressure is called the *normal melting point.* Just as freezing and melting points are the same, condensation points and boiling points are the same temperature. For this reason published tables are of freezing points and boiling points.

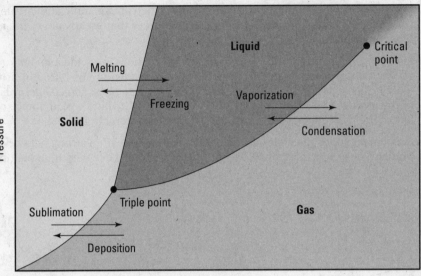

Figure 11-3:
A phase
diagram
shows how
a substance
moves
through
states as
temperature
and pres-
sure vary.

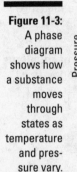

At the surface of a liquid, molecules can enter the gas phase more easily than elsewhere within the liquid because the motions of those molecules aren't as constrained by the molecules around them. So, these surface molecules can enter the gas phase at temperatures below the liquid's characteristic boiling point. This low-temperature phase change is called *evaporation* and is very sensitive to pressure. Low pressures allow for greater evaporation, while high pressures encourage molecules to re-enter the liquid phase in a process called *condensation*. The pressure of the gas over the surface of a liquid is called the *vapor pressure*. It is important not to confuse this with atmospheric pressure due to other gases in the air. For example a sample of liquid water at 20°C has a vapor pressure of 0.023atm while the atmosphere has a pressure of 1.0atm. Understandably, liquids with low boiling points tend to have high vapor pressures; particles in liquids with low boiling points are weakly attracted to each other. At the surface of a liquid, weakly interacting particles have more of a chance to escape into vapor phase, thereby increasing the vapor pressure. See how kinetic theory helps make sense of things?

In addition to having high vapor pressure and low boiling points, substances with weakly interacting molecules tend to have low surface tension and low cohesion. *Surface tension* is a measure of the amount of energy it takes to spread out a substance over a larger surface; the weaker the interactions between molecules, the easier this is to do. *Cohesion* is the tendency of the molecules of a substance to attract one another. *Adhesion* is the tendency of those molecules to bond to the molecules of another substance. Substances that are both adhesive and cohesive display *capillary action,* the ability to pull themselves through narrow tubes.

At the right combination of pressure and temperature, matter can move directly from solid to a gas or vapor. This type of phase change is called *sublimation,* and is the kind of phase change responsible for the white mist that emanates from dry ice, the common name for solid carbon dioxide. Movement in the opposite direction, from gas directly into solid phase, is called *deposition.*

For any given type of matter there is a unique combination of pressure and temperature at the nexus of all three states. This pressure-temperature combination is called the *triple point.* At the triple point, all three phases coexist. In the case of good old H_2O, going to the triple point produces ice-water vapor. Take a moment to bask in the weirdness.

Other weird phases include the following:

- ✔ ***Plasma*** is a gaslike state in which electrons pop off gaseous atoms to produce a mixture of free electrons and cations (which are atoms or molecules with positive charge). For most types of matter, achieving the plasma state requires very high temperatures, very low pressures, or both. Matter at the surface of the sun, for example, exists as plasma.

- ✔ ***Supercritical fluids*** (SCF) exist under high temperature-high pressure conditions. For a given type of matter, there is a unique combination of temperature and pressure called the *critical point.* At temperatures and pressures higher than those at this point, the phase boundary between liquid and gas disappears, and the matter exists as a kind of liquidy gas or gassy liquid. Supercritical fluids can diffuse through solids like gases do, but can also dissolve things like liquids do. SCFs are being used in some areas as extraction agents in dry cleaning.

Changing phase and temperature along heating curves

Starting from the solid phase, you can add heat energy to a sample, causing it to progress through liquid and vapor phases. If you measure the temperature of the sample as you do this, you'll find that it has a staircase pattern. This pattern is called a *heating curve,* and results from the fact that it takes energy simply to move particles from solid to liquid and from liquid to vapor (moving in the opposite direction releases energy). When a substance is at its melting point or freezing point, added heat energy goes toward disrupting attractive forces between the molecules instead of increasing the temperature (the average kinetic energy) of the molecules. Like phase diagrams, the exact shape of a heating curve varies from one substance to another. The heating curve for water is shown in Figure 11-4. Heating curves usually assume a constant rate of energy input.

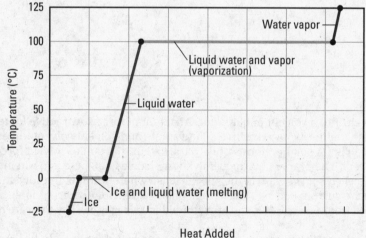

Figure 11-4: A heating curve for water.

Dissolving with Distinction: Solubility and Types of Solutions

Compounds can form mixtures. When compounds mix completely, right down to the level of individual molecules, we call the mixture a *solution*. Each type of compound in a solution is called a *component*. The component of which there is the most is usually called the *solvent*. The other components are called *solutes*. Although most people think "liquid" when they think of solutions, a solution can be a solid, liquid, or gas. The only criterion is that the components are completely intermixed. We explain what you need to know in this section. By far the most important solutions (99% on the AP exam) are those where water is the solvent. Master those before worrying about other solutions.

Forces in solvation

For gases, forming a solution is a straightforward process. Gases simply diffuse into a common volume (see Chapter 9 for more about diffusion). Things are a bit more complicated for condensed states like liquids and solids. In liquids and solids, molecules or ions are crammed so closely together that *intermolecular forces* (forces between molecules) are very important. Examples of these kinds of forces include dipole-dipole, hydrogen bonding, and London dispersion forces as discussed previously. In addition, ion-dipole forces can be important in solutions, as shown for water in Figure 11-5.

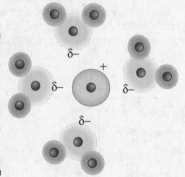

Figure 11-5:
Water molecules participate in ion-dipole interactions with a cation.

Introducing a solute into a solvent initiates a tournament of forces. Attractive forces between solute and solvent compete with attractive solute-solute and solvent-solvent forces, as depicted in Figure 11-6. A solution forms only to the extent that solute-solvent forces dominate over the others. The process in which solvent molecules compete and win in the tournament of forces is called *solvation* or, in the specific case where water is the solvent, *hydration*. Solvated solutes are surrounded by solvent molecules. When solute ions or molecules become separated from one another and surrounded in this way, we say they are *dissolved*.

Imagine that the members of a ridiculously popular band exit their hotel to be greeted by an assembled throng of fans and the media. The band members attempt to cling to each other, but are soon overwhelmed by the crowd's ceaseless, repeated attempts to get closer. Soon, each member of the band is surrounded by his own attending shell of reporters and hyperventilating fans. So it is with dissolution.

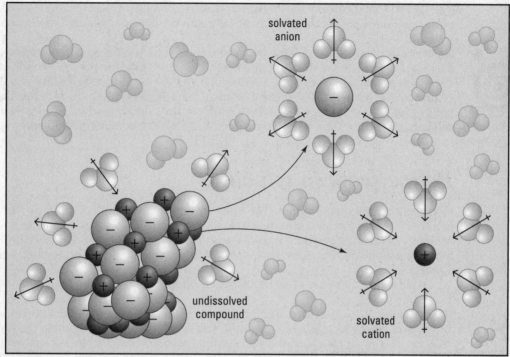

Figure 11-6:
An ionic
compound
dissolves in
water.

The tournament of forces plays out differently among different combinations of components. In mixtures where solute and solvent are strongly attracted to one another, more solute can be dissolved. One factor that always tends to favor solvation is *entropy,* a kind of disorder or "randomness" within a system. Dissolved solutes are less ordered than undissolved solutes. Beyond a certain point, however, adding more solute to a solution doesn't result in a greater amount of solvation. At this point, the solution is in dynamic equilibrium; the rate at which solute becomes solvated equals the rate at which dissolved solute *crystallizes,* or falls out of solution. A solution in this state is *saturated.* By contrast, an *unsaturated* solution is one that can accommodate more solute. A *supersaturated* solution is a temporary one in which more solute is dissolved than is necessary to make a saturated solution. A supersaturated solution is unstable; solute molecules may crash out of solution given the slightest perturbation. The situation is like that of Wile E. Coyote, who runs off a cliff and remains suspended in the air until he looks down — at which point he inevitably falls.

To dissolve or not to dissolve: Solubility

The concentration of solute (the amount of solute relative to the amount of solvent or the total amount of solution) required to make a saturated solution is the *solubility* of that solute. Solubility varies with the conditions of the solution. The same solute may have different solubility in different solvents, and at different temperatures, and so on.

When one liquid is added to another, the extent to which they intermix is called *miscibility.* Typically, liquids that have similar properties mix well — they are *miscible.* Liquids with dissimilar properties often don't mix well — they are *immiscible.* This pattern is summarized by the phrase, "like dissolves like." Alternately, you may understand lack of miscibility in terms of the Italian Salad Dressing Principle. Inspect a bottle of Italian salad dressing that has been sitting in your refrigerator. Observe the following: The dressing consists of two distinct layers, an oily layer and a watery layer. Before using, you must shake the bottle to temporarily mix the layers. Eventually, they will separate again because water is polar and oil is nonpolar.

(See Chapter 5 if the distinction between polar and nonpolar is lost on you.) Polar and non-polar liquids mix poorly, though occasionally with positive gastronomic consequences.

Similarity or difference in polarity between components is often a good predictor of solubility, regardless of whether those components are liquid, solid, or gas. This rule is often described as "like dissolves like." Why is polarity such a good predictor? Because polarity is central to the tournament of forces that underlies solubility. So, solids held together by ionic bonds (the most polar type of bond) or polar covalent bonds tend to dissolve well in polar solvents, like water. For a refresher on ionic and covalent bonding, visit Chapter 5.

Heat effects on solubility

Increasing temperature magnifies the effects of entropy on a system. Because the entropy of a solute is usually increased when it dissolves, increasing temperature usually increases solubility — for solid and liquid solutes, anyway. Another way to understand the effect of temperature on solubility is to think about heat as a reactant in the dissolution reaction:

Solid solute + solvent + heat → Dissolved solute

Heat is usually absorbed when a solute dissolves. Increasing temperature corresponds to added heat. So, by increasing temperature you supply a needed reactant in the dissolution reaction. (In those rare cases where dissolution releases heat, increasing temperature can decrease solubility.) NaCl is an example where the solubility changes very little with temperature change.

Gaseous solutes behave differently than do solid or liquid solutes with respect to temperature. Increasing the temperature tends to decrease the solubility of gas in liquid. To understand this pattern, recall the concept of vapor pressure from earlier in the chapter. Increasing temperature increases vapor pressure because added heat increases the kinetic energy of the particles in solution. With added energy, these particles stand a greater chance of breaking free from the intermolecular forces that hold them in solution. A classic, real-life example of temperature's effect on gas solubility is carbonated soda. Which goes flat (loses its dissolved carbon dioxide gas) more quickly: warm soda or cold soda?

Pressure effects on gas solubility

The comparison of gas solubility in liquids with the concept of vapor pressure highlights another important pattern: Increasing pressure increases the solubility of a gas in liquid. Just as high pressures make it more difficult for surface-dwelling liquid molecules to escape into vapor phase, high pressures inhibit the escape of gases dissolved in solvent, as highlighted by Figure 11-7. The relationship between pressure and gas solubility is summarized by *Henry's law:*

$$S_A = k \times P_A$$

where S_A is the solubility of A, P_A is the partial pressure of A in the vapor over the solution, and k is *Henry's constant*. The value of Henry's constant depends on the gas, solvent, and temperature, and is accurate for small concentrations of solute. A particularly useful form of Henry's law relates the change in solubility (S) that accompanies a change in pressure (P) between two different states:

$$S_1 / P_1 = S_2 / P_2$$

According to this relationship, tripling the pressure triples the gas solubility, for example.

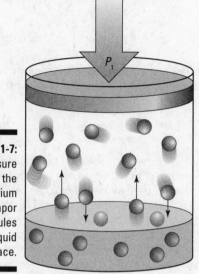
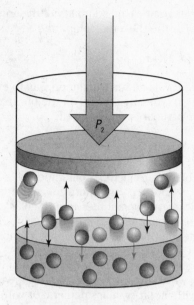

Figure 11-7:
Pressure alters the equilibrium of vapor molecules at a liquid interface.

Measuring solute concentration

It seems that different solutes dissolve to different extents in different solvents in different conditions. How can anybody keep track of all these differences? Chemists do so by measuring *concentration*. Qualitatively, a solution with a large amount of solute is said to be *concentrated*. A solution with only a small amount of solute is said to be *dilute*. As you may suspect, simply describing a solution as concentrated or dilute is usually about as useful as calling it "pretty" or naming it "Fifi." We need numbers. Two important ways to measure concentration are *molarity* and *percent solution*.

Molarity relates the amount of solute to the volume of the solution:

$$\text{Molarity}\ (M) = \frac{\text{moles solute}}{\text{liters solution}}$$

In order to calculate molarity, you may have to use conversion factors to move between units. For example, if you are given the mass of a solute in grams, use the molar mass of that solute to convert the given mass into moles. If you are given the volume of solution in cm^3 or some other unit, you'll need to convert that volume into liters.

The units of molarity are always mol L^{-1}. These units are often abbreviated as M and referred to as "molar." Thus, $0.25M$ KOH(aq) is described as "Point two-five molar potassium hydroxide" and contains 0.25 moles of KOH per liter of solution. Note that this does *not* mean that there are 0.25 moles KOH per liter of *solvent* (water, in this case) — only the final volume of the solution (solute plus solvent) is important in molarity. Why? Because often volumes just don't add up when two substances are mixed. (It is kinetic theory again!) Molar concentrations of a substance are often denoted by brackets, as in [KOH] = 0.25. Like other units, the unit of molarity can be modified by standard prefixes, as in millimolar (mM, 10^{-3} mol L^{-1}) and micromolar (μM, 10^{-6} mol L^{-1}).

One important quantity that is measured in units of molarity is the *solubility product constant*, K_{sp}. The solubility product is useful for measuring the dynamic equilibrium of ionic compounds in any given solvent. Once a saturated solution of the compound has been made, further addition of that compound has no effect on the concentration of dissolved solute. The concentrations of the component ions of the compound therefore remain constant and reflect the characteristic solubility of the compound in that solvent. The K_{sp} measures this solubility. For the the ionic compound X_AY_B,

$$K_{sp} = [X]^A \times [Y]^B$$

Percent solution is another common way to express concentration. The precise units of percent solution typically depend on the phase of each component. For solids dissolved in liquids, mass percent is usually used:

$$\text{Mass \%} = 100\% \times \frac{\text{mass solute}}{\text{total mass solution}}$$

This kind of measurement is sometimes called a mass-mass percent solution because one mass is divided by another. Very dilute concentrations (as in the concentration of a contaminant in drinking water) are sometimes expressed as a special mass percent called *parts per million (ppm)* or *parts per billion (ppb)*. In these metrics, the mass of the solute is divided by the total mass of the solution, and the resulting fraction is multiplied by 10^6 (ppm) or by 10^9 (ppb).

Clearly, it's important to pay attention to units when working with concentration. Only by observing which units are attached to a measurement can you determine whether you are working with molarity, mass percent, or with mass-mass, mass-volume, or volume-volume percent solution.

Real-life chemists in real-life labs don't make every solution from scratch. Instead, they make concentrated *stock solutions* (starting solutions) and then make *dilutions* (solutions in which solvent is added to stock solution) of those stocks as necessary for a given experiment.

To make a dilution, you simply add a small quantity of a concentrated stock solution to an amount of pure solvent. The resulting solution contains the amount of solute originally taken from the stock solution, but disperses that solute throughout a greater volume. So, the final concentration is lower; the final solution is less concentrated and more dilute.

But how do you know how much of the stock solution to use, and how much of the pure solvent to use? It depends on the concentration of the stock and on the concentration and volume of the final solution you want. You can answer these kinds of pressing questions by using the dilution equation, which relates concentration (C) and volume (V) between initial and final states:

$$C_1 \times V_1 = C_2 \times V_2$$

This equation can be used with any units of concentration, provided the same units are used throughout the calculation. Because molarity is such a common way to express concentration, the dilution equation is sometimes expressed in the following way, where M_1 and M_2 refer to the initial and final molarity, respectively:

$$M_1 \times V_1 = M_2 \times V_2$$

Dissolving with Perfection: Ideal Solutions and Colligative Properties

If you've read the rest of this chapter, you may consider yourself a recently minted expert in solubility and molarity, ready to write off solutions as another chemistry topic mastered. Don't. You, as a chemist worth your salt, must be aware of another piece to the puzzle: *colligative properties*. Colligative properties are the properties of a solution compared to a pure solvent that change as a function of the number of solute particles in solution, regardless of what kind of particles. The presence of extra particles in a formerly pure solvent has a significant impact on some of that solvent's characteristic properties, such as vapor pressure, freezing point, and boiling point.

Understanding ideal solutions

Understanding how solute particles affect the properties of a solution requires you to know first whether you're dealing with an "ideal solution." An *ideal solution* is one in which properties change proportionally (that is, in a linear way) with the amount of added solute. Thankfully, only ideal solutions are on the AP test!

Two kinds of solutions tend to approach ideal behavior:

- Very dilute solutions
- Solutions in which solute-solvent interactions are about the same strength as solvent-solvent and solute-solute interactions

Ideal solutions obey Raoult's law. Raoult's law states that the vapor pressure over the surface of an ideal solution should be the sum of the vapor pressures of the pure components multiplied by their mole fraction in the solution. In other words, each solution component should contribute exactly its fair share to the total vapor pressure, no more, no less. For a two-component solution

$$P_{total} = P_A \times X_A + P_B \times X_B$$

where P_{total} is the total vapor pressure over the solution, P_A and P_B are the vapor pressures of pure samples of components A and B, and X_A and X_B are the mole fractions of components A and B in the solution.

Mole fraction is the ratio of the number of moles of one component in a solution to the number of moles of all the components in the solution.

In general, the mole fractions of a two-component solution are expressed as

$$X_A = \frac{n_A}{n_A + n_B} \ and \ X_B = \frac{n_B}{n_A + n_B}$$

where n_A is the number of moles of component A (like a solute) and n_B is the number of moles of component B (like a solvent).

Raoult's law makes a prediction: If you add a nonvolatile solute (one that contributes no vapor pressure of its own) to a solvent, the vapor pressure of the resulting solution should be lower than the vapor pressure of the pure solvent. If you've got a solution that seems to obey Raoult's law, then you've got a solution for which you can make useful predictions about colligative properties.

Using molality to predict colligative properties

To correctly account for the effects of solute particles on some colligative properties, you need a new way to measure solution concentration: *molality*. No, that's not a typo. Molality is different from molarity.

Like the difference in their names, the difference between molarity and molality is subtle. Whereas molarity measures the moles of solute per liter of solution, molality measures the *moles of solute particles per kilogram of solvent:*

$$\text{Molality } (m) = \frac{\text{moles solute particles}}{\text{kilogram solvent}}$$

Notice that the numerator of the fraction for calculating molality includes "solute particles" and not just "solute." What's the difference? When one mole of the ionic compound NaCl dissolves into one liter of aqueous solution, it produces 1 molar sodium chloride, $1M$ NaCl(aq). But when NaCl dissolves, it becomes one mole of Na$^+$ cation and one mole of Cl$^-$ anion — two moles of solute particles. So, when one mole of NaCl dissolves into one kilogram of water, it produces 2 molal sodium chloride, $2m$ NaCl(aq).

Calculating molality is no more or less difficult than calculating molarity, so you may be asking yourself, "Why all the fuss?" Is it even worth adding another quantity and another variable to memorize? Yes! Although molarity is exceptionally convenient for calculating concentrations and working out how to make dilutions in the most efficient way, molality is useful for predicting important colligative properties, including the boiling point of a solution. When solute particles are added to a solvent, the boiling point of the resulting solution tends to increase relative to the boiling point of the pure solvent. This phenomenon is called *boiling point elevation.* The more solute you add, the greater you elevate the boiling point.

 Boiling point elevation is directly proportional to the molality of a solution, but chemists have found that some solvents are more susceptible to this change than others. The formula for the change in the boiling point (T_b) of a solution therefore contains a proportionality constant, K_b (not to be confused with an equilibrium constant!). The K_b is determined by experiment and in practice you usually look it up in a table such as Table 11-1. The formula for the boiling point elevation is

$$\Delta T_b = K_b \times m$$

Note the use of the Greek letter delta (Δ) in the formula to indicate that you are calculating a *change* in boiling point, not the boiling point itself. You'll need to add this number to the boiling point of the pure solvent to get the boiling point of the solution. The units of K_b are given in degrees Celsius per molality ($^\circ C \ m^{-1}$).

Table 11-1	Common K_b Values	
Solvent	*K_b ($^\circ C \ m^{-1}$)*	*T_b of pure solvent ($^\circ C$)*
Acetic acid	3.07	118.1
Benzene	2.53	80.1
Camphor	5.95	204.0
Carbon tetrachloride	4.95	76.7
Cyclohexane	2.79	80.7
Ethanol	1.19	78.4
Phenol	3.56	181.7
Water	0.512	100.0

Boiling point elevations are a result of the attraction between solvent and solute particles in a solution. Adding solute particles increases these intermolecular attractions because there are more particles around to attract one another. Solvent particles must therefore achieve a greater kinetic energy to overcome this extra attractive force and boil off the surface. Greater kinetic energy means a higher temperature, and therefore a higher boiling point. An alternative explanation is that there are simply fewer solvent molecules on the surface to escape as some surface locations are occupied by solute particles.

The second of the important colligative properties you can calculate by using molality is the freezing point (T_f) of a solution. When solute particles are added to a solvent, the freezing point of the solution tends to decrease relative to that of the pure solvent. This phenomenon is called *freezing point depression*. The more solute you add, the more you decrease the freezing point. This is the reason, for example, that you sprinkle salt on icy sidewalks. The salt mixes with the ice and lowers its freezing point. If this new freezing point is lower than the outside temperature, the ice melts, eliminating the spectacular wipeouts so common on salt-free sidewalks. The colder it is outside, the more salt is needed to melt the ice and lower the freezing point to below the ambient temperature.

Like boiling point elevation, freezing point depression is directly proportional to the molality of the solution. So, the formula for freezing point depression contains a constant of proportionality, K_f, that depends on the solvent in question. The formula for freezing point depression is

$$T_f = K_f \times m$$

To calculate the new freezing point of a compound, you must *subtract* the change in freezing point from the freezing point of the pure solvent. Table 11-2 lists several common K_f values.

Table 11-2	Common K_f Values	
Solvent	*K_f (°C m^{-1})*	*T_f of pure solvent (°C)*
Acetic acid	3.90	16.6
Benzene	5.12	5.5
Camphor	37.7	179
Carbon tetrachloride	30.0	−22.3
Cyclohexane	20.2	6.4
Ethanol	1.99	−114.6
Phenol	7.40	41
Water	1.86	0.0

Freezing point depressions are the result of solute particles interrupting the crystalline order of a frozen solid. In order to reach a solid, frozen state, the solution must achieve an even lower average kinetic energy. Lower kinetic energy means lower temperature, and therefore a lower freezing point.

In summary, adding solute particles to a solvent increases the stability of the liquid phase. Boiling point elevation and freezing point depression mean that greater changes in energy are required to shift the solution out of liquid phase into solid or vapor phase. This effect can be seen in a phase diagram that overlays the behavior of a solution with that of the pure solvent, as shown in Figure 11-8.

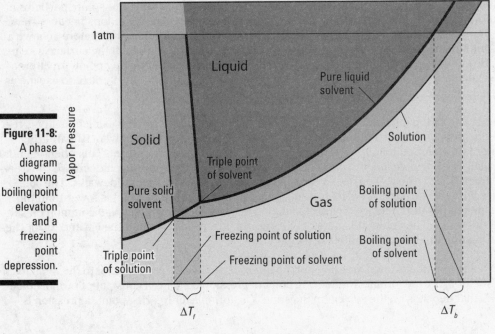

Figure 11-8:
A phase diagram showing boiling point elevation and a freezing point depression.

By carefully measuring boiling point elevations and freezing point depressions, you can determine the molar mass of a mystery compound that is added to a known quantity of solvent. To pull off this trick you must know the total mass of the pile of mystery solute that has been dissolved, the mass of solvent into which the compound was dissolved, and either the change in the freezing or boiling point or the new freezing or boiling point itself. From this information you then follow this set of simple steps to determine the molecular mass:

1. **If you know the boiling point of the solution, calculate the ΔT_b by subtracting the boiling point of the pure solvent from that value. If you know the freezing point of the solution, subtract the freezing point of the pure solvent to that value to get the ΔT_f.**

2. **Look up the K_b or K_f of the solvent (in a place like Table 11-1 or 11-2).**

3. **Solve for the molality of the solution using the equation for ΔT_b or ΔT_f.**

4. **Calculate the number of moles of solute in the solution by multiplying the molality calculated in Step 3 by the mass of solvent, in kilograms.**

5. **Divide the mass of the pile of mystery solute that has been dissolved by the number of moles calculated in Step 4. The result of this calculation is the molar mass (number of grams per mole) of the mystery compound. From this value, you can often make an educated guess about the identity of the compound.**

Osmotic pressure is another property of a solution that depends on the number of solute particles. To understand osmotic pressure, it helps to understand *osmosis*, the movement of solvent molecules through a semipermeable barrier (one which lets some things pass, but not others) from areas of low solute concentration to areas of high solute concentration. Osmosis is a result of the more general process of *diffusion*, the net movement of a substance from where it is more concentrated to where it is less concentrated. In solutions with higher solute concentration, solvent is less concentrated. In solutions with lower solute concentration, solvent is more concentrated. So, given the opportunity, solvent diffuses toward solutions with higher solute concentration.

If solvent and a solution made with that solvent are separated by a semipermeable membrane in a setup like the one shown in Figure 11-9, solvent molecules will move by osmosis into the solute-containing chamber. This movement of solute will alter the surface height within each chamber until the difference in height causes enough pressure to prevent a further net movement of solvent into the solute-containing chamber. This pressure difference is equal to the osmotic pressure, Π, of the solution. Osmotic pressure depends on the moles of solute particles per unit volume of solution; that is, in contrast to boiling points and freezing points, osmotic pressure depends on molarity, not molality:

$$\pi = \left(\frac{n_{solute}}{V_{solution}} \right) RT = MRT$$

where n_{solute} is the moles of solute particles, $V_{solution}$ is the volume of solution, M is molarity of particles, R is the gas constant and T is temperature.

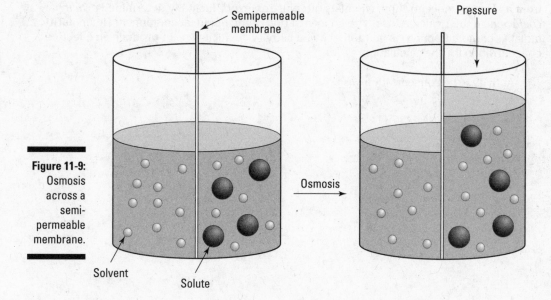

Figure 11-9:
Osmosis across a semipermeable membrane.

Osmosis and osmotic pressure are important in biology because cell membranes are semipermeable, allowing transport of solvent water molecules while restricting transport of many solutes. So, if a cell finds itself in a *hypertonic* environment (one with higher solute concentration than that of the cell), osmosis may cause the cell to shrivel as water diffuses outward. If the cell finds itself in a *hypotonic* environment (one with lower solute concentration than that of the cell), osmosis may cause the cell to swell (or explode!) as water diffuses inward. So, even if you yawn at osmotic pressure, the cells that compose you consider it a life-or-death kind of thing.

Dissolving with Reality: Nonideal Solutions

Many solutions come very close to ideal behavior. But some solutions deviate pretty significantly from ideal behavior. And for sensitive applications, sometimes even very close isn't close enough. Situations like these force us to deal with *nonideal solutions,* ones that don't obey Raoult's law or Henry's law, and whose properties aren't proportional to the amount of added solute.

Nonideal solutions occur when solute concentrations are very high and/or when solute-solvent interactions are significantly more attractive or repulsive than solute-solute interactions and solvent-solvent interactions.

✔ When solute-solvent interactions are especially attractive, the solute is effectively more dissolved (that is, more intermixed) than an ideal solute. The vapor pressure of the solvent is lower than predicted by Raoult's law. The partial pressure of a dilute solute is higher than that predicted by Henry's law.

✔ When solute-solvent interactions are especially repulsive, the solute is effectively less intermixed than an ideal solute. The vapor pressure of the solution is higher than predicted by Raoult's law. The partial pressure of a dilute solute is lower than that predicted by Henry's law.

To account for nonideal behavior, chemists use the concept of *activity,* the effective concentration of a component in solution. When the activity of a component differs significantly from its formal concentration, chemists often use an experimentally determined *activity coefficient* in their calculations. The concentration of the nonideal component (its molarity, molality or mole fraction) is multiplied by an activity coefficient, γ, to produce an effective concentration for the calculation:

Activity = $\gamma \times$ Concentration

Chapter 12

Answering Questions on Solids, Liquids, and Solutions

. . . .

In This Chapter

▶ Summarizing what you need to remember

▶ Testing your knowledge with some practice questions

▶ Finding explanations for the answers

. . . .

Chapter 11 describes the different kinds of interactions between particles within different ent phases of matter and how changes in temperature and pressure can cause matter to move between phases. Chapter 11 also explores solutions — what they are, how to measure their concentration, and how their behavior changes with added solute. Just to be certain you remember the key points of Chapter 11, we've included those important concepts for you to review in this chapter before you take a stab at answering questions about solids, liquids, and solutions.

Getting a Grip on What Not to Forget

Here is a collection of the most important points to remember for the AP exam about how particles behave in different states of matter, how samples move between states, and how solute particles interact with solvent within solutions.

Kinetic theory

Just as kinetic theory can explain gas behavior, the following list shows how kinetic theory can also help explain the properties and behavior of the *condensed states* — liquids and solids, in case you haven't read Chapter 11:

✔ Particles within liquids are close together, but can slide past one another (that is, *translate*), so that liquids are fluid, meaning liquids have *short-range order*.

✔ Particles within a solid cannot translate past one another, but vibrate about an average position. So solids have *long-range order*.

✔ In condensed states, interparticle attractions and repulsions are critical in determining the properties of the substance.

Solid states

Solid states include several categories with distinct characteristics:

- ✔ The particles within *ionic solids* are strongly attracted to one another, and these solids have high melting points. *Lattice energy* measures the strength of the ion-ion interactions within these solids.

- ✔ *Molecular solids* have weaker attractions between particles and typically have lower melting points.

- ✔ *Covalent (network) solids* are bound together by multiple covalent bonds and tend to be exceptionally strong.

- ✔ *Metallic solids* consist of a lattice of positively charged atoms within a sea of mobile electrons coming from valence shells, which accounts for the high electrical conductivity of metals.

- ✔ Particles within *crystalline solids* pack into highly ordered arrangements of a repeating *unit cell* and tend to have well-defined melting temperatures. Particles within *amorphous solids* exhibit far less long-range order and tend to melt over broad temperature ranges.

Condensed states

Within condensed states, a collection of "weak" forces (that is generally, but not always, weaker than covalent or ionic bonds) can play important roles:

- ✔ *Hydrogen bonds* can occur when two electronegative atoms (like oxygen, nitrogen, or fluorine) essentially "share" an electropositive hydrogen atom. Formally, the hydrogen is covalently bound to one of the electronegative atoms.

- ✔ *Dipole-dipole* forces can occur between the partially charged parts of polar molecules. Opposite partial charges attract and same-sign partial charges repel. Dipoles can also interact with ions according the the same basic concept.

- ✔ *London dispersion forces* are attractive forces that are generally weaker than hydrogen bonds, dipole-dipole, and ion-dipole forces at long distances. London forces occur between *induced dipoles,* partial charges that flicker into and out of existence as electron clouds interact with one another.

Phase diagrams

Phase diagrams reveal the effects of temperature and pressure on the *phase* (physical state) of a substance. The following list describes how to interpret the diagrams:

- ✔ A field of the phase diagram represents boundaries between phases. Crossing a line indicates a *phase change*, such as melting, freezing, boiling, condensing, subliming, or desubliming (depositing).

- ✔ The *triple point* represents a substance-specific combination of temperature and pressure at which the solid, liquid, and gas phases of a substance coexist.

- ✔ At very high temperatures and/or very low pressures, gases can move into *plasma phase,* in which ionized particles intermingle with free electrons.

- ✔ At temperatures and pressures above a substance-specific condition called the *critical point,* the distinction between liquid and gas phase disappears. A substance in this condition is called a *supercritical fluid.*

✔ *Heating curves* show how the temperature of a substance varies as heat is added to it. These curves tend to exhibit a staircase-pattern in which temperature increases as heat is added at a constant rate, and then levels off during phase changes because heat must be added simply to shift the substance from one phase to another.

Solutions

Solutions are mixtures in which different components are completely and evenly mixed down to the level of individual molecules, ions, or atoms.

✔ The major component of a solution is the *solvent*. Minor components are *solutes*.

✔ Substances *dissolve* if and when solvent-solute attractions dominate over solvent-solvent and solute-solute attractions.

✔ *Saturated* solutions are those that hold the maximum amount of a dissolved solute. *Unsaturated* solutions can accommodate further solute additions.

✔ The *solubility* of a solute is the concentration of the substance required to produce a saturated solution. So, substances with large solubility can be dissolved to a large degree, while smaller amounts of substances with low solubility will dissolve.

✔ Solubility can vary in the following ways, depending on temperature, pressure, and the physical/chemical properties (especially *polarity*) of the solute and solvent:

  • Increasing temperature increases the solubility of most (but not all) solid solutes in liquid solutions.

  • Increasing temperature decreases the solubility of gases in liquid solutions.

  • Increasing pressure increases the solubility of gases in liquid solutions, as expressed by *Henry's law*: $S_1 / P_1 = S_2 / P_2$.

  • Polar solutes dissolve best in polar solvents. Nonpolar solutes dissolve best in nonpolar solvents. In other words, "like dissolves like."

Ideal solutions have some important points to remember:

✔ Ideal solutions obey Raoult's law. This law states that in the vapor over a liquid solution, each solution component should have a partial pressure in proportion to its mole fraction within the solution. Raoult's law implies that adding a solute to a solvent should reduce the partial pressure of the solvent within the vapor over the solution.

✔ In *ideal solutions* (those which are dilute and in which all intermolecular forces are about the same strength), adding solute particles has predictable effects on *colligative properties* like boiling point, freezing point, and osmotic pressure. Most AP questions involve only nonvolatile solutes — those that have essentially no vapor pressure of their own to add to that of the solvent. Thus only the solvent properties are affected.

  • Adding solute particles raises the boiling point of a solvent in a phenomenon called *boiling point elevation*: $\Delta T_b = K_b \times m$.

  • Adding solute particles lowers the freezing point of a solvent in a phenomenon called *freezing point depression*: $\Delta T_f = K_f \times m$.

  • Adding solute particles increases the *osmotic pressure*, π, of a solution.

Many real-life solutions are nonideal solutions, especially at high concentrations or when one kind of intermolecular force dominates over others in solution. (Fortunately, AP examinations never ask about these.) We give you examples of some nonideal solution effects in the following list:

✔ Strong solute-solvent interactions lead to vapor pressures lower than those predicted by Raoult's law.

✔ Weak solute-solvent interactions lead to vapor pressures higher than those predicted by Raoult's law.

✔ The activity coefficient, γ, is used to correct for nonideal solution effects on solute concentration: Activity = $\gamma \times$ Concentration.

More concentrated solutions can be diluted with solvent to yield less-concentrated solutions, as expressed by the *dilution equation*: $C_1 \times V_1 = C_2 \times V_2$.

Reporting the concentration of a solute

Different methods for reporting the concentration of a solute include molarity, molality, percent solution, mole fraction, and partial pressure. The following list reminds you how those methods are expressed:

✔ **Molarity:** M = moles solute / liters solution

✔ **Molality:** m = moles solute particles / kilograms solution

✔ **Percent solution** can refer to mass percent, volume percent, or mass/volume percent:

- Mass % = 100% $\times$ (mass solute / total mass solution)
- Volume % = 100% $\times$ (volume solute / total volume solution)
- Mass/Volume % = 100% $\times$ (grams solute / 100 milliliters solution)

✔ **Mole fraction:** χ = moles component / total moles of all components

✔ **Partial pressure** of A, P_A = mole fraction of A $\times$ total pressure

Testing Your Knowledge

No matter how we summarize it, clearly you need to know a lot about solids, liquids, gases, and solutions. Test your mastery of the material in Chapter 11 by giving these questions a try.

Questions 1 through 5 refer to the phase diagram below.

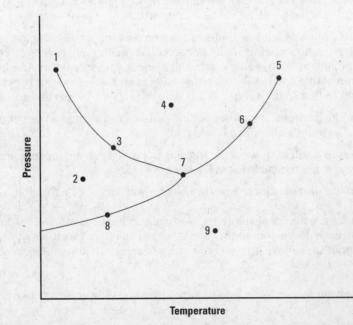

1. Which points correspond to a melting/freezing equilibrium?

    (A) 1 and 5

    (B) 1 and 3

    (C) 2 and 4

    (D) 6 and 7

    (E) 7 and 8

2. Which point(s) correspond(s) to homogenous phase(s)?

    I. 1

    II. 4

    III. 7

    (A) I only

    (B) II only

    (C) III only

    (D) I and II only

    (E) I, II and III

3. Which point corresponds to a sublimation/deposition equilibrium?

    (A) 1

    (B) 3

    (C) 5

    (D) 6

    (E) 8

4. Which point corresponds to the critical point?

    (A) 1

    (B) 2

    (C) 5

    (D) 7

    (E) 9

5. Which point corresponds to the triple point?

    (A) 1

    (B) 2

    (C) 5

    (D) 7

    (E) 9

6. How much water must be added to 150mL of 0.500M KCl to make 0.150M KCl?

(A) 45mL

(B) 150.mL

(C) 350.mL

(D) 500.mL

(E) 650.mL

7. The K_{sp} of lead (II) chloride is 1.7×10^{-5}. What is the molar concentration of a chloride ion in 1.0L of a saturated $PbCl_2$ solution?

(A) 5.7×10^{-6}

(B) 1.1×10^{-5}

(C) 1.6×10^{-2}

(D) 2.6×10^{-2}

(E) 3.2×10^{-2}

8. You can prepare 0.75 molal NaCl by dissolving 15g NaCl in what amount of water?

(A) 0.40kg

(B) 0.34kg

(C) 0.27kg

(D) 0.20kg

(E) 0.26kg

9. If 1.6×10^{-3} moles of an ideal gas are dissolved in 2L of a saturated aqueous solution at 1atm pressure, what will be the molar concentration of the gas under 2.5atm pressure?

(A) 1.6×10^{-3}

(B) 0.8×10^{-3}

(C) 4.0×10^{-3}

(D) 2.0×10^{-3}

(E) 3.2×10^{-3}

Questions 10 through 14 refer to the following set of choices:

(A) Particles vibrate about average positions.

(B) Particles are ordered and occur within a sea of mobile electrons.

(C) Particles are ionized, disordered and highly energetic.

(D) Particles diffuse rapidly and can dissolve many solutes.

(E) Particles do not translate but lack long-range order.

10. Supercritical fluid

11. Metallic solid

12. Amorphous solid

13. Plasma

14. Crystalline solid

15. One mole of glucose, $C_6H_{12}O_6$, dissolved in 1.00kg water results in a solution that boils at 100.52 °C and freezes at −1.86 °C. If 300. grams of ribose, $C_5H_{10}O_5$, are dissolved in 4.00kg water, what will be the boiling and freezing points of the resulting solution?

    (A) 100.26 °C, −0.93 °C

    (B) 100.13 °C, −0.47 °C

    (C) 100.52 °C, −1.86 °C

    (D) 101.04 °C, −3.72 °C

    (E) 101.56 °C, −5.58 °C

16. Which solution has the lowest boiling point?

    (A) 0.5M $FeCl_3$

    (B) 1.0M $MgCl_2$

    (C) 0.5M $CaCl_2$

    (D) 1.0M $NaNO_3$

    (E) 1.5M NaCl

17. Ethanol, CH_3CH_2OH, engages in which of the following types of intermolecular interactions?

    I. Hydrogen bonding

    II. Dipole-dipole interactions

    III. London forces

    (A) I only

    (B) II only

    (C) III only

    (D) I and II only

    (E) I, II, and III

Questions 18 through 22 refer to the following diagram and the accompanying set of choices with regard to the heat being transferred:

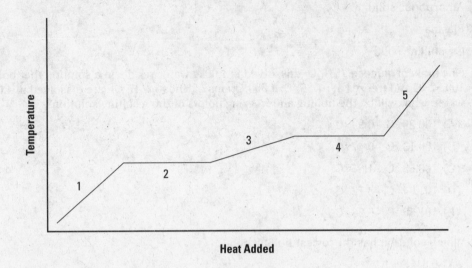

Heat Added

(A) Liquid phase

(B) Increase in average kinetic energy of particles

(C) Decrease in average kinetic energy of particles

(D) Heat of fusion

(E) Heat of vaporization

18. Segment 2 corresponds to this phase or phase change.

19. Moving left to right on segment 1 represents this.

20. Moving right to left on segment 5 represents this.

21. Segment 4 corresponds to this phase or phase change.

22. Segment 3 corresponds to this phase or phase change..

Checking Your Work

Dissolve any lingering uncertainties you have about your answers. Check them here.

The following diagram refers to items explained in the answers to questions 1 through 5:

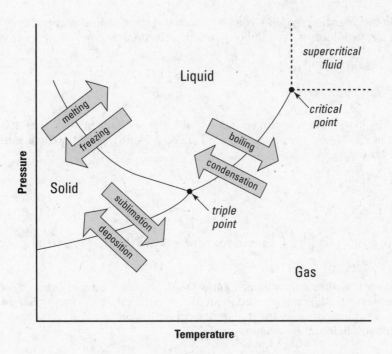

1. (B). The melting/freezing equilibrium occurs at the boundary line between solid and liquid phases. Although different substances have differently shaped phase diagrams, it is typically the case that low temperature–high pressure conditions favor solid phase; high temperature–low pressure conditions favor gas phase; and liquid phases occur at intermediate temperatures and pressures.

2. (B). Homogenous phases are those that are uniform throughout and not in equilibrium between multiple phases. So, on a phase diagram, homogenous phases are found in the open spaces between phase boundary lines.

3. (E). Sublimation is direct movement from solid to gas phase, and deposition is the reverse of that process. So, a sublimation/deposition equilibrium occurs at the boundary line between solid and gas phases.

4. (C). The critical point for a substance is the unque combination of pressure and temperature beyond which liquid and gas phases cease to be distinct. Any combination of temperature and pressure that is simultaneously above both the critical temperature and critical pressure causes the substance to become a supercritical fluid.

5. (D). The triple point for a substance is a unique combination of temperature and pressure at which all three major phases (solid, liquid, and gas) are in simultaneous equilibrium. The triple point occurs at the convergence of the three phase boundary lines.

6. (C). Attack this problem by using the dilution equation: $C_1 \times V_1 = C_2 \times V_2$.

$$(0.500M) \times (150\text{mL}) = (0.150M) \times V_2$$

Solving for V_2 gives you 500mL. So, the final solution should have a volume of 500mL. Because the initial solution had a volume of 150mL, you must add 350mL of water to achieve the desired final volume.

7. (E). Solving this question requires you to understand the dissolution reaction of lead (II) chloride and the definition of the solubility product constant, K_{sp}.

$$PbCl_2(s) + H_2O(l) \leftrightarrow Pb^{2+}(aq) + 2Cl^-(aq)$$

$$K_{sp} = 1.7 \times 10^{-5} = [Pb^{2+}][Cl^-]^2$$

Notice from the dissolution reaction equation that for every mole of $PbCl_2$ that dissolves, you get one mole of Pb^{2+} cation and *two* moles of Cl^- anions. Using this fact, we can substitute into the equation for the K_{sp}:

$$1.7 \times 10^{-5} = [x][2x]^2$$

$$1.7 \times 10^{-5} = 4x^3$$

Solving for x, you get 1.6×10^{-2}. However, the concentration of Cl^- is 2x, so the answer is 3.2×10^{-2}.

8. (B). To answer this question, you must know the definition of molality, m = (moles solute) / (kilograms solvent). Next, you must find the moles of NaCl solute by converting from the given mass of 15 grams. Because the molar mass of NaCl is 58.45 g mol^{-1}, you have 0.257mol NaCl. Next, substitute into the definition of molality:

$$0.75m = (0.257mol\ NaCl) / (X\ kilograms\ water)$$

Solving for X, you get 0.34kg water.

9. (D). This question requires you to apply a result of Henry's law, stating that the solubility of an ideal gas in a liquid solvent is directly proportional to the pressure: $S_1 / P_1 = S_2 / P_2$.

Before substituting into the proportion, however, you should calculate the initial molar solubility:

$$Molarity = (1.6 \times 10^{-3}mol) / (2L) = 0.80 \times 10^{-3}\ M$$

Substituting into the proportion, you get $(0.80 \times 10^{-3}\ M) / (1atm) = S_2 / (2.5atm)$

Solving for S_2 gives you $2.0 \times 10^{-3}\ M$.

10. (D). Supercritical fluids occur in high temperature–high pressure conditions beyond the thermodynamic critical point of a substance. Under these conditions, the substance has properties of both a gas (high diffusivity) and a liquid (ability to dissolve solute).

11. (B). The atoms of a metallic solid are packed within an ordered lattice, but are so electropositive that electrons move freely about the lattice, which accounts for the high electrical conductivity of metals.

12. (E). In an amorphous solid, particles — often polymers — are packed tightly enough that the substance isn't fluid, but lack the extreme order and structural regularity of a crystalline solid.

13. (C). Provided enough energy, the particles of a gas (often, a superheated gas) can ionize to create plasma, a state of matter that is diffuse and fluid like gas, but has unusual properties due to its ionic nature.

14. (A). Within a crystalline solid, particles are tightly packed within highly ordered, regularly repeating structural units called unit cells. So little translational freedom is available to the particles within a crystalline solid that all they can usually do is to vibrate in place. Such is the price of perfection.

15. (A). This question tests your knowledge of two colligative properties, boiling point elevation (ΔT_b) and freezing point depression, (ΔT_f). You are required to know that colligative properties, which apply to ideal solutions, depend solely on the number of solute particles in solution, and not on the physical or chemical properties of those particles. The number of solute particles is measured by molality, m. The effects of added solute on boiling and freezing points depend on the solvent. By giving you data on the boiling point and freezing point of a $1M$ glucose solution, the question essentuially gives you the boiling point and freezing point proportionality constants, $K_b = 0.52\ ^\circ C\ m^{-1}$ and $K_f = -1.86\ ^\circ C\ m^{-1}$. Next, you must determine the molality of the ribose solution. To do so, you must calculate the molar mass of ribose, 150 g mol^{-1}. Because 300g ribose are dissolved, there are 2.0 moles ribose in the solution.

    molality = (2.0mol ribose) / (4.0kg solvent) = 0.50*m*

Next, substitute into the equations for boiling point elevation and freezing point depression:

    $\Delta T_b = (0.52\ ^\circ C\ m^{-1}) \times (0.50m) = 0.26\ ^\circ C$

    $\Delta T_f = (^\circ 1.86\ ^\circ C\ m^{-1}) \times (0.50m) = -0.93\ ^\circ C$

Adding these quantites to the standard boiling point and freezing point of water gives $T_b = 100.26\ ^\circ C$ and $T_f = -0.93\ ^\circ C$.

16. (D). This question requires to understand that boiling point, as a colligative property, depends on the number of moles of solute *particles* in solution, not on the number of moles of solute compound. For example, one mole of dissolved KCl contributes 2 moles of particles (one mole K$^+$ and one mole Cl$^-$). Taking this factor into account, it is clear that in $1.0M$ NaNO$_3$, the NaNO$_3$ contributes the fewest particles to solution (one mole Na$^+$ and one mole of the polyatomic ion NO$_3^-$), and therefore elevates the boiling point the least.

17. (E). With its hydroxyl functional group, ethanol is quite capable of forming hydrogen bonds, which require a hydrogen covalently bonded to an electronegative atom (like oxygen) and also require a second electronegative atom (like the oxygen of a second ethanol molecule). Hydrogen bonds are themselves a kind of dipole-dipole interaction, employing the strong dipoles set up by the electronegative atoms. Finally, all condensed phase molecules engage in London dispersion interactions. So, ethanol participates in all three kinds of intermolecular interactions.

18. (D). The heat added within this segment clearly does not go toward rasing the temperature of the substance. Instead, the added heat goes toward rearranging particles at a given temperature from solid phase into liquid phase. We know this as it is the first of the phase changes to be shown (moving left to right on the diagram).The heat required to accomplish this phase change is called the heat of fusion, ΔH_{fus}.

19. (B). Heat added in this segment clearly goes toward increasing the temperature of the substance. Temperature is a measure of the average kinetic energy of the particles within a sample.

20. (C). Moving from right to left within the diagram corresponds to heat transferred out of the system, and doing so within this segement clearly results in a decrease in the temperature of the substance. Falling temperature corresponds to a decrease in the average kinetic energy of the particles in a system.

21. (E). The heat added within this segment clearly does not go toward raising the temperature of the substance. Instead the added heat goes toward rearranging particles at a given temperature from liquid phase to gas phase (the second phase change shown in the diagram). The heat required to accomplish this phase change is called the heat of vaporization, ΔH_{vap}.

22. (A). This segment lies beween the plateaus corresponding to fusion (melting) and vaporization (boiling). The phase that sits between melting and boiling is liquid. Note that the average kinetic energy of the particles is increasing during this segment.

Part IV
Transforming Matter in Reactions

The 5th Wave   By Rich Tennant

"Okay—now that the paramedic is here with the defibrillator and smelling salts, prepare to learn about covalent bonds."

In this part . . .

When matter transforms, we call it chemistry. Breaking and making bonds, swapping protons, swapping electrons . . . these kinds of simple transactions underlie all of chemistry and are the subject of this part. We show you how to recognize the major categories of reactions, and how to keep track of which bits of matter do what to which other bits. Once you know which reaction is going on, you can predict all sorts of things, like how much product will be made, and how much reactant will be left over. In this part, you'll learn how acidity and charge can take center stage in a chemical reaction, without changing the basic rules by which chemical change takes place. Finally, you'll learn to discern the difference between how far a reaction goes, and how quickly it gets where it's going. Each question deals with energy, but the two questions are independent. We show you a handful of equations that allow you to calculate answers to both questions. This part is the beating heart of the book.

Chapter 13

Reacting by the Numbers: Reactions and Stoichiometry

Stoichiometry. Such a complicated word for such a simple idea. The Greek roots of the word mean "measuring elements," which doesn't sound nearly as intimidating. Moreover, the ancient Greeks couldn't tell an ionic bond from an ionic column, so just how technical and scary could stoichiometry really be? Simply stated, stoichiometry (stoh-ick-eee-AH-muh-tree) is the quantitative relationship between atoms in chemical substances.

We will begin this chapter by walking you through each of the types of chemical reaction you will encounter on the AP chemistry exam. Several of them will be discussed in greater detail in later chapters. Then, we will show you how to use your understanding of chemical equations to do AP-type Chemistry problems, and Chapter 14 will allow you to practice your new-found skills.

The first half of the chapter describes the basics of stoichiometry, then delves into the seven types of reaction that you'd do well to recognize (notice how their names tell you what happens in each reaction). By recognizing the patterns of these seven types of reaction, you can often predict reaction products given only a set of reactants. There are no perfect guidelines, and predicting reaction products can take what is called "chemical intuition," a sense of what reaction is likely to occur based on knowing the outcomes of similar reactions. Still, if you are given both reactants and products, you should be able to tell what kind of reaction connects them, and if you are given reactants and the type of reaction, you should be able to predict likely products. Figuring out the formulas of products often requires you to apply knowledge about how ionic and molecular compounds are put together. To review these concepts, see Chapter 5.

The remainder of the chapter covers how to use your knowledge of these reaction types to calculate the quantities that allow you to quantify the success of a reaction.

Stoichiometry Lingo: How to Decipher Chemical Reactions

Chemists have a nice, ordered set of rules for describing chemical reactions. Reactants are always on the lefthand side of the equation, and products on the right, for example. Chemists would rather stick their hands in a vat of concentrated hydrochloric acid than reverse this order. There are also a nice, ordered set of chemical symbols that chemists use to describe reactions, and each symbol has one precise meaning. These fundamentals of stoichiometry are described in this section and, since they form the basis of all of stoichiometry, which is a huge point-gainer on the AP exam, it would behoove you to memorize them before moving on.

In compound formulas and reaction equations, you express stoichiometry by using subscripted numbers on atoms and coefficients in front of groups of atoms. Just like an accountant can't afford to make a mistake when tracking the dollars and cents in your parents' bank account, you can't afford to lose track of any atoms on the AP exam. The road to poor scores in AP Chemistry is littered with students who couldn't master stoichiometry!

In general, all chemical equations are written in the basic form:

Reactants → Products

where the arrow in the middle means *yields* or "turns into." The basic idea is that the reactants react, and the reaction produces products. By *reacting*, we simply mean that bonds within the reactants are broken, to be replaced by new and different bonds within the products.

Chemists fill chemical equations with symbols because they think it looks cool and, more important, because the symbols help pack a lot of meaning into a small space. Table 13-1 summarizes the most important symbols you'll find in chemical equations.

Table 13-1	Symbols Commonly Used in Chemical Equations
Symbol	*Explanation*
+	Separates two reactants or products
→	The "yields" symbol separates the reactants from the products. The single arrowhead suggests the reaction occurs in only one direction.
↔	A two-way yield symbol means the reaction can occur in both the forward and reverse directions. You may also see this symbol written as two stacked arrows with opposing arrowheads.
(s)	A compound followed by this symbol exists as a solid.
(l)	A compound followed by this symbol exists as a liquid.
(g)	A compound followed by this symbol exists as a gas.
(aq)	A compound followed by this symbol exists in aqueous solution, dissolved in water.
Δ	This symbol, usually written above the yields symbol, signifies that heat is added to the reactants.
Ni, LiCl	Sometimes a chemical symbol (such as those for nickel or lithium chloride here) is written above the yields symbol. This means that the indicated chemical was added as a catalyst. Catalysts speed up reactions but do not otherwise participate in them.

After you understand how to interpret chemical symbols, compound names (see Chapter 5), and the symbols in Table 13-1, there's not a lot you can't understand. You're equipped, for example, to decode a chemical equation into an English sentence describing a reaction. Conversely, you can translate an English sentence into the chemical equation it describes. When you're fluent in this language, you regrettably won't be able to talk to the animals; you will, however, be able to describe their metabolism in great detail.

Building and Breaking: Synthesis and Decomposition Reactions

The two simplest types of chemical reactions involve the joining of two compounds to form one compound or the breaking apart of one compound to form two. We describe both of these simple reactions in the sections that follow.

Synthesis

In synthesis (or combination) reactions, two or more reactants combine to form a single product, following the general pattern

$$A + B \rightarrow C$$

For example,

$$2Na(s) + Cl_2(g) \rightarrow 2NaCl(s)$$

The combining of elements to form compounds (like NaCl) is a particularly common kind of combination reaction. Here is another such example:

$$2Ca(s) + O_2(g) \rightarrow 2CaO(s)$$

Compounds can also combine to form new compounds, such as in the combination of sodium oxide with water to form sodium hydroxide:

$$Na_2O(s) + H_2O(l) \rightarrow 2NaOH(aq)$$

Decomposition

In a decomposition reaction, a single reactant breaks down (decomposes) into two or more products, following the general pattern

$$A \rightarrow B + C$$

For example,

$$2H_2O(l) \rightarrow 2H_2(g) + O_2(g)$$

Notice that combination and decomposition reactions are the same reaction in opposite directions.

Many decomposition reactions produce gaseous products, such as in the decomposition of carbonic acid into water and carbon dioxide:

$$H_2CO_3(aq) \rightarrow H_2O(l) + CO_2(g)$$

Swapping Spots: Displacement Reactions

Many common reactions involve the swapping of one or more atoms or polyatomic ions in a compound. You can check these reactions out in the following sections.

Single replacement

In a single replacement reaction, a single, more-reactive element or group replaces a less-reactive element or group, following the general pattern

$$A + BC \rightarrow AC + B$$

For example,

$$Zn(s) + CuSO_4(aq) \rightarrow ZnSO_4(aq) + Cu(s)$$

 Single replacement reactions in which metals replace other metals are especially common. You can determine which metals are likely to replace which others by using the *metal activity series,* a ranked list of metals in which ones higher on the list tend to replace ones lower on the list. Table 13-2 presents the metal activity series.

Table 13-2	Metal Activity Series in Order of Decreasing Reactivity
Metal	*Notes*
Lithium, potassium, strontium, calcium, sodium	Most reactive metals; react with cold water to form hydroxide and hydrogen gas
Magnesium, aluminum, zinc, chromium	React with hot water/acid to form oxides and hydrogen gas
Iron, cadmium, cobalt, nickel, tin, lead	Replace hydrogen ion from dilute strong acids
Hydrogen	Nonmetal, listed in reactive order
Antimony, arsenic, bismuth, copper	Combine directly with oxygen to form oxides
Mercury, silver, palladium, platinum, gold	Least reactive metals; often found as free metals; oxides decompose easily

Double replacement

Double replacement is a special form of *metathesis reaction* (that is, a reaction in which two reacting species exchange bonds). Double replacement reactions tend to occur between ionic compounds in solution. In these reactions, cations (atoms or groups with positive charge) from each reactant swap places to form ionic compounds with the opposing anions (atoms or groups with negative charge), following the general pattern

$$AB + CD \rightarrow AD + CB$$

For example,

$$KCl(aq) + AgNO_3(aq) \rightarrow AgCl(s) + KNO_3(aq)$$

Of course, ions dissolved in solution move about freely, not as part of cation-anion complexes. So, to allow double replacement reactions to progress, one of several things must occur:

✔ One of the product compounds must be insoluble, so it precipitates (forms an insoluble solid) out of solution after it forms.

✔ One of the products must be a gas that bubbles out of solution after it forms.

✔ One of the products must be a solvent molecule, such as H_2O, that separates from the ionic compounds after it forms.

Burning Up: Combustion Reactions

Oxygen is always a reactant in combustion reactions, which often release heat and light as they occur. Combustion reactions frequently involve hydrocarbon reactants (like propane, $C_3H_8(g)$, the gas used to fire up backyard grills), and yield carbon dioxide and water as products. For example,

$$C_3H_8(g) + 5O_2(g) \rightarrow 3CO_2(g) + 4H_2O(l)$$

Combustion reactions also include combination reactions between elements and oxygen, such as:

$$S(s) + O_2(g) \rightarrow SO_2(g)$$

So, if the reactants include oxygen (O_2) and a hydrocarbon or an element, you're probably dealing with a combustion reaction. If the products are carbon dioxide and water, you're almost certainly dealing with a combustion reaction.

Getting Sour: Acid-Base Reactions

Acids and bases are complicated enough to deserve their own chapter, so flip forward to Chapter 17 for the nitty-gritty details of this reaction type. In the meantime, and so you have a comprehensive list of reaction types which the AP chemistry exam may quiz you on in one place, we'll provide a summary here.

Neutralization reactions, which occur when an acid and a base are mixed together to form a salt and water, are the bread and butter of acid-base reactions. The best way to spot a reaction of this type is to look for water among the products of a reaction. If the water came from a double replacement reaction in which a hydrogen from one reactant (the acid) joined up with a hydroxide from the other (the base), then you are definitely dealing with an acid-base reaction.

Not all acid-base reactions are so simple, however. Bases do not necessarily contain hydroxide, and the products are not always water and a salt. See Chapter 17 for the details of the myriad of acid-base reactions.

Getting Charged: Oxidation-Reduction Reactions

Oxidation-reduction (or redox) reactions, like acid-base reactions, have an entire chapter (Chapter 19) devoted to them. Redox reactions are concerned with the transfer of electrons between reactants. Because they involve the transfer of charged particles, ions are a staple of redox reactions and are the things to look out for when trying to spot one. The species in a redox reaction that loses electrons to the other reactant is *oxidized* in the reaction, while the species that gains the electrons is *reduced*. There are two easy mnemonics commonly used to remember this, so take your pick or make up your own:

- ✔ **LEO and GER:** Lose electrons, oxidation. Gain electrons, reduction.
- ✔ **OIL RIG:** Oxidation is lost, reduction is gained.

Oxidation and reduction reactions always occur together, but may be written separately as half reactions in a chemistry problem. In these half reactions, electrons being transferred are written explicitly, so if you see them, take it as a dead giveaway that you are dealing with a redox reaction.

The Stoichiometry Underground: Moles

Chemical equations are all well and good, but what do they actually mean? For translating chemical equations into useful chemical data, you can find no better or more fundamental tool than the mole. As an AP chemistry student, you're probably already exhaustively familiar with the mole, but in the following sections, we provide you with a comprehensive review of the mole and its usefulness in case you forgot any of the basics along the way.

Counting your particles: The mole

Chemists routinely deal with hunks of material containing trillions of trillions of atoms, but ridiculously large numbers can induce migraines. For this reason, chemists count particles (like atoms and molecules) in multiples of a quantity called the *mole*. Initially, counting particles in moles can be counterintuitive, similar to the weirdness of Dustin Hoffman's character in the movie *Rainman,* when he informs Tom Cruise's character that he has spilled 0.41 fraction of a box of 200 toothpicks onto the floor. Instead of referring to 82 individual toothpick particles, he refers to a fraction of a larger unit, the 200-toothpick box. A mole is a very big box of toothpicks — 6.022×10^{23} toothpicks, to be precise.

If 6.022×10^{23} strikes you as an unfathomably large number, then you're thinking about it correctly. It's larger, in fact, than the number of stars in the sky or the number of fish in the sea, and is many, many times more than the number of people who have ever been born throughout all of human history. When you think about the number of particles in something as simple as, say, a cup of water, all your previous conceptions of "big numbers" are blown out of the water, as it were.

The number 6.022×10^{23}, known as *Avogadro's number,* is named after the 19th century Italian scientist Amedeo Avogadro. Posthumously, Avogadro really pulled one off in giving his name to this number, because he never actually thought of it. The real brain behind Avogadro's number was that of a French scientist named Jean Baptiste Perrin. Nearly 100 years after Avogadro had his final pasta, Perrin named the number after him as an homage. Ironically, this humble act of tribute has misdirected the resentment of countless hordes of high school chemistry students to Avogadro instead of Perrin.

Avogadro's number is the conversion factor used to move between particle counts and numbers of moles:

1 mole / 6.022×10^{23} particles

Like all conversion factors, you can invert it to move in the other direction, from moles to particles.

Assigning mass and volume to your particles

Chemists always begin a discussion about moles with Avogadro's number. They do this for two reasons. First, it makes sense to start the discussion with the way the mole was originally defined. Second, it's a sufficiently large number to intimidate the unworthy.

Still, for all its primacy and intimidating size, Avogadro's number quickly grows tedious in everyday use. More interesting is the fact that one mole of a pure *monatomic* substance (in other words, a substance that always appears as a single atom) turns out to possess exactly its atomic mass's worth of grams. In other words, one mole of monatomic hydrogen weighs about 1 gram. One mole of monatomic helium weighs about 4 grams. The same is true no matter where you wander through the corridors of the periodic table. If we define *molar mass* as the mass of one mole of any substance, then the number listed as the atomic mass of an element equals that element's molar mass *if the element is monatomic*. Guess what — chemists actually set up the value of the mole to make that happen, though it took years of arguments and an international commission to get the final details down to the last decimal place just right.

Of course, chemistry involves the making and breaking of bonds, so talk of pure monatomic substances gets you only so far. How lucky, then, that calculating the mass per mole of a complex molecule is essentially no different than finding the mass per mole of a monatomic element. For example, one molecule of glucose ($C_6H_{12}O_6$) is assembled from 6 carbon atoms, 12 hydrogen atoms, and 6 oxygen atoms. To calculate the number of grams per mole of a complex molecule (such as glucose), simply do the following:

1. **Multiply the number of atoms per mole of the first element by its atomic mass.**

   In this case, the first element is carbon, and you'd multiply its atomic mass, 12, by the number of atoms, 6.

2. **Multiply the number of atoms per mole of the second element by its atomic mass.**

   Here, you multiply hydrogen's atomic mass of 1 by the number of hydrogen atoms, 12.

3. **Multiply the number of atoms per mole of the third element by its atomic mass. Keep going until you've covered all the elements in the molecule.**

   The third element in glucose is oxygen, so you multiply 16, the atomic mass by 6, the number of oxygen atoms.

4. **Finally, add the masses together.**

   In this example, $(12\text{g mol}^{-1} \times 6) + (1\text{g mol}^{-1} \times 12) + (16\text{g mol}^{-1} \times 6) = 180\text{g mol}^{-1}$.

This kind of quantity, called the *gram molecular mass,* is exceptionally convenient for chemists, who are much more inclined to measure the mass of a substance than to count all of millions or billions of individual particles that make it up.

If chemists don't try to intimidate you with large numbers, they may attempt to do so by throwing around big words. For example, older chemists may distinguish between the molar masses of pure elements, molecular compounds, and ionic compounds by referring to them as the *gram atomic mass, gram molecular mass,* and *gram formula mass,* respectively. Don't be fooled! The basic concept behind each is the same: molar mass. Fortunately these other terms are never used by those modern folks who write AP chemistry exams.

It's all very good to find the mass of a solid or liquid and then go about calculating the number of moles in that sample. But what about gases? Let's not engage in phase discrimination; gases are made of matter, too, and their moles have the right to stand and be counted. Fortunately, there's a convenient way to convert between the moles of gaseous particles and their *volume.* Unlike gram atomic/molecular/formula masses, this conversion factor is constant *no matter what kinds of molecules make up the gas.* Every gas assumed to behave ideally has a volume of 22.4 liters per mole, regardless of the size of the gaseous molecules.

Before you start your hooray-chemistry-is-finally-getting-simple dance, however, understand that certain conditions apply to this conversion factor. For example, it's only true at *standard temperature and pressure (STP),* or 0° Celsius and 1atm. Also, the figure of $22.4L \ mol^{-1}$ applies only to the extent that a gas resembles an ideal gas, one whose particles have zero volume and neither attract nor repel one another. Ultimately, no gas is truly ideal, but many are so close to being so that the $22.4L \ mol^{-1}$ conversion is very useful.

Giving credit where it's due: Percent composition

Chemists are often concerned with precisely what percentage of a compound's mass consists of one particular element. Lying awake at night, uttering prayers to Avogadro, they fret over this quantity, called *percent composition.* Percent composition was mentioned in Chapter 5, but now that you have reviewed the basics of stoichiometry, we will explain in detail how to calculate it.

Follow these three simple steps:

1. **Calculate the molar mass of the compound, as we explain in the previous section.**

   Percent compositions are completely irrelevant to molar masses for pure monatomic substances, which by definition have 100 percent composition of a given element.

2. **Multiply the atomic mass of each element present in the compound by the number of atoms of that element present in one molecule.**

3. **Divide each of the masses calculated in Step 2 by the total mass calculated in Step 1. Multiply each fractional quotient by 100%. Voilà! You have the *percent composition* by mass of each element in the compound.**

Checking out empirical formulas

What if you don't know the formula of a compound? Chemists sometimes find themselves in this disconcerting scenario. Rather than curse Avogadro (or perhaps *after* doing so), they analyze samples of the frustrating unknown to identify the percent composition. From there, they calculate the ratios of different types of atoms in the compound. They express these ratios as an *empirical formula,* the lowest whole-number ratio of elements in a compound.

To find an empirical formula given percent composition, use the following procedure:

1. **Assume that you have 100g of the unknown compound.**

   The beauty of this little trick is that you conveniently gift yourself with the same number of grams of each elemental component as its contribution to the percent composition. For example, if you assume that you have 100g of a compound composed of 60.3 percent magnesium and 39.7 percent oxygen, you know that you have 60.3g of magnesium and 39.7g of oxygen.

2. **Convert the assumed masses from Step 1 into moles by using gram atomic masses.**

3. **Divide each of the element-by-element mole quantities from Step 2 by the lowest among them.**

   This division yields the mole ratios of the elements of the compound.

4. **If any of your mole ratios aren't whole numbers, multiply all numbers by the smallest possible factor that produces whole-number mole ratios for all of the elements.**

   For example, if there is one nitrogen atom for every 0.5 oxygen atoms in a compound, the empirical formula is not $N_1O_{0.5}$. Such a formula casually suggests that an oxygen atom has been split, something that would create a small-scale nuclear explosion. Though impressive-sounding, this scenario is almost certainly false. Far more likely is that the atoms of nitrogen to oxygen are combining in a 1:0.5 *ratio,* but do so in groups of $2 \times (1{:}0.5) = 2{:}1$. The empirical formula is thus N_2O.

   Because the original percent composition data is typically experimental, expect to see a bit of error in the numbers. For example, 2.03 is probably within experimental error of 2.

5. **Write the empirical formula by attaching these whole-number mole ratios as subscripts to the chemical symbol of each element. Order the elements according to the general rules for naming ionic and molecular compounds (described in Chapter 5).**

Differentiating between empirical formulas and molecular formulas

Many compounds in nature, particularly compounds made of carbon, hydrogen, and oxygen, are composed of atoms that occur in numbers that are multiples of their empirical formula. In other words, their empirical formulas don't reflect the actual numbers of atoms within them, but only the ratios of those atoms. What a nuisance! Fortunately, this is an old nuisance, so chemists have devised a means to deal with it. To account for these annoying types of compounds, chemists are careful to differentiate between an empirical formula and a *molecular formula*. A molecular formula uses subscripts that report the actual number of atoms of each type in a molecule of the compound (a *formula unit* accomplishes the same thing for ionic compounds).

Molecular formulas are associated with molar masses that are simple whole-number multiples of the corresponding *empirical formula mass*. In other words, a molecule with the empirical formula CH_2O has an empirical formula mass of $30.0g \, mol^{-1}$ (12 for the carbon + 2 for the two hydrogens + 16 for the oxygen). The molecule may have a molecular formula of CH_2O, $C_2H_4O_2$, $C_3H_6O_3$, and so on. As a result, the compound may have a molar mass of $30.0g \, mol^{-1}$, $60.0g \, mol^{-1}$, $90.0g \, mol^{-1}$, and so on.

You can't calculate a molecular formula based on percent composition alone. If you attempt to do so, Avogadro and Perrin will rise from their graves, find you, and slap you 6.022×10^{23} times per cheek (Ooh, that smarts!). The folly of such an approach is made clear by comparing formaldehyde with glucose. The two compounds have the same empirical formula, CH_2O, but different molecular formulas, CH_2O and $C_6H_{12}O_6$, respectively. Glucose is a simple sugar, the one made by photosynthesis and the one broken down during cellular respiration. You can dissolve it into your coffee with pleasant results. Formaldehyde is a carcinogenic component of smog. Solutions of formaldehyde have historically been used to embalm dead bodies. You are not advised to dissolve formaldehyde into your coffee. In other words, molecular formulas differ from empirical formulas, and the difference is important.

To determine a molecular formula, you must know the molar mass of the compound as well as the empirical formula (or enough information to calculate it yourself from the percent composition; see the previous section for details). With these tools in hand, calculating the molecular formula involves a three-step process:

1. **Calculate the empirical formula mass.**

2. **Divide the molar mass by the empirical formula mass.**

3. **Multiply each of the subscripts within the empirical formula by the number calculated in Step 2.**

Keeping the See-Saw Straight: Balancing Reaction Equations

Equations that simply show reactants on one side of the reaction arrow and products on the other are *skeleton equations,* and are perfectly adequate for a qualitative description of the reaction: who are the reactants and who are the products. But if you look closely, you'll see that those equations just don't add up quantitatively. As written, the mass of one mole of each of the reactants doesn't equal the mass of one mole of each of the products. The skeleton equations break the *Law of Conservation of Mass,* which states that all the mass present at the beginning of a reaction must be present at the end. To be quantitatively accurate, these equations must be *balanced* so the masses of reactants and products are equal.

In chemistry this is easy — the requirement is met as mentioned earlier by assuring that the numbers of the same types of atoms are equal on each side. All the equations we have so far written have met this balance requirement. (The crime of losing or splitting atoms is punishable by an Avogadro number of lashes!).

How to use coefficients

To balance an equation, you use *coefficients* to alter the number of moles of reactants and/or products so the mass on one side of the equation equals the mass on the other side. A *coefficient* is simply a number that precedes the symbol of an element or compound, multiplying the number of moles of that *entire* compound within the equation. Coefficients are different from *subscripts,* which multiply the number of atoms or groups within a compound. Consider the following:

$$4Cu(NO_3)_2$$

The number 4 that precedes the compound is a coefficient, indicating that there are four moles of copper (II) nitrate. The subscripted 3 and 2 within the compound indicate that each nitrate contains three oxygen atoms, and that there are two nitrate groups per atom of copper. Coefficients and subscripts multiply to yield the total number of moles of each atom:

$4\text{mol Cu}(NO_3)_2 \times (1\text{mol Cu/mol Cu}(NO_3)_2) = \textbf{4mol Cu}$

$4\text{mol Cu}(NO_3)_2 \times (2\text{mol } NO_3/\text{mol Cu}(NO_3)_2) \times 1\text{mol N/mol } NO_3 = \textbf{8mol N}$

$4\text{mol Cu}(NO_3)_2 \times (2\text{mol } NO_3/\text{mol Cu}(NO_3)_2) \times 3\text{mol O/mol } NO_3 = \textbf{24mol O}$

When you balance an equation, *you change only the coefficients.* Changing subscripts alters the chemical compounds themselves. If your pencil were equipped with an electrical shocking device, that device would activate the moment you attempted to change a subscript while balancing an equation.

Here's a simple recipe for balancing equations:

1. **Given a skeleton equation (one that includes formulas for reactants and products), count up the number of each kind of atom on each side of the equation.**

   If you recognize any polyatomic ions, you can count these as one whole group (as if they were their own form of element). See Chapter 5 for information on recognizing polyatomic ions.

2. **Use coefficients to balance the elements or polyatomic ions, one at a time.**

   For simplicity, start with those elements or ions that appear only once on each side.

3. **Check the equation to ensure that each element or ion is balanced.**

   Checking is important because you may have "ping-ponged" several times from reactants to products and back — there's plenty of opportunity for error.

4. **When you're sure the reaction is balanced, check to make sure it's in lowest terms.**

   For example,

   $$4H_2(g) + 2O_2(g) \rightarrow 4H_2O(l)$$

   should be reduced to

   $$2H_2(g) + O_2(g) \rightarrow 2H_2O(l)$$

You can't begin to wrap your brain around the unimaginably large number of possible chemical reactions. It's good that so many reactions can occur, because they make things like life and the universe possible. From the perspective of a mere human brain trying to grok all these reactions, we have another bit of good news: A few categories of reactions pop up over and over again. After you see the very basic patterns in these categories, you'll be able to make sense of the majority of reactions out there.

Conversions

Mass and energy are conserved. It's the law. Unfortunately, this means that there's no such thing as a free lunch, or any other type of free meal. Ever. On the other hand, the conservation of mass makes it possible to predict how chemical reactions will turn out.

Balancing equations can seem like a chore, like taking out the trash. But a balanced equation is far better than any collection of coffee grounds and orange peels because such an equation is a useful tool. After you've got a balanced equation, you can use the coefficients to build *mole-mole conversion factors.* These kinds of conversion factors tell you how much of any given product you get by reacting any given amount of reactant. This is one of those calculations that makes chemists particularly useful, so they needn't get by on looks and charm alone.

Consider the following balanced equation for generating ammonia from nitrogen and hydrogen gases:

$$N_2(g) + 3H_2(g) \rightarrow 2NH_3(g)$$

For every mole of nitrogen molecules reactanting, a chemist expects two moles of ammonia product. Similarly, for every three moles of hydrogen molecules reacting, the chemist expects two moles of ammonia product. These expectations are based on the coefficients of the balanced equation and are expressed as mole-mole conversion factors as shown in Figure 13-1.

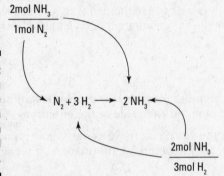

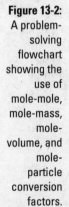

Figure 13-1:
Building mole-mole conversion factors from a balanced equation.

The mole is the beating heart of stoichiometry, the central unit through which other quantities flow. Real-life chemists don't have magic mole vision, however. A chemist can't look at a pile of potassium chloride crystals, squint her eyes, and proclaim: "That's 0.539 moles of salt." Well, she could proclaim such a thing, but she wouldn't bet her pocket protector on it. Real reagents (reactants) tend to be measured in units of mass or volume, and occasionally even in actual numbers of particles. Real products are measured in the same way. So, you need to be able to use *mole-mass, mole-volume,* and *mole-particle conversion factors* to translate between these different dialects of counting. Figure 13-2 summarizes the interrelationship between all these things and serves as a flowchart for problem solving. All roads lead to and from the mole.

Figure 13-2:
A problem-solving flowchart showing the use of mole-mole, mole-mass, mole-volume, and mole-particle conversion factors.

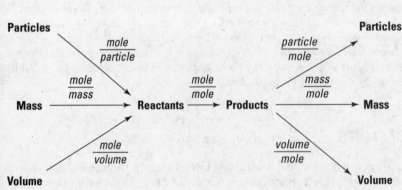

Getting rid of mere spectators: Net ionic equations

Chemistry is often conducted in aqueous solutions. Soluble ionic compounds dissolve into their component ions, and these ions can react to form new products. In these kinds of reactions, sometimes only the cation or anion of a dissolved compound reacts. The other ion merely watches the whole affair, twiddling its charged thumbs in electrostatic boredom. These uninvolved ions are called *spectator ions.*

Because spectator ions don't actually participate in the chemistry of a reaction, you don't need to include them in a chemical equation. Doing so leads to a needlessly complicated reaction equation. So, chemists prefer to write *net ionic equations,* which omit the spectator ions. A net ionic equation doesn't include every component that might be present in a given beaker. Rather, it includes only those components that actually react.

Here is a simple recipe for making net ionic equations of your own:

1. **Examine the starting equation to determine which ionic compounds are dissolved, as indicated by the (*aq*) symbol following the compound name.**

   $Zn(s) + HCl(aq) \rightarrow ZnCl_2(aq) + H_2(g)$

2. **Rewrite the equation, explicitly separating dissolved ionic compounds into their component ions.**

   This step requires you to recognize common polyatomic ions, so be sure to familiarize yourself with them.

   $Zn(s) + H^+(aq) + \cancel{Cl^-}(aq) \rightarrow Zn^{2+}(aq) + \cancel{2Cl^-}(aq) + H_2(g)$

3. **Compare the reactant and product sides of the rewritten reaction. Any dissolved ions that appear in the same form on both sides are spectator ions. Cross out the spectator ions to produce a net reaction.**

   $Zn(s) + H^+(aq) + Cl^-(\cancel{aq}) \rightarrow Zn^{2+}(aq) + 2Cl^-(\cancel{aq}) + H_2(g)$

   Net reaction:

   $Zn(s) + H^+(aq) \rightarrow Zn^{2+}(aq) + H_2(g)$

   As written, the preceding reaction is imbalanced with respect to the number of hydrogen atoms and the amount of positive charge.

4. **Balance the net reaction for mass and charge.**

   $Zn(s) + 2H^+(aq) \rightarrow Zn^{2+}(aq) + H_2(g)$

If you wish, you can balance the equation for atoms and charge first (at Step 1). This way, when you cross out spectator ions at Step 3, you'll be crossing out equivalent numbers of ions. Either method produces the same net ionic equation in the end. Some people prefer to balance the starting reaction equation, but others prefer to balance the net reaction because it is a simpler equation.

Running Out Early: Limiting Reagents

In real-life chemistry, not all of the reactants present convert into product. That would be perfect and convenient. Does that sound like real life to you? More typically, one reagent is completely used up, and others are left behind, perhaps to react another day.

The situation resembles that of a horde of Hollywood hopefuls lined up for a limited number of slots as extras in a film. Only so many eager faces react with an available slot to produce a happily (albeit pitifully) employed actor. The remaining actors are in excess, muttering quietly all the way back to their jobs as waiters. In this scenario, the slots are the limiting reagent.

Those standing in line demand to know, how many slots are there? With this key piece of data, they can deduce how many of their huddled mass will end up with a gig. Or, they can figure out how many will continue to waste their film school degrees serving penne with basil and goat cheese to chemists on vacation.

Chemists demand to know, which reactant will run out first? In other words, which reactant is the _limiting reagent?_ Knowing that information allows them to calculate how much product they can expect, based on how much of the limiting reagent they've put into the reaction. Also, identifying the limiting reagent allows them to calculate how much of the excess reagent they'll have left over when all the smoke clears. Either way, the first step is to figure out which is the limiting reagent.

In any chemical reaction, you can simply pick one reagent as a candidate for the limiting reagent, calculate how many moles of that reagent you have, and then calculate how many grams of the other reagent you'd need to react both to completion. You'll discover one of two things. Either you have an excess of the first reagent or you have an excess of the second reagent. The one you have in excess the the excess reagent. The one that is not in excess is the limiting reagent.

Having Something to Show for Yourself: Percent Yield

In a way, reactants have it easy. Maybe they'll make something of themselves and actually react. Or maybe they'll just lean against the inside of the beaker, flip through a back issue of _People_ magazine, and sip a caramel macchiato.

Chemists don't have it so easy. Someone is paying them to do reactions. That someone doesn't have time or money for excuses about loitering reactants. So you, as a fresh-faced chemist, have to be concerned with just how completely your reactants react to form products. To compare the amount of product obtained from a reaction with the amount that should have been obtained, chemists use _percent yield._ You determine percent yield with the following formula:

Percent yield = 100% × (actual yield) / (theoretical yield)

Lovely, but what is an actual yield and what is a theoretical yield? An _actual yield_ is . . . well . . . the amount of product actually produced by the reaction. A _theoretical yield_ is the amount of product that could have been produced had everything gone perfectly, as described by theory — in other words, as predicted by your painstaking calculations.

Things never go perfectly. Reagents stick to the sides of flasks. Impurities sabotage reactions. Chemists attempt to dance. None of these ghastly things are _supposed_ to occur, but they do. So, actual yields fall short of theoretical yields.

Chapter 14

Answering Questions on Reactions and Stoichiometry

In This Chapter
▶ Going over important points to remember
▶ Putting in some practice with example questions
▶ Seeing how your answers stack up

*I*f we had to name one single thing that is key to acing the AP chemistry exam, a solid understanding of stoichiometry would be it. Stoichiometry is chemistry's bread and butter. Make sure that you understand all of the concepts outlined in Chapter 13 in detail and you will score some major points on exam day. To ensure you nail the concepts of stoichiometry, we not only include the major points that are essential to your success, but we also include practice questions near the end of the chapter so you can sharpen your test-taking skills on stoichiometry.

Making Sure You React to Questions with the Right Answers

Stoichiometry and the skills that surround it might seem mundane, but they are central skills that let you grapple with the very heart of chemistry: reactions. It's well worth your time to make sure you know these basics back and forth.

Knowing reactions when you see them

These tips and reminders are all about recognition — knowing how to decode chemical formulas and reaction equations to understand what reaction is taking place, and even how to predict the products of a reaction when all you're given is the reactants.

✔ Reaction equations use symbols to indicate details of the reaction. Review the symbols in Chapter 13 in Table 13-1 and make sure that you're familiar with all of them.

✔ Make sure that you're familiar with the seven common reaction types and are able to identify them and predict products if given only reactants. It is not so important to memeorize the names of the types as it is to be able to predict products and balance equations. Many reactions on tests expect "familiar" compounds as products, not truly exotic ones, so don't get hung up on memorizing lots of exotic compounds!

- Synthesis/Combination reactions: A + B → C. Two or more reactants combine to form a single product.

- Decomposition reactions: A → B + C. Four types of decomposition reaction are seen quite commonly on the AP chemistry exam. These are

  metal carbonate → metal oxide + $CO_2(g)$

  metal chlorate → metal chloride + $O_2(g)$

  metal hydroxide → metal oxide + H_2O

  acid → nonmetal oxide + H_2O

- Single replacement reactions: A + BC → B + AC. Single replacement reactions are all redox reactions! The likelihood of single replacement reactions occurring relies on the reactivity series of metals outlined in Table 13-2. A will only displace B if it is a more-reactive species, which means it appears higher in the reactivity series.

- Double replacement reactions: AB + CD → AD + CB. A and D will not necessarily combine in the same ratios as A and B did. Take careful account of the charges of each species and make sure to balance the equation in the end.

- Combustion reactions: These contain oxygen among the reactants. The most common types of combustion reactions are those where hydrocarbons are burned to form water and carbon dioxide as products, but a familiar trap is to forget that substances that already contain oxygen, such as ethanol (C_2H_6O), also "burn" to form carbon dioxide and water. All combustion reactions are also redox reactions.

- Neutralization, or acid/base reactions: Acid + Base → Water + Salt. Chapter 17 explains these in greater detail.

- There are also a number of acid-base reaction types in addition to neutralization reactions. These are

  Nonmetal oxide + H_2O → acid

  Metal oxide + H_2O → base

  Metal + $H_2O(l)$ → base + $H_2(g)$

  Active metal + acid → salt + $H_2(g)$

  Carbonate + acid → $CO_2(g)$ + $H_2O(l)$ + salt

  Base + salt → metal hydroxide precipitate + salt

  Strong base + salt containing NH_4 → NH_3 + H_2O + salt

- Redox reactions are outlined in detail in Chapter 19. They typically involve the transfer of electrons between one reactant and the other. The species that loses the electrons is oxidized and that which gains the electrons is reduced. The mnemonics "LEO and GER" or "OIL RIG" will help you remember this information.

Dealing with the numbers that flow from stoichiometry

Coming up with accurate formulas and reaction equation isn't the end of the story — it's usually just the beginning. Once you've got an equation, you've got to make sure that it's balanced, and then use the correct stoichiometry to solve problems. These tips deal with that process.

✔ Avogadro's number, 6.022×10^{23}, is equivalent to the number of particles in one mole of a substance. By "particles" we mean atoms, ions, or molecules — whichever is relevant for the circumstance.

✔ The gram atomic, molecular, or formula mass of a substance is calculated by adding up the atomic masses of all of its components (multiplying by the number of atoms of each substance present where applicable). It's equivalent to the mass in grams of one mole of particles of that substance.

✔ At standard temperature and pressure, one mole of an ideal gas would occupy 22.4 liters volume. No gases are truly ideal (see Chapter 9 on gas laws for further details). However, this approximation is a very good one in many cases and is very useful in converting volume to moles of a gas and vice versa.

✔ The percent composition of a substance takes into account the mass ratios of the atoms that make it up as well as the ratios in which they combine. To calculate a percent composition of a molecular or ionic compound, calculate the gram molecular or gram formula mass of the compound. Then, calculate the mass of each type of atom in the substance (by multiplying the mass by the number of atoms present in each molecule) and divide by the total mass of the compound.

✔ You can also use percent compositions to find the formula of a compound, though it is a somewhat complicated process. If you follow the five following steps, you'll get it every time.

1. **Assume that you have 100g of the unknown compound.**

2. **Convert the assumed masses from Step 1 into moles by using gram atomic masses.**

3. **Divide each of the element-by-element mole quantities from Step 2 by the lowest among them.**

4. **If any of your mole ratios aren't whole numbers, multiply all numbers by the smallest possible factor that produces whole-number mole ratios for all of the elements.**

5. **Write the empirical formula by attaching these whole-number mole ratios as subscripts to the chemical symbol of each element. Order the elements according to the general rules for naming ionic and molecular compounds (described in Chapter 5).**

✔ Given an empirical formula and either a formula or molar mass for a compound, you can also determine its molecular formula by dividing the gram formula mass by the empirical formula mass and then by taking this whole number value and multiplying the subscripts for each atom in the empirical formula by that number.

✔ Before you take the AP exam, make sure you've mastered the method for balancing equations. Unless the AP exam tells you that an equation is balanced, always check. To check that an equation is properly balanced, simply count the total number of each type of atom on each side of the equation and make sure that they match. Also be sure to check that any total charges on each side are balanced.

✔ Always consider the possibility of limiting reagents in reactions. Limiting reagents determine theoretical yields. Percent yield is 100 percent times the actual yield divided by the theoretical yield.

Testing Your Knowledge

Now that you've grounded yourself in a solid knowledge of reactions and stoichiometry, see if you can keep your balance through these practice questions.

Multiple choice

1. Which of the following represents a process in which a species is reduced?

   (A) $Mg \rightarrow Mg^{2+}$

   (B) $Cl_2 \rightarrow Cl^-$

   (C) $Ni^{3+} \rightarrow Ni^{4+}$

   (D) $CO \rightarrow CO_2$

   (E) $NO_2^- \rightarrow NO_3^{-5\%}$

2. A sample of Li_2SO_4 (molar mass 110g) is reported to be 10.2% lithium. Assuming that none of the impurities contain any lithium, what percentage of the sample is pure lithium sulfate?

   (A) 65%

   (B) 70%

   (C) 75%

   (D) 80%

   (E) 85%

3. When the following equation for the acid base reaction above is balanced and all of the coefficients are reduced to lowest whole-number terms, the coefficient on the H_2O is

   $$\_\_H_2SO_4 + \_\_Ca(OH)_2 \rightarrow \_\_CaSO_4 + \_\_H_2O$$

   (A) 1.

   (B) 2.

   (C) 3.

   (D) 4.

   (E) 5.

4. What is the simplest formula for a compound containing only carbon and hydrogen and containing 16.67% H?

   (A) CH_4

   (B) C_3H_8

   (C) C_5H_{12}

   (D) C_6H_{14}

   (E) C_7H_{16}

5. A substance has an empirical formula of CH_2O and a molar mass of 180.0. What is its molecular formula?

   (A) $C_2H_4O_2$

   (B) $C_3H_6O_3$

   (C) $C_4H_8O_4$

   (D) $C_5H_{10}O_5$

   (E) $C_6H_{12}O_6$

6. In the reaction

$$2Fe_2O_3 + 3C \rightarrow 4Fe + 3CO_2$$

if 500.g of Fe_2O_3 reacts with 75.0g of C, how many grams of Fe will be produced?

(A) 56.4g

(B) 75.0g

(C) 316.g

(D) 349.g

(E) 465.g

7. The mass of 5 atoms of lead is

(A) 3.44×10^{-20}g.

(B) 1.72×10^{-21}g.

(C) 8.60×10^{-21}g.

(D) 1.20×10^{-23}g.

(E) 6.02×10^{-23}g.

8. $AgNO_3(aq) + KCl(aq) \rightarrow$ _____ + _____

What are the missing products?

(A) $AgNO_3(aq) + KCl(aq)$ (no reaction)

(B) $AgCl_2(s) + K_2NO_3(aq)$

(C) $AgCl(s) + KNO_3(aq)$

(D) $Ag^{2+}(aq) + Cl^-(aq) + KNO_3 (aq)$

(E) $AgCl_2(s) + K^+(aq) + Cl^- (aq)$

9. $HCl + NaOH \rightarrow NaCl + H_2O$

If 30.0g of HCl is mixed with excess NaOH according to the reaction above and 41.0g of NaCl is produced, what is the difference between the actual and the theoretical yields for this reaction?

(A) 7.1g

(B) 8.3g

(C) 15.5g

(D) 41.0g

(E) 48.1g

Free response

10. Do each of the following conversions based on the reaction below.

SiO_2 (s) + 6HF(g) → H_2SiF_6(aq) + 2H_2O(l)

(a) 5.25g HF to moles

(b) 6.00 moles SiO_2 to grams of H_2SiF_6

(c) 10.0 moles HF to liters HF (at STP)

(d) 3.20 mol SiO_2 to molecules H_2SiF_6

11. Write the formulas to show the reactants and the products for the reactions described below. Assume that the reaction occurs and that solutions are aqueous unless otherwise indicated. Represent substances in solution as ions if the substances are extensively ionized. Omit spectator ions or molecules. You do not need to balance the equations.

(a) Iron (II) sulfide is added to a solution of hydrochloric acid.

(b) The combustion of methane in large excess of air.

(c) Calcium carbonate is strongly heated (hint: carbon dioxide is one of the products).

(d) A solution of potassium hydroxide is added to solid ammonium chloride.

(e) A strip of iron is added to a solution of copper (II) sulfate.

(f) Chlorine gas is bubbled into a solution of sodium fluoride.

(g) Dilute sulfuric acid is added to a solution of strontium nitrate.

(h) Copper ribbon is burned in oxygen.

12. For each of the following three reactions, in part one for each reaction write a balanced equation for the reaction and in part two for each reaction answer the question about the reaction. In the balanced equation, coefficients should be in terms of lowest whole numbers. Assume that solutions are aqueous unless otherwise indicated. Represent substances in solution as ions if the substances are extensively ionized. Omit formulas for any ions or molecules that are unchanged by the reaction.

(a) Excess nitric acid is added to solid sodium bicarbonate.

(i) Balanced Equation:

(ii) What is the minimum mass of sodium bicarbonate that must be added in order to react completely with 25.0g of nitric acid?

(b) A solution containing the magnesium ion (an oxidizing agent) is mixed with a solution containing lead(II) (a reducing agent).

(i) Balanced Equation:

(ii) If the contents of the reaction mixture described above are filtered, what substance(s), if any, would remain on the filter paper?

(c) A solution of sodium hydroxide is added to a solution of copper (II) sulfate.

(i) Balanced Equation:

(ii) What is the percent composition of oxygen in sodium hydroxide?

13. For each of the following, use appropriate chemical principles to explain the observation.

    (a) Lithium metal will react with water, but copper will not.

    (b) The human body is made of 60% to 70% water, but hydrogen is only the third most abundant element in the body by mass.

    (c) Water can act as both an acid and a base.

14. A compound is 36.5% sodium, 25.4% sulfur, and 38.1% oxygen.

    (a) What is its empirical formula?

    (b) If the compound in (a) has a molar mass of 366.3g, what is its molecular formula?

15. Calculate the percent composition of each element in caffeine $C_8H_{10}N_4$.

Checking Your Work

Still on your feet? Good. Now check with the judges and receive your score. If you faltered, find out why. If you triumphed, enjoy it.

1. (B). This reaction is the only one where electrons are gained by the reactant to form the product, a sure sign of a reduction reaction. The negative sign on the chlorine is an indicator that extra electrons are present, and the fact that the reactant is of neutral charge cements the reaction as reduction.

2. (D). Begin by finding the percent composition of lithium in pure lithium sulfate, 14.0g ÷ 110g = 12.7%. Impurities, however, have driven the lithium concentration down to 10.2% lithium. Dividing this percentage by the first will give you the percent of the pure compound in the sample. 10.2 ÷ 12.7 = 80.3%, which is closest to answer choice D.

3. (B). There are two tricks to balancing an equation like this. The first is to begin by balancing the polyatomic ions, in this case SO_4 and OH^-. But where is the OH^- among the products? This point is the second important one. In balancing equations, particularly neutralization reactions like this one, it is often extremely useful to write H_2O as HOH. The fully balanced equation here is $H_2SO_4 + Ca(OH)_2 \rightarrow CaSO_4 + 2H_2O$. That's right. Water is the only component that takes a subscript greater than 1, and that subscript is 2.

4. (C). Here you're given a percent composition and are asked to derive an empirical formula. Follow the steps outlined in Chapter 13 to accomplish this task. Or, since the answers are right in front of you and percent composition is really a fairly easy calculation to make, you can simply calculate the percent composition of hydrogen in each of the compounds among the answer choices and see which one matches the question. You will find that this answer is choice C, pentane.

5. (E). Calculate the empirical formula mass of CH_2O (30g per mole) and divide the molar mass by it, giving you 180 ÷ 30 = 6. Multiply each of the subscripts in the empirical formula to give you the molecular formula $C_6H_{12}O_6$, answer choice E.

6. (D). First, identify the limiting reagent in the reaction. Converting grams of iron (III) oxide to grams of carbon reveals that 56.4g of carbon are needed to react completely with 500.g of iron (III) oxide.

$$\frac{500g\ Fe_2O_3}{1} \times \frac{1mol\ Fe_2O_3}{159.7g\ Fe_2O_3} \times \frac{3mol\ C}{2mol\ Fe_2O_3} \times \frac{12.0g\ C}{1mol\ C} = 56.4g\ C$$

This discovery means that there is excess carbon among the reactants and iron oxide should be used as the limiting reagent. With this established, all that remains is to convert grams of iron (III) oxide to grams of the iron product according to the calculation

$$\frac{500g\ Fe_2O_3}{1} \times \frac{1mol\ Fe_2O_3}{159.7g\ Fe_2O_3} \times \frac{4mol\ Fe}{2mol\ Fe_2O_3} \times \frac{55.8g\ Fe}{1mol\ Fe} = 349.4g\ Fe$$

This matches answer choice D.

7. (B). Remember that you must convert to moles first and cannot convert directly from particles to grams. Set up your three conversion factors as follows and you get an answer matching answer choice B.

$$\frac{5atoms\ Pb}{1} \times \frac{1mol\ Pb}{6.022 \times 10^{23}\ atoms\ Pb} \times \frac{207.2g\ Pb}{1mol\ Pb} = 1.72 \times 10^{-21}g$$

8. (C). This choice represents a double replacement reaction. The first thing that you will need to do is deduce the charge on the silver. As a diligent AP chemistry student, you should already have memorized the charges on the common polyatomic ions and know that NO_3 carries a charge of -1. Your excellent powers of deduction should then lead you to the conclusion that you are dealing with Ag^+, whose charge balances that of NO_3^- in a 1:1 ratio. When silver and potassium switch places, the silver and chlorine and potassium and nitrate will all combine in 1:1 ratios, giving the products $AgCl$ and KNO_3, which is consistent with answer choice C.

9. (A). By telling you that NaOH is present in excess, the problem saves you from having to do a limiting reagent calculation. All you need to do now is to convert grams of HCl, your limiting reagent, to grams of the NaCl product as in the calculation

$$\frac{30.0g\ HCl}{1} \times \frac{1mol\ HCl}{36.5g\ HCl} \times \frac{1mol\ NaCl}{1mol\ HCl} \times \frac{58.5g\ NaCl}{1mol\ NaCl} = 48.1g\ NaCl$$

Be careful, though, that you've read the problem carefully. It does not ask for the theoretical yield or the percent yield, but the *difference* between the theoretical and actual yields. Subtract the two to get 48.1g – 41.0g = 7.1g, answer choice A.

10. For all of the conversions in this problem, be careful to set up your conversion factors so that the units cancel correctly.

Here is the answer for (a):

$$\frac{5.25g\ HF}{1} \times \frac{1mol\ HF}{20.0g\ HF} = 0.263mol\ HF$$

Here is the answer for (b):

$$\frac{6.00mol\ SiO_2}{1} \times \frac{1mol\ H_2SiF_6}{1mol\ SiO_2} \times \frac{144.1g\ H_2SiF_6}{1mol\ H_2SiF_6} = 865.g\ H_2SiF_6$$

Here is the answer for (c):

$$\frac{10.0mol\ HF}{1} \times \frac{22.4L\ HF}{1mol\ HF} = 224L\ HF$$

Here is the answer for (d):

$$\frac{3.20mol\ SiO_2}{1} \times \frac{1mol\ H_2SiF_6}{1mol\ SiO_2} \times \frac{6.022 \times 10^{23}\ molecules\ H_2SiF_6}{1mol\ H_2SiF_6} = 1.93 \times 10^{24}\ molecules\ H_2SiF_6$$

11. (a) The complete, balanced reaction here is $FeS(s) + 2HCl(aq) \rightarrow FeCl_2(aq) + H_2S(g)$. However, the problem has asked you to eliminate spectators and also allows you to ignore balancing. Chlorine is the only spectator here, so eliminating it from both sides leaves you with $FeS(s) + H^+(aq) \rightarrow Fe^{2+}(aq) + H_2S(g)$.

(b) This reaction does not take place in solution, so there will not be any spectators. Recall that all combustion reactions require the reactant O_2 and that the products are always water and carbon dioxide. With this knowledge, you are left with $CH_4 + O_2 \rightarrow CO_2 + H_2O$, which you do not need to balance. In minimal air, some CO will form.

(c) Since this substance is "strongly heated" you should be able to identify it as a decomposition reaction immediately. If one of the products is CO_2, the remaining product must be CaO, giving you the reaction $CaCO_3 \rightarrow CaO + CO_2$.

(d) This is a double replacement reaction of the pattern. $KOH (aq) + NH_4Cl(s) \rightarrow NH_4OH(aq) + KCl(aq)$. You know that both of the products are soluble because one contains ammonium and the other contains an alkali metal, salts of both of which are always soluble. Eliminating spectators, this leaves you with the equation $NH_4Cl(s) \rightarrow NH_4^+(aq) + Cl^-(aq)$.

(e) This is both a single replacement reaction and a redox reaction. Iron is more reactive than copper and so displaces it in solution. (Recall that you are to assume that a reaction always occurs in this type of problem, so even if you didn't remember that iron was higher than copper on the reactivity series, you could still treat it as such.) Eliminating spectators, your net equation is $Fe(s) + Cu^{2+}(aq) \rightarrow Cu(s) + Fe^{2+}(aq)$.

(f) Nothing happens here — fluorine is a stronger oxidizing agent than chlorine.

(g) This double replacement reaction follows the equation $H_2SO_4(aq) + Sr(NO_3)_2(s) \rightarrow HNO_3(aq) + SrSO_4(s)$, where you know that strontium sulfate is a precipitate because it is insoluble. Eliminating spectators, you are left with the equation $SO_4^{2-}(aq) + Sr(NO_3)_2(s) \rightarrow NO_3^-(aq) + SrSO_4(s)$.

(h) This is a combustion/combination reaction in which a metal is burned to create a metal oxide according to the equation. $Cu(s) + O_2(g) \rightarrow CuO(s)$.

12. (a) (i) This is one of those seven acid/base reactions outlined earlier in the chapter in which an acid and carbonate mix to form three products: water, carbon dioxide, and a salt. In this case, the reaction is therefore $NaHCO_3(s) + HNO_3(aq) \rightarrow CO_2(g) + H_2O(l) + NaNO_3(aq)$. Both the nitric acid and the sodium nitrate product will be significantly dissociated in water, leaving you with the equation $NaHCO_3(s) + H^+(aq) + NO_3^-(aq) \rightarrow CO_2(g) + H_2O(l) + Na^+(aq) + NO_3^-(aq)$. NO_3^- is the only spectator ion here, so the final equation is $NaHCO_3 + H^+ \rightarrow CO_2 + H_2O + Na^+$. Note that you do not need to include phases in your final answer.

(ii) Convert mass of nitric acid to mass of sodium bicarbonate as follows.

$$\frac{25.0g \ HNO_3}{1} \times \frac{1mol \ HNO_3}{63.0g \ HNO_3} \times \frac{1mol \ NaHCO_3}{1mol \ HNO_3} \times \frac{84.0g \ NaHCO_3}{1mol \ NaHCO_3} = 33.3g \ NaHCO_3$$

(b) (i) Magnesium is the oxidizing agent, so it must be reduced (gain electrons) in the reaction. Lead (II) is the reducing agent, so it must be oxidized (lose electrons) in the reaction. This leaves you with the net reaction. $Mg^{2+} + Pb^{2+} \rightarrow Mg + Pb^{4+}$.

(ii) As a solid, magnesium will be on the paper. If the lead salt formed is insoluble then it may also be on the paper. Either way, $Mg(s)$ will precipitate.

(c) (i) This reaction also follows one of the seven acid/base reaction types listed earlier, this time the base + salt → metal hydroxide precipitate + salt. In other words, this is a relatively simple double replacement reaction. The complete, balanced reaction is $2NaOH(aq)$ + $CuSO_4(aq) \rightarrow Cu(OH)_2(s) + Na_2SO_4(aq)$. Sodium and sulfate both appear in aqueous solution on both sides of the equation, so eliminating them as spectators, you are left with $2OH^- + Cu^{2+} \rightarrow Cu(OH)_2$.

(ii) 40%. The percent composition of oxygen can be calculated by dividing the total number of grams per mole of oxygen by the molecular mass of sodium hydroxide. This results in $16g \div 40g \times 100 = 40\%$.

13. (a) Lithium is a very reactive metal and is significantly higher than hydrogen on the activity series and is therefore bound to displace it in a reaction. Copper, however, is lower than hydrogen on the activity series and therefore will not replace it.

(b) Both carbon and oxygen are more abundant in the human body than hydrogen by mass because even though there are many more hydrogen atoms than either carbon or oxygen atoms, hydrogen is the lightest of the elements. Carbon is 12 times more massive and oxygen is 16 times more massive than hydrogen.

(c) Simply speaking, water can act as both an acid and a base because it contains both an acid's characteristic hydrogen and the hydroxide ion common to most bases. Another way to look at it is that water can accept or donate a proton.

14. (a) Na_2SO_3. To find the empirical formula from the percent composition, follow the steps outlined in Chapter 13. Begin by assuming that you have a total of 100g of the substance, which means that you have 36.5g sodium, 25.4g sulfur and 38.1g oxygen. Divide each of these by their gram atomic masses as follows:

$$\frac{36.5g\ Na}{1} \times \frac{1mol\ Na}{23.0g\ Na} = 1.59mol\ Na$$

$$\frac{25.4g\ S}{1} \times \frac{1mol\ S}{32.1g\ S} = 0.79mol\ S$$

$$\frac{38.1g\ O}{1} \times \frac{1mol\ O}{16.0g\ O} = 2.38mol\ O$$

Next, divide each of these by the lowest among them (1.12), giving 1.59mol Na ÷ 0.79 = 2.0mol Na, 0.79mol S ÷ 0.79 = 1.0mol S, and 2.38mol O ÷ 0.79 = 3.0mol O. Attach each of these as subscripts on your empirical formula Na_2SO_3. This is your empirical formula.

(b) $Na_6S_3O_9$. You are given the molar mass and will need to divide it by the empirical formula mass in order to find your conversion factor from the empirical to molecular formulae. The empirical formula mass of Na_2SO_3 is 122.1g per mole. 366.3 divided by 122.1 is 3. Multiply each of the subscripts in the empirical formula by this number to arrive at the molecular formula of $Na_6S_3O_9$.

15. 59.3% C, 6.2%H and 34.6%N. Begin all percent composition calculations by finding the mass per mole of each type of atom in the substance by multiplying the number of each atom present by its gram atomic mass. In this case, you get $8 \times 12.0g = 96g$ C, $10 \times 1.0g = 10.0g$ H, and $4 \times 14.0g = 56.0g$ N. Next calculate the molar mass of the compound by adding these three values together (96.0 + 10.0 + 56.0 = 162.0g). Finally, divide each of the individual masses by this molar mass and multiply by 100 to give you the percent composition. You should get 59.3% C, 6.2% H, and 34.6% N.

Chapter 15

Going Nowhere: Equilibrium

. . . .

In This Chapter

▶ Settling down with the concept of chemical equilibrium

▶ Using Le Chatelier to find your easy chair

▶ Measuring reactions at rest with equilibrium constants

. . . .

*E*quilibrium . . . it's a pleasant-sounding word. What does it mean? In everyday language, equilibrium connotes calmness and composure, or some state of balance. The chemical definition of equilibrium most closely mirrors the last of these meanings: balance. In this chapter, we describe how the idea of balance applies to chemical reactions.

Shifting Matter: An Overview of Chemical Equilibrium

On the microscopic scale, most reactions can occur in both the "forward" and "reverse" directions. When we say that a reaction is "going forward," what we mean is that the rate of the forward reaction is larger than the rate of the reverse reaction (see Figure 15-1); in other words, at any given moment more reactant mass converts into product than there is product mass converting back into reactant. Remember, chemical reactions simply convert a given amount of mass between states.

Given enough time, a closed chemical system (one protected from outside influences) reaches a state in which the rate of the forward reaction equals the rate of the reverse reaction, as shown in Figure 15-2. On the microscopic scale, chemistry occurs constantly in both directions, but on the larger, "macroscopic" scale, the reaction appears to be "at rest." We call this state of chemical balance *dynamic equilibrium*.

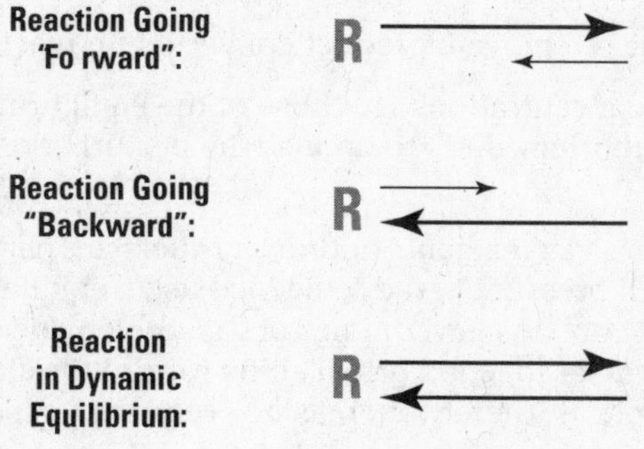

Figure 15-1:
Equilibrium occurs when forward and backward rates are equal.

Reaction Going "Fo rward":

R ⟶ ⟵ P

Reaction Going "Backward":

R ⟶ ⟵ P

Reaction in Dynamic Equilibrium:

R ⟶ ⟵ P

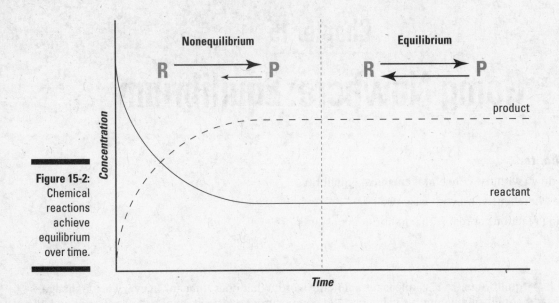

Figure 15-2:
Chemical
reactions
achieve
equilibrium
over time.

Once a chemical system achieves equilibrium, the concentrations of reactants and products no longer change (macroscopically, at least). The resulting combination of concentrations is called an *equilibrium mixture,* and can be used to characterize the reaction, as will be shown in a later section. Whether the equilibrium mixture contains more products or more reactants (and to what extent) depends on differences in energy between reactants and products. The details of this energy-equilibrium relationship are described in detail in Chapter 21.

Predicting Shifts with Le Chatelier's Principle

After a chemical system has reached equilibrium, that equilibrium can be disrupted, or *perturbed.* Think of systems at equilibrium as people who have finally found their easy chair at the end of a long day. You may rouse them to take out the trash, but they'll return to the easy chair at the first opportunity. This concept is more or less the idea behind *Le Chatelier's Principle:* The equilibrium of a perturbed system shifts in the direction that opposes the perturbation. Perturbations include changes in the following:

✔ **Concentration:** If a system is at equilibrium, adding or removing reactant or product disrupts the equilibrium. Roused from its easy chair, the equilibrium reasserts itself in response.

 • If reactant is added or product is removed, reactant converts into product.

 • If product is added or reactant is removed, product converts into reactant.

Either way, chemistry occurs until concentrations are those of the equilibrium mixture once more. In other words, the equilibrium shifts to oppose the perturbation, as shown in Figure 15-3.

✔ **Pressure:** Reactions that include gases as reactants and/or products are particularly sensitive to pressure perturbation. If pressure is suddenly increased, equilibrium shifts toward the side of the reaction that contains fewer moles of gas, thereby decreasing pressure. If pressure is suddenly decreased, equilibrium shifts toward the side of the reaction that contains more moles of gas, thereby increasing pressure. Consider the following reaction:

$$N_2(g) + 3H_2(g) \leftrightarrow 2NH_3(g)$$

A given amount of mass on the reactant side of the equation (as N_2 and H_2) corresponds to double the moles of gas as the same mass on the product side (as NH_3). Imagine that the system is at equilibrium at a low pressure. Now imagine that the pressure suddenly increases, perturbing that equilibrium. Reactant (N_2 and H_2) converts to product (NH_3) so the total moles of gas decrease, thereby lowering the pressure.

If the system suddenly shifts to lower pressure, NH_3 converts to N_2 and H_2 so the total moles of gas increase, thereby raising the pressure. The equilibrium shifts to oppose the perturbation. These responses are summarized in Figure 15-4.

✔ **Temperature:** Reactions that absorb or give off heat (that is, most reactions) can be perturbed from equilibrium by changes in temperature. The easiest way to understand this behavior is to explicitly include heat as a reactant or product in the reaction equation:

$$A + B \leftrightarrow C + \text{heat}$$

Imagine that this reaction is at equilibrium. Now imagine that the temperature suddenly increases. Product C absorbs heat, converting to reactants A and B. Because the heat "product" has been decreased, the temperature of the system decreases.

If the temperature suddenly shifts down from equilibrium, reactants A and B convert to product C, releasing heat. The released heat increases the temperature of the system. The equilibrium shifts to oppose the perturbation. These responses of the equilibrium to changes in temperature are reviewed in Figure 15-5.

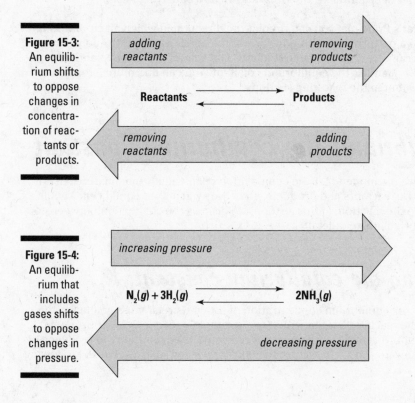

Figure 15-3:
An equilibrium shifts to oppose changes in concentration of reactants or products.

adding reactants   removing products

Reactants ⟶ **Products**

removing reactants   adding products

Figure 15-4:
An equilibrium that includes gases shifts to oppose changes in pressure.

increasing pressure

$N_2(g) + 3H_2(g)$ ⟶ $2NH_3(g)$

decreasing pressure

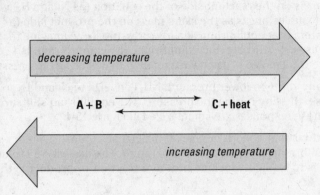

Figure 15-5:
An equilibrium that includes heat as a reactant or (as shown) as a product shifts to oppose changes in temperature.

What is the effect of adding a catalyst on an equilibrium mixture? Precisely nothing. Catalysts may help speed the journey to the equilibrium mixture, but they have no effect on its composition.

What if the reaction starts on one "side" of the equilibrium mixture as opposed to the other side? In other words, what if the same reaction starts from an "all reactants" state versus an "all products" state? Given time, each reaction will find its way to the same equilibrium mixture. In science, that kind of beautiful symmetry is known as a Satisfying Result.

So, by using Le Chatelier's Principle, you can predict the direction in which an equilibrium will shift in response to a perturbation. But how far will the equilibrium shift? And where exactly was the equilibrium "resting" in the first place? The answers to these quantitative questions involve a new quantity: the equilibrium constant. You can find more information on the equilibrium constant in the following section.

Measuring Equilibrium: The Equilibrium Constant

In a previous section, we mentioned that the composition of an equilibrium mixture (that is, the concentrations of the reactants and products when the system is at equilibrium) could be used to characterize the reaction. In this section, we discuss the connection between concentrations and equilibrium in great detail.

Understanding the equilibrium constant, K_{eq}

In essence, the greater the equilibrium concentration of products relative to reactants, the more spontaneous is the reaction. Spontaneous (or "favorable") reactions favor products. Nonspontaneous (or "unfavorable") reactions favor reactants. To quantitate the tendency of a reaction to proceed toward products, we use the *equilibrium constant*, K_{eq}. For the following reaction

$$aA + bB \leftrightarrow cC + dD$$

the equilibrium constant is calculated as follows:

$$K_{eq} = \frac{[C]^c [D]^d}{[A]^a [B]^b}$$

In general, concentrations of the products are multiplied together in the numerator. Concentrations of the reactants are multiplied together in the denominator. Stoichiometric coefficients show up as exponents on their corresponding reactants or products. Once actual values (that is, concentrations from an equilibrium mixture) are entered for each concentration and coefficient, the whole expression reduces to a single number, one ranging from zero to infinity. Real-life values always occupy the middle ground between these two extremes. Whatever it is, that final number has meaning.

Very favorable reactions produce a lot of product, so they have K_{eq} values much larger than 1. Very unfavorable reactions convert very little reactant into product, so they have K_{eq} values between 0 and 1. In a reaction with $K_{eq} = 1$, the amount of product equals the amount of reactant at equilibrium.

Note that you can only calculate K_{eq} by using concentrations measured at equilibrium. Concentrations measured before a reaction reaches equilibrium can be used to calculate a *reaction quotient, Q:*

$$Q = \frac{[C]^c [D]^d}{[A]^a [B]^b} \ (nonequilibrium)$$

If $Q < K_{eq}$, the reaction will progress "to the right," making more product. If $Q > K_{eq}$, the reaction will shift "to the left," converting product into reactant. If $Q = K_{eq}$, the reaction is at equilibrium. Because equilibria are temperature dependent, the K_{eq} at one temperature may be the Q at another temperature.

You may have noticed that the expression given for calculating the K_{eq} lists each reactant or product within brackets, as in [X]. If those brackets look unfamiliar, you'll want to review Chapter 11, where you'll learn that they indicate the molar concentration of a solute, $M = mol \ L^{-1}$.

But what if you're dealing with a gas phase reaction, like the one pictured in Figure 15-4? Well, in that case you'll work with partial pressures of the reactant and product gases (see Chapter 9), and the expression for K_{eq} is:

$$K_{eq} = \frac{(P_C)^c (P_D)^d}{(P_A)^a (P_B)^b}$$

Fine, you say, but what if my reaction includes some combination of liquids, gases, and liquid or gas solutions? And what about solids? Relax, there's an answer—but it comes at the price of some new vocabulary:

Homogenous equilbria contain reactants and products that are all of the same phase. These are the easiest cases because you can simply apply the K_{eq} expression directly, as already described.

Heterogeneous equilibria contain reactants and products of different phases. You can still calculate K_{eq} for such reactions, but you've got to follow a few simple rules:

- Pure liquids and pure solids are omitted from the K_{eq} expression of a heterogeneous equilibrium; these concentrations simply equal 1. This includes the solvent (such as H_2O in the case of aqueous solutions), even if solvent molecules are directly involved in the chemistry of the reaction. The solvent is present in such high concentration that its involvement as a reactant or product doesn't change its concentration.

- Partial pressures (P) of reactant or product gases and molar concentrations (M) of reactant or product solutes can appear together in the K_{eq} expression, because their concentrations may change significantly during the course of a reaction.

Using K_{eq} to solve problems

Some equilibrium problems don't simply list final, equilibrium concentrations, but instead give you a combination of initial concentrations, final concentrations, and possibly the K_{eq} itself. At this point, you may have developed that sinking feeling in your stomach that suggests that all this can get pretty complicated. Not to worry; you just need to be organized. Follow these steps for dealing with equilibrium problems that confuse you:

1. **Start with a balanced equation for the reaction, including the phases of the reactants and products.**

   For example

   $NH_4Cl(s) \leftrightarrow NH_3(g) + HCl(g)$

2. **Check the phases to see if you're dealing with a homogenous equilibrium or a heterogeneous equilibrium.** If you've got a heterogeneous equilibrium on your hands, cross out any pure (unmixed) liquid or solid substances to ensure that you omit them from the final K_{eq} expression. Continuing with the example introduced in step 1:

   ~~$NH_4Cl(s)$~~ $\leftrightarrow NH_3(g) + HCl(g)$

3. **Set up the K_{eq} expression, including the actual K_{eq} value, if known (0.036 in this case):**

   $K_{eq} = (P\,NH_3)(P\,HCl) = 0.036$

4. **Underneath each remaining reactant or product, list the following quantities, in order: initial concentration/partial pressure, change in concentration/partial pressure and equilibrium concentration/partial pressure. Leave unknown quantities blank.**

   The resulting table might look like this:

	$NH_4Cl(s)$	$\leftrightarrow$	$NH_3(g)$	+	$HCl(g)$
Initial:	1		0atm		0.072atm
Change:	1		+xatm		+xatm
Equilibrium:	1		xatm		0.072 + xatm

This kind of table is known as an ICEbox, drawing on the initials of the quantities in the three rows: Initial, Change, and Equilibrium. How exactly you fill in the ICEbox depends on which values are known and which are unknown. In the given example, the partial pressures of both product gases were known. By examining the balanced equation, it is clear that any changes in NH_3 and HCl products occur in a one-to-one ratio, because each product has a stoichiometric coefficient of 1. So, the unknown change in partial pressure (x) experienced by the gases en route to equilibrium is the same for each gas. Having filled in the blanks for Initial and Change, the Equilibrium values are calculated as the sum of the two. Just substitute the "Equilibrium" values into the K_{eq} expression and solve for the unknown:

$K_{eq} = (x)(0.072 + x) = 0.036$

$x^2 + 0.072x - 0.036 = 0$ (a quadratic equation)

$x = 0.16$ OR $x = -0.23$

Solving the quadratic equation gives two possible values for x. We can exclude the –0.23 solution because it results in negative values for the partial pressures, which is a physical impossibility. So, x = 0.16atm is the solution. Therefore, the equilibrium mixture of the reaction contains 0.16atm NH_3 and 0.072 + 0.16 = 0.23atm of HCl.

Here are a few additional tips for working with equilibrium constants:

✔ Equilibrium constants for the "forward" and "reverse" directions of the same reaction are simply inverses of each other:

> If . . . K_{eq} for Reactants ↔ Products is $K_{forward}$
>
> And . . . K_{eq} for Products ↔ Reactants is $K_{reverse}$
>
> Then . . . $K_{forward} = (K_{reverse})^{-1}$

✔ Multiple equilibria can be connected or "coupled" when a product from one equilibrium is the reactant for another. The equilibrium constant for the overall, coupled reaction is the product of the equilibrium constants for the component reactions:

> If . . . K_{eq} for A ↔ B is K_{AB}
>
> And . . . K_{eq} for B ↔ C is K_{BC}
>
> Then . . . K_{eq} for A ↔ C is $K_{AC} = (K_{AB})(K_{BC})$

Specializing Equilibrium: Constants for Different Reaction Types

In previous sections of this chapter, we discussed how to calculate the K_{eq} for gas phase reactions by using the partial pressures of the gases in an equilibrium mixture. That's great stuff, as far as it goes. But consider this: In a constant-volume system, increasing the total concentration of a gas also increases its partial pressure. So what? The upshot of this fact is that calculating the equilibrium constant in terms of pressures can sometimes give a value of K_{eq} different from the one you'd get if you calculated in terms of molar concentrations. To account for this discrepancy, chemists sometimes use the quantities K_P and K_C to make clear whether a gas phase equilibrium constant was calculated using pressure or molar concentration, repectively.

Is there any reliable relationship between K_P and K_C? Yes:

$K_P = K_C \leftrightarrow (RT)^{\Delta n}$

In this equation, the exponent (n corresponds to the difference in the moles of gas between the product and reactant sides of the reaction equation. For example, in the reaction

$$2NH_3(g) \leftrightarrow N_2(g) + 3H_2(g)$$

the exponent $\Delta n = n_{product} - n_{reactant} = 4 - 2 = 2$.

Other kinds of specialized equilibrium constants you'll encounter are *acid and base dissociation constants*, K_a and K_b, and the *solubility product constant*, K_{sp}.

K_a and K_b are discussed at length in Chapter 17. Here, you need only understand that they are the equilibrium constants that correspond to acid dissociation and base dissociation reactions.

✔ Strong acids possess large values of K_a.

✔ Strong bases possess large values of K_b.

K_{sp} is the equilibrium constant for the dissociation reaction of an ionic solid as it dissolves. This constant is discussed in the context of solutions and solute concentrations in Chapter 11. Here, we simply show how K_{sp} fits under the umbrella of "equilibrium constant" and discuss how K_{sp} relates to a phenomenon called the common ion effect.

When an ionic solid (such as silver bomide) dissolves, its cation and anion components separate:

$$AgBr(s) + H_2O(l) \leftrightarrow Ag^+(aq) + Br^-(aq)$$

Because AgBr is a pure solid and H_2O is the solvent, those two compounds are assigned concentrations of 1, and effectively fall out of the equilibrium constant expression, such that:

$$K_{sp} = [Ag^+][Br^-]$$

This equation shows that

- ✔ The K_{sp} is the *product* of the concentrations of the dissolved ions.
- ✔ Very soluble compounds possess large K_{sp} values.
- ✔ Very insoluble compounds possess small K_{sp} values.

The solubility product governs the *common ion effect,* a shift in the solubility equilibrium of a dissolved ionic solid upon the addition of a second compound, one that contains an ion in common with the first compound. For example, if sodium bromide, NaBr, is added to a solution of silver bromide, AgBr, a certain amount of dissolved silver bromide shifts back into solid form and precipitates from solution.

Why? Adding sodium bromide results in the addition of Br⁻ to solution. This addition perturbs the silver bromide equilibrium, effectively adding Br⁻ "product." To oppose this perturbation, the equilibrium shifts mass toward the AgBr "reactant" side, as shown in Figure 15-6. How far does the equilibrium shift? How much AgBr precipitates? Answer: AgBr precipitates until the concentrations of Ag⁺ and Br⁻ in solution once again match the K_{sp} for silver bromide.

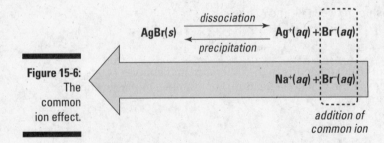

Figure 15-6: The common ion effect.

Chapter 16

Answering Questions on Equilibrium

. . . .

In This Chapter

▶ Keeping the important points in mind

▶ Trying your hand at some practice questions

▶ Understanding the answers with explanations

. . . .

Chapter 15 explained the teeterings and totterings of chemical equilibrium. This topic is a conceptual octopus, wrapping its long tentacles around a variety of other topics. In other words, knowing equilibrium well helps you to understand other things well. So, embrace the octopus, and give these practice questions a try.

Leveling Out the Facts on Equilibrium

Understanding equilibrium helps you to predict how a reaction at rest will respond to a change in conditions. Making these predictions requires you to understand some things about the reactants and products, such as what phase they are in, and whether heat acts as a reactant or product in the reaction. Different categories of reactions sometimes receive the high honor of an equilibrium constant with its own special name. The following points summarize the most important concepts and details about equilibrium.

Defining equilibrium

Like other concepts in chemistry, equilibrium is an important idea that is often expressed by using equations. These points describe equilibrium both verbally and mathematically.

✔ Chemical equilibrium is "dynamic equilibrium," in which forward and reverse reactions occur at the same rate.

✔ Equilibrium constants for forward and reverse reactions are inverses of one another.

✔ The equilibrium constant for a coupled reaction is equal to the product of the equilibrium constants for each individual reaction.

✔ The equilibrium constant K_{eq} quantifies the tendency of a reaction to favor products. K_{eq} is calculated by taking the concentrations of products raised to the power of their coefficients divided by the concentrations of reactants raised to their coefficients. Thus a reaction of the form $aA + bB \leftrightarrow cC + dD$ will have a $K_{eq} = [A]^a[B]^b \div ([C]^c[D]^d)$.

✔ If you are dealing with a reaction between gases, then you will use the expression

$$K_{eq} = \frac{\left(P_C\right)^c \left(P_D\right)^d}{\left(P_A\right)^a \left(P_B\right)^b}$$

✔ If a K_{eq} value is significantly greater than 1, that is an indication that the reaction favors the product heavily. K_{eq} values between 0 and 1 indicate that reactants are favored. A K_{eq} value equal to 1 indicates that at equilibrium, the product of the concentrations of reactants (raised to the appropriate powers) is equal to the product of concentrations of products (raised to the appropriate powers). Note the individual concentrations themselves do not have to be equal for this to be true!

✔ Reactions in which all of the products and reactants are in the same phase are subject to homogeneous equilibrium and K_{eq} is calculated normally. If, however, the products and reactants are in more than one phase, you will have to apply the rules of heterogeneous equilibrium, which are as follows:

  • Pure liquids (including solvents) and solids are omitted from the K_{eq} expression.

  • Partial pressures of gases take the place of concentrations in the K_{eq} expression and both concentrations and partial pressures may appear in the same expression.

Predicting the shifting

Equilibria are all about rest. If you rouse them, they oppose you, shifting toward reactants or products — whatever it takes — to get back to their bean bag chairs. Here's how they do it:

✔ Le Chatelier's Principle describes the tendency of reactions to act to reverse perturbations:

  • **Concentration:** If reactant is added or product is removed, the forward reaction will be favored. If product is added or reactant is removed, the reverse reaction will be favored.

  • **Pressure:** If pressure is increased, the reaction will favor the side with fewer moles of gas. If the pressure is decreased, the reaction will favor the side with more moles of gas.

  • **Temperature:** A decrease in temperature will cause the reaction to favor the side that releases more heat. An increase in temperature will cause the reaction to favor the side that does not release heat.

  • **Catalysts:** Catalysts speed reactions but do not affect the concentration of products or reactants at equilibrium.

✔ If you are given a mixture of initial and final concentrations, use the following five steps to avoid confusion in solving the problem:

  • **Step 1:** Start with a balanced equation for the reaction, including the phases of the reactants and products.

  • **Step 2:** Check the phases to see if you're dealing with a homogenous equilibrium or a heterogeneous equilibrium. If you've got a heterogeneous equilibrium on your hands, cross out any pure (unmixed) liquid or solid substances to ensure that you omit them from the final K_{eq} expression.

  • **Step 3:** Set up the K_{eq} expression, including the actual K_{eq} value, if known.

- **Step 4:** Make an ICEbox. Underneath each remaining reactant or product, list the following quantities, in order: initial concentration/partial pressure, change in concentration/partial pressure, and equilibrium concentration/partial pressure. For the reaction A + B ↔ C,

	[A]	[B]	[C]

Initial:

Change:

Equilibrium:

Use known values to fill in the empty slots in the ICEbox.

- **Step 5:** Substitute the "Equilibrium" values into the K_{eq} expression and solve for the unknown.

Special constants

Several different quantities are all calculated just like K_{eq}, but go by other names. In most cases, these quantities are in fact K_{eq} values, putting on airs because they think they're special enough to deserve their own titles. In another case (the reaction quotient, Q), the special title occurs because the reaction quotient is not an equilibrium quantity at all.

✔ The expressions K_P and K_c are used to distinguish between gaseous phase equilibrium constants in which partial pressures were used (K_P) or molar concentrations (K_c). The relationship between the two is $K_P = K_C \times (RT)^{\Delta n}$, where Δn corresponds to the difference in the moles of gas between the product and reactant sides of the reaction equation.

✔ Other specialized equilibrium constants are discussed in Chapter 11 (K_{sp}) and 17 (K_a and K_b). K_{sp} is simply the equilibrium constant for a dissolution reaction. K_a and K_b are simply the equilibrium constants for acid or base dissociation.

✔ K_{eq} is only valid for a reaction that has reached equilibrium. A similar quantity called the reaction quotient Q is used when the reaction is not at equilibrium. If $Q < K_{eq}$, the reaction will progress "to the right," making more product. If $Q > K_{eq}$, the reaction will shift "to the left," converting product into reactant. If $Q = K_{eq}$, the reaction is at equilibrium.

Testing Your Knowledge

Shifting back and forth with equilibria is dizzying. Steady yourself for a moment and give these practice questions a shot.

Multiple choice

Questions 1 through 4 refer to the following choices.

(A) K_a

(B) K_{sp}

(C) K_{eq}

(D) Q

(E) K_p

1. This is used for a reaction that has not yet reached equilibrium.

2. This equilibrium constant is specific to gases.

3. This equilibrium constant is specific to dissociated solutions of ionic solids.

4. This equilibrium constant can be used to calculate pH.

$$CO_2(g) + H_2(g) \leftrightarrow CO(g) + H_2O(l)$$

5. According to the balanced equation above, which of the following will favor the reverse reaction?

   (I) Decreasing the pressure

   (II) Adding CO to the reaction mixture

   (III) Removing CO_2 from the reaction mixture

   (A) I only

   (B) II and III

   (C) I and III

   (D) I, II, and III

   (E) None of the above

$$SnO_2(s) + 2CO(g) \leftrightarrow Sn(s) + 2CO_2(g)$$

6. Which of the following is a correct equilibrium constant expression for the above reaction?

   (A) $$K_{eq} = \frac{[Sn][CO_2]^2}{[SnO_2][CO]^2}$$

   (B) $$K_{eq} = \frac{[SnO_2][CO]^2}{[Sn][CO_2]^2}$$

   (C) $$K_{eq} = \frac{[Sn][CO_2]}{[SnO_2][CO]}$$

   (D) $$K_{eq} = \frac{[CO_2]^2}{[CO]^2}$$

   (E) $$K_{eq} = \frac{[Sn]}{[SnO_2]}$$

7. If the K_{eq} of a forward reaction is 8.3×10^{-7}, what is the K_{eq} for the reverse reaction?

   (A) 8.3×10^{-7}

   (B) 8.3×10^{7}

   (C) 1.2×10^{6}

   (D) 1.2×10^{-6}

   (E) Cannot be determined from the given information

8. What is the value of the equilibrium constant for the reaction $H_2(g) + I_2(g) \leftrightarrow 2HI$ if the equilibrium mixture contains $0.092M$ H_2, $0.023M$ I_2, and $0.056M$ HI?

    (A) 0.015

    (B) 0.15

    (C) 1.5

    (D) 15

    (E) 150

9. If the K_{sp} for the dissociation of nickel(II) hydroxide is 6.0×10^{-16}, what is the concentration of nickel ions in solution?

    (A) $5.3 \times 10\text{-}6$

    (B) $1.1 \times 10\text{-}5$

    (C) $3.0 \times 10\text{-}16$

    (D) $6.0 \times 10\text{-}16$

    (E) Cannot be determined from the given information

$$2NO_2Cl(g) \leftrightarrow 2NO_2(g) + Cl_2(g)$$

10. If the K_c of the above reaction is 8.90 at 350(C, what is its K_p?

    (A) 0.174

    (B) 0.310

    (C) 256

    (D) 350.

    (E) 455

Free response

11. $CH_3COOH(aq) + H_2O(l) \leftrightarrow CH_3COO^-(aq) + H_3O^+(aq)$ $K_a = 1.8\times10^{-5}$

    (a) Write the equilibrium constant expression K_a for the reaction above.

    (b) If the initial concentration of acetic acid is 0.25M, what is the equilibrium concentration of CH_3COO^-?

Checking Your Work

It feels good to be done with the practice questions, doesn't it? It feels even better to know that you answered them correctly. Make yourself feel good — check your work.

1. (D). The reaction quotient Q is used in place of K_{eq} when a reaction has not yet reached equilibrium.

2. (E). K_p is used for gases.

3. (B). The solubility product constant K_{sp} is used for the dissociation of ionic solids.

4. (A). K_a can be used to calculate pH in a solution of an acid or a base.

5. (D). Decreasing the pressure will favor the side with the greater number of moles of gas (the reactants here). Adding products and removing reactants will also act to favor the production of reactants (the reverse reaction) so all three are valid.

6. (D). You eliminate all pure solids and liquids from heterogeneous equilibrium constant expressions, so D is the correct answer.

7. (C). The K_{eq} of a reverse reaction can be calculated by taking the inverse of the forward reaction. $1 \div (8.3 \times 10^{-7}) = 1.2 \times 10^6$.

8. (C). The equilibrium constant expression is

$$K_{eq} = \frac{\left[HI \right]^2}{\left[H_2 \right]\left[I_2 \right]} = \frac{0.056^2}{0.092 \times 0.023} = 1.5.$$

9. (A). The K_{sp} expression for the dissociation of $Ni(OH)_2$ is $K_{sp} = [Ni^{2+}][OH^-]^2$. You also know that for every one nickel ion that dissolves in solution, there must be 2 OH^- ions so $[Ni^{2+}] = 0.50[OH^-]$. Call this concentration X and plug it into the K_{sp} expression and you have $6.0 \times 10^{-16} = 0.50X \times X^2$. Solve this expression for X and you should get $1.06 \times 10^{-5} = X$, but as stated before, $[Ni^{2+}]$ is half of that value, which is 5.3×10^{-6}.

10. (E). You will need to use the expression $K_P = K_C \times (RT)^{\Delta n}$. The only tricky parts to applying this equation are to remember to convert your temperature to K (623K) and to correctly solve for Δn. Do this by subtracting the number of moles of gaseous *reactants* from the number of moles of gaseous *products*, giving you $3 - 2 = 1$. Your expression then becomes $K_P = 8.90 \times (0.0821 \times 623)^1 = 455$.

11. (a) Neglecting the pure liquid H_2O, the equilibrium constant expression becomes

$$K_a = \frac{\left[CH_3COO^- \right]\left[H_3O^+ \right]}{\left[CH_3COOH \right]}$$

(b) 2.1×10^{-3}. You need to make an ICEbox for this

	$CH_3COOH(aq)$	$\rightarrow$	$H_3O^+(aq)$	$+$	$CH_3COO^-(aq)$
Initial	0.25M		0M		0M
Change	–xM		+xM		+xM
Equilibrium	(0.25 – x)M		xM		xM

Then, use the equilibrium constant expression you developed in (a) to solve for x through the following equation

$$K_a = \frac{\left[x \right]\left[x \right]}{\left[0.25 - x \right]} = 1.8 \times 10^{-5}$$

In such problems, we always make the approximation that x is very, very small relative to 0.25. So, we can neglect the $-x$ in the denominator, turning the expression into the much simpler

$$K_a = \frac{x^2}{\left[0.25 \right]} = 1.8 \times 10^{-5}$$

Solving the expression for x^2 gives you $x^2 = 4.5 \times 10^{-6}$. Taking the square root of both sides gives you $x = 2.1 \times 10^{-3}$.

Chapter 17

Giving and Taking in Solution: Acids and Bases

. . . .

In This Chapter

▶ Defining Arrhenius, Brønsted-Lowry and Lewis acids and bases

▶ Titrating

▶ Using buffered solutions

▶ Defining the solubility product constant

▶ Working with acid and base dissociation constants

. . . .

*I*f you've read any comic books, watched any superhero flicks, or even tuned in to one of those fictional solve-the-crime-in-50-minutes shows on TV, you've likely come across a reference to acids being dangerous substances. Acids are generally thought of as something that evil villains intend to spray in the face of a hero or heroine, but somehow usually manage to spill on themselves. However, you encounter and even ingest a wide variety of fairly innocuous acids in everyday life. Citric acid, present in citrus fruits, such as lemons and oranges, is very ingestible, as is acetic acid, a.k.a. vinegar.

Strong acids can indeed burn the skin and must be handled with care in the laboratory. However, strong bases can burn skin as well. Chemists must have a more sophisticated understanding of the difference between an acid and a base and their relative strengths than simply their propensity to burn. This chapter focuses on how you can identify acids and bases, as well as several ways to determine their strengths, all of which are fair game on the AP chemistry exam.

Thinking Acidly: Three Different Ways to Define Acid-Base Behavior

As chemists came to understand acids and bases as more than just "stuff that burns," their understanding of how to define them evolved as well. It's often said that acids taste sour, while bases taste bitter, but we do *not* recommend that you go around tasting chemicals in the laboratory to identify them as acids or bases. In the following sections, we explain three much safer methods you can use to tell the difference between the two.

Method 1: Arrhenius sticks to the basics

Svante Arrhenius was a Swedish chemist who is credited not only with the acid-base determination method that is named after him, but with an even more fundamental chemical concept: that of *dissociation*. Arrhenius proposed an explanation in his PhD thesis for a phenomenon that, at the time, had chemists all over the world scratching their heads. What had them perplexed was this: Although neither pure salt nor pure water are good conductors of electricity, solutions in which salts are dissolved in water tend to be excellent conductors of electricity.

Arrhenius proposed that aqueous solutions of salts conducted electricity because the bonds between atoms in the salts had been broken simply by mixing them into the water, forming ions. Although the ion had been defined several decades earlier by Michael Faraday, chemists generally believed at the time that ions could only form through *electrolysis,* or the breaking of chemical bonds using electric currents, so Arrhenius's theory was met with some skepticism. Ironically, although his thesis committee wasn't overly impressed and gave him a grade just barely sufficient to pass, Arrhenius eventually managed to win over the scientific community with his research. He was awarded the Nobel Prize in Chemistry in 1903 for the same ideas that nearly cost him his doctorate.

Arrhenius subsequently expanded his theories to form one of the most widely used and straightforward definitions of acids and bases. Arrhenius said that acids are substances that form hydrogen (H^+) ions when they dissociate in water, while bases are substances that form hydroxide (OH^-) ions when they dissociate in water.

Peruse Table 17-1 for a list of common acids and bases and note that all of the acids in the list contain a hydrogen and most of the bases contain a hydroxide. The Arrhenius definition of acids and bases is very straightforward and works for many common acids and bases, but it's limited by its narrow definition of bases.

Table 17-1	Common Acids and Bases		
Acid Name	*Chemical Formula*	*Base Name*	*Chemical Formula*
Hydrochloric Acid	HCl	Potassium hydroxide	KOH
Hydrofluoric Acid	HF	Sodium hydroxide	NaOH
Nitric Acid	HNO_3	Calcium hydroxide	$Ca(OH)_2$
Nitrous Acid	HNO_2	Magnesium hydroxide	$Mg(OH)_2$
Sulfuric Acid	H_2SO_4	Sodium carbonate	Na_2CO_3
Citric Acid	$C_5H_7O_5COOH$	Ammonia	NH_3
Acetic Acid	CH_3COOH		

Method 2: Brønsted-Lowry tackles bases without a hydroxide ion

You no doubt noticed that some of the bases in Table 17-1 do not contain a hydroxide ion, which means that the Arrhenius definition of acids and bases can't apply. When chemists realized that several substances without a hydroxide ion still behaved like bases, they reluctantly acknowledged that another determination method was needed. Independently proposed by Johannes Brønsted and Thomas Lowry in 1923 and therefore named after both of them, the Brønsted-Lowry method for determining acids and bases accounts for those pesky non-hydroxide containing bases.

Under the Brønsted-Lowry definition, an acid is a substance that donates a hydrogen-ion (H^+) in an acid-base reaction, while a base is a substance that accepts that hydrogen ion from the acid. When ionized to form a hydrogen cation (an atom with a positive charge), hydrogen loses its one and only electron and is left with only a single proton. For this reason, Brønsted-Lowry acids are often called *proton donors* and Brønsted-Lowry bases are called *proton acceptors*.

The best way to spot Brønsted-Lowry acids and bases is to keep careful track of hydrogen ions in a chemical equation. Consider, for example, the dissociation of the base sodium carbonate in water. Note that although sodium carbonate is a base, it doesn't contain a hydroxide ion.

$$NaCO_3 + H_2O \rightarrow H_2CO_3 + NaOH$$

This is a simple double replacement reaction (see Chapter 13 for an introduction to these types of reactions). A hydrogen ion from water switches places with the sodium of sodium carbonate to form the products: carbonic acid and sodium hydroxide. By the Brønsted-Lowry definition, water is the acid because it gave up its hydrogen to $NaCO_3$. This makes $NaCO_3$ the base because it accepted the hydrogen ion from H_2O.

On the AP exam, if you ever see carbonic acid formed as a product, you will be expected to know that it breaks down into CO_2 and H_2O, making this reaction $NaCO_3 + H_2O \rightarrow H_2O + CO_2 + NaOH$.

Almost (but not quite) *all* the acids and bases you will encounter on the AP test will be Bronsted-Lowry type so master these first.

What about the substances on the righthand side of the equation? Brønsted-Lowry theory calls the products of an acid-base reaction the *conjugate acid* and *conjugate base*. The conjugate acid is produced when the base accepts a proton (in this case, H_2CO_3), while the conjugate base is formed when the acid loses its hydrogen (in this case, NaOH). This reaction also brings up a very important point about the strength of each of these acids and bases. Although sodium carbonate is a very strong base, its conjugate acid, carbonic acid, is a very weak acid. Weak acids always form strong conjugate bases and vice versa. The same is true of strong acids.

Method 3: Lewis relies on electron pairs

In the same year that Brønsted and Lowry proposed their definition of acids and bases, an American chemist named Gilbert Lewis proposed an alternate definition that not only encompassed the Brønsted-Lowry theory, but also accounted for acid-base reactions in which a hydrogen ion is not exchanged. Lewis's definition relies on tracking lone pairs of electrons. Under his theory, a base is any substance that donates a pair of electrons to form a coordinate covalent bond with another substance, while an acid is a substance that accepts that electron pair in such a reaction. Recall that a coordinate covalent bond is a covalent bond in which both of the bonding electrons are donated by one of the atoms forming the bond.

All Brønsted-Lowry acids are Lewis acids, but in practice, the term *Lewis acid* is generally reserved for Lewis acids that don't also fit the Brønsted-Lowry definition. The best way to spot a Lewis acid/base pair is to draw a Lewis dot structure of the reacting substances, noting the presence of lone pairs of electrons. (We introduce Lewis structures in Chapter 7.) For example, consider the reaction between ammonia (NH_3) and boron trifluoride (BF_3):

$$NH_3 + BF_3 \rightarrow NH_3BF_3$$

At first glance, neither the reactants nor the product appear to be acids or bases, but the reactants are revealed as a Lewis acid-base pair when drawn as Lewis dot structures as in Figure 17-1. Ammonia donates its lone pair of electrons to the bond with boron trifluoride, making ammonia the Lewis base and boron trifluoride the Lewis acid.

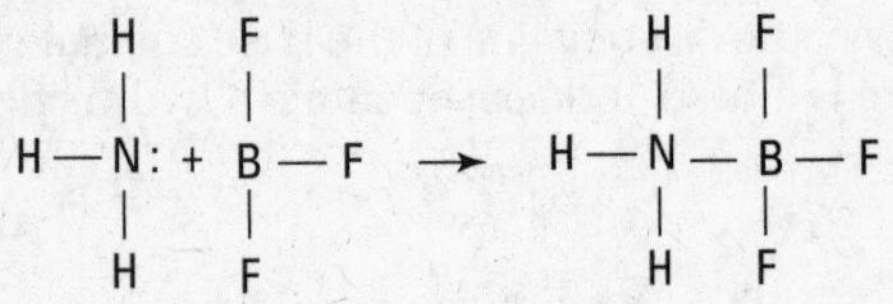

Figure 17-1:
The Lewis dot structures of ammonia and boron trifluoride.

$$\begin{array}{ccc} & H & F \\ & | & | \\ H-N: \ + & B-F & \rightarrow \\ & | & | \\ & H & F \end{array} \qquad \begin{array}{ccc} H & F \\ | & | \\ H-N-B-F \\ | & | \\ H & F \end{array}$$

Sometimes you can identify the Lewis acid and base in a compound without drawing the Lewis dot structure. You can do this by identifying reactants that are electron rich (bases) or electron poor (acids). Metal cations, for example, are electron poor and tend to act as a Lewis acid in a reaction, accepting a pair of electrons.

In practice, it's much simpler to use the Arrhenius or Brønsted-Lowry definitions of acid and base, but you'll need to use the Lewis definition when hydrogen ions aren't being exchanged. You can pick and choose among the definitions when you're asked to identify acid and base in a reaction.

Describing Acid-Base Equilibria with Constants

Arrhenius's concept of dissociation gives us another convenient way of measuring the strength of an acid or base. Although water tends to dissociate all acids and bases, the degree to which they dissociate depends on their strength. Strong acids such as HCl, HNO_3, and H_2SO_4 dissociate completely in water, while weak acids dissociate only partially. Practically speaking, a weak acid is any acid that doesn't dissociate completely in water. Note that while acid-base reactions do occur in other solvents, the AP exam will only deal with aqueous (water) solutions.

To measure the amount of dissociation occurring when a weak acid is in aqueous solution, chemists use a constant called the *acid dissociation constant* (K_a). K_a is a special variety of the equilibrium constant introduced in Chapter 15. As we explain in Chapter 15, the equilibrium constant of a chemical reaction is the concentration of products divided by concentration of reactants and indicates the balance between products and reactants in a reaction. The acid dissociation constant is simply the equilibrium constant of a reaction in which an acid is mixed with water and from which the water concentration has been removed. This is done because the concentration of water is a constant in dilute solutions, and a better indicator of acidity is the concentration of the dissociated products divided by the concentration of the acid reactant. The general form of the acid dissociation constant is therefore

$$K_a = \frac{\left[H_3O^+\right] \times \left[A^-\right]}{\left[HA\right]}$$

Where [HA] is the concentration of the acid before it loses its hydrogen, and [A⁻] is the concentration of its conjugate base. Notice that the concentration of the hydronium ion (H_3O^+) is used in place of the concentration of H^+, which we use to describe acids earlier in this chapter. In truth, they are one and the same and this is just shorthand for this species, since "bare" protons never exist in water. Generally speaking, H^+ ions in aqueous solution will be caught up by atoms of water in solution, making H_3O^+ ions.

A similar situation exists for bases. Strong bases such as KOH, NaOH, and $Ca(OH)_2$ dissociate completely in water. Weak bases don't dissociate completely in water, and their strength is measured by the *base dissociation constant,* or K_b.

$$K_b = \frac{\left[OH^-\right] \times \left[B^+\right]}{\left[BOH\right]}$$

Here, BOH is the base, and B^+ is its conjugate acid. This can also be written in terms of the acid and base:

$$K_b = \frac{\left[OH^-\right] \times \left[HA\right]}{\left[A^-\right]}$$

In problems where you are asked to calculate K_a or K_b, you'll generally be given the concentration, or molarity, of the original acid or base and the concentration of its conjugate *or* hydronium/hydroxide, but rarely both. Dissociation of a single molecule of acid involves the splitting of that acid into one molecule of its conjugate base and one hydrogen ion, while dissociation of a base always involves the splitting of that base into one molecule of the conjugate acid and one hydroxide ion. For this reason, the concentration of the conjugate and the concentration of the hydronium or hydroxide ion in any dissociation are equal, so you only need one to know the other. You may also be given the pH and be asked to figure out the [H⁺] (equivalent to [H₃O⁺]) or [OH⁻]. K_a and K_b are also constants at constant temperature. Remember the above only works if you can assume that *only* the acid (or base) + water is initaially present. What solutions can be acidic or basic without an acid or base present? Salt solutions!

In chemistry, a "salt" is not necessarily the substance you sprinkle on French fries, but has a much broader definition. Rather, a *salt* is any substance that is a combination of an anion (an atom with a negative charge) and a cation (an atom with a positive charge) and is created in a neutralization reaction. Salts, therefore, have a tendency to dissociate in water. The degree of dissociation possible — in other words, the solubility of the salt — varies greatly from one salt to another.

Chemists use a quantity called the *solubility product constant* or K_{sp} to compare the solubilities of salts. K_{sp} is calculated in much the same way as an equilibrium constant (K_{eq}). The product concentrations are multiplied together, each raised to the power of its coefficient in the balanced dissociation equation. There is one key difference, however, between a K_{sp} and a K_{eq}. K_{sp} is a quantity specific to a *saturated* solution of salt, so the concentration of the undissociated salt reactant has absolutely no bearing on its value. If the solution is saturated, then the amount of possible dissociation is at its maximum, and any additional solute added merely settles on the bottom.

Measuring Acidity and Basicity with pH

A substance's identity as an acid or a base is only one of many things that a chemist may need to know about it. Sulfuric acid and water, for example, can both act as acids, but using sulfuric acid to wash your face in the morning would be a grave error indeed. Sulfuric acid and water differ greatly in *acidity*, a measurement of an acid's strength. A similar quantity, called *basicity*, measures a base's strength.

Acidity and basicity are measured in terms of quantities called *pH* and *pOH*, respectively. Both are simple scales ranging from 0 to 14, with low numbers on the pH scale representing a higher acidity and therefore a stronger acid. On both scales, a measurement of 7 indicates a *neutral solution*. On the pH scale, any number lower than 7 indicates that the solution is acidic, with acidity increasing as pH decreases, while any number higher than 7 indicates a basic solution, with basicity increasing as pH increases. In other words, the farther the pH gets away from 7, the more acidic or basic a substance gets. pOH shows exactly the same relationship between distance from 7 and acidity or basicity, only this time low numbers indicate very basic solutions, while high numbers indicate very acidic solutions. Both the pH and the pOH scales are logarithmic, meaning that every time you increase them by 1, the concentration of H^+ or OH^- in the solution increases by 10. An acid with a pH of 1 has 100 times more H^+ ions than and acid with a pH of 3.

pH is calculated using the formula $pH = -\log[H^+]$, where the brackets around H^+ indicate that it's a measurement of the concentration of hydrogen ions in moles per liter (or molarity; see Chapter 11). pOH is calculated using a similar formula, with OH^- concentration replacing the H^+ concentration: $pOH = -\log[OH^-]$. (The word *log* in each formula stands for base 10 *logarithm*.)

Because a substance with high acidity must have low basicity, a low pH (high acidity) indicates a high pOH for a substance and vice versa. In fact, a very convenient relationship between pH and pOH allows you to solve for one when you have the other: $pH + pOH = 14$.

You'll often be given a pH or pOH and be asked to solve for the H^+ or OH^- concentrations instead of the other way around. The logarithms in the pH and pOH equations make it tricky to solve for $[H^+]$ or $[OH^-]$, but if you remember that a log is inversed by raising 10 to the power of each side of an equation, you will quickly arrive at a convenient formula for $[H^+]$, namely $[H^+] = 10^{-pH}$. Similarly, $[OH^-]$ can be calculated using the formula $[OH^-] = 10^{-pOH}$. As with pH and pOH, a convenient relationship exists between $[H^+]$ and $[OH^-]$, which multiply together to equal a constant. This constant, called the *ion product constant for water*, or K_W, is calculated as follows: $K_W = [H^+] \times [OH^-] = 1 \times 10^{-14}$.

Taking It One Proton at a Time: Titration

So far we have treated acids and bases separately, but chemists are awfully fond of mixing things. In the real world of chemistry, acids and bases often meet in solution and when they do, they're drawn to one another. These unions of acid and base are called *neutralization reactions* because the low pH of the acid and the high pH of the base neutralize one another. This does not, however, mean that the final solution will be neutral (pH=7) because not all acids and bases have equal abilities to neutralize nor do they have to be mixed in equal amounts or at equal concentrations to create a neutralization reaction.

Although strong acids and bases have their uses, the prolonged presence of a strong acid or base in an environment not equipped to handle it can be very damaging. In the laboratory, for example, you need to handle strong acids and bases carefully, and deliberately perform neutralization reactions when appropriate. A lazy chemistry student will get a brow-beating from her hawk-eyed teacher if she attempts to dump a concentrated acid or base down the laboratory sink. Doing so can damage the pipes and is generally unsafe. Instead, a responsible chemist will neutralize acidic laboratory waste with a base such as baking soda, and will neutralize basic waste with an acid. Doing so makes most solutions perfectly safe to dump down the drain and often results in the creation of a satisfyingly sizzly solution while the reaction is occurring.

Neutralizing equivalents

At heart, neutralization reactions in which the base contains a hydroxide ion are simple double replacement reactions of the form $HA + BOH \rightarrow BA + H_2O$ (or in other words, an acid reacts with a base to form a salt and water). You're asked to write a number of such reactions in this chapter, so be sure to review double replacement reactions and balancing equations in Chapter 13 before you delve into the new and exciting world of neutralization.

Unlike people, not all acids and bases are created equally. Some have an innate ability to neutralize more effectively than others. Consider hydrochloric acid (HCl) and sulfuric acid (H_2SO_4), for example. If you mixed 1M sodium hydroxide (NaOH) together with 1M hydrochloric acid, you'd need to add equal amounts of each to create a neutral solution. If you mixed sodium hydroxide with sulfuric acid, however, you'd need to add twice as much sodium hydroxide as sulfuric acid. Why this blatant inequality of acids? The answer lies in the balanced neutralization reactions for both acid/base pairs.

$$HCl + NaOH \rightarrow NaCl + H_2O$$

$$H_2SO_4 + 2NaOH \rightarrow Na_2SO_4 + 2H_2O$$

The coefficients in the balanced equations are the key to understanding this inequality. To balance the equation, the coefficient 2 needs to be added to sodium hydroxide, indicating that two moles of it must be present to neutralize one mole of sulfuric acid. On a molecular level, this happens because sulfuric acid has two acidic hydrogen atoms to give up, and the single hydroxide in a molecule of sodium hydroxide can only neutralize one of those two acidic hydrogens to form water. Therefore, two moles of sodium hydroxide are needed for every one mole of sulfuric acid. Hydrochloric acid, on the other hand, has only one acidic hydrogen to contribute, so it can be neutralized by an equal amount of sodium hydroxide, which has only one hydroxide to contribute to neutralization.

The number of moles of an acid or base multiplied by the number of hydrogens or hydroxides that a molecule has to contribute in a neutralization reaction is called the number of *equivalents* of that substance. Basically, the number of *effective* neutralizing moles available determines the ratio of acid to base in a neutralization reaction.

Titrating with equivalents

Imagine you're a newly hired laboratory assistant who's been asked to alphabetize the chemicals on the shelves of a chemistry laboratory during a lull in experimenting. As you reach for the bottle of sulfuric acid, your first-day jitters get the better of you, and you knock over the bottle. Some careless chemist failed to screw the cap on tightly! You quickly neutralize the acid with a splash of baking soda and wipe up the now-nicely neutral solution. As you pick up the bottle, however, you notice that the spilled acid burned away most of the label! You know it's sulfuric acid, but there are several different concentrations of sulfuric acid on the shelves, and you don't know the molarity of the solution in this bottle. Knowing that your boss will surely blame you if she sees the damaged bottle, and not wanting to get sacked on your very first day, you quickly come up with a way to determine the molarity of the solution and save your job.

You know that the bottle contains sulfuric acid of a mystery concentration, and you notice bottles of 1M sodium hydroxide, a strong base, and phenolphthalein, a pH indicator, among the chemicals on the shelves. You measure a small amount of the mystery acid into a beaker and add a little phenolphthalein. You reason that if you drop small amounts of sodium hydroxide into the solution until the phenolphthalein indicates that the solution is neutral by turning the appropriate color, you'll be able to figure out the acid's concentration.

You can do this by making a simple calculation of the number of moles of sodium hydroxide you have added, and then reasoning that the mystery acid must have an equal number of equivalents to have been neutralized. This then leads to the number of moles of acid, and that, in turn, can be divided by the volume of acid you added to the beaker to get the molarity. Whew! You relabel the bottle and rejoice in the fact that you can come in and do menial labor in the lab again tomorrow.

This process is called a *titration,* and it's often used by chemists to determine the molarity of acids and bases. In a titration calculation, you generally know the identity of an acid or base of unknown concentration, and the identity and molarity of the acid or base which you're going to use to neutralize it. Given this information, you then follow six simple steps:

1. **Measure out a small volume of the mystery acid or base.**

2. **Add a pH indicator such as phenolphthalein.**

   Be sure to take note of the color that will indicate that the solution has reached neutrality. Again, this does not necessarily mean pH=7. Rather, the solution will reach neutrality when the reaction between the acid and base is complete, which may leave the solution with a pH other than 7.

3. **Neutralize.**

   Drop the acid or base of known concentration into the solution until the indicator shows that it's neutral, keeping careful track of the volume added.

4. **Calculate the number of moles added.**

   Multiply the number of liters of acid or base added by the molarity of that acid or base to get the number of moles added.

5. **Account for equivalents.**

   Determine how many moles of the mystery substance being neutralized are present using equivalents. How many moles of your acid do you need to neutralize one mole of base or vice versa?

6. **Solve for molarity.**

   Divide the number of moles of the mystery acid or base by the number of liters measured out in Step 1, giving you the molarity.

The titration process is often visualized using a graph showing concentration of base on one axis and concentration of acid on the other as in Figure 17-2. The interaction of the two traces out a *titration curve,* which has a characteristic *s* shape.

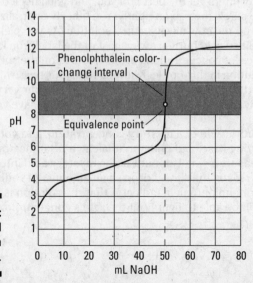

Figure 17-2:
A typical
titration
curve.

Keeping Steady: Buffers

You may have noticed that the titration curve shown in Figure 17-2 has a flattened area in the middle where pH doesn't change significantly, even when you add a conspicuous amount of base. This region is called a *buffer region.*

Certain solutions, called *buffered solutions,* resist changes in pH like a stubborn child resists eating her Brussels sprouts: steadfastly at first, but choking them down reluctantly if enough pressure is applied (such as the threat of no dessert). Although buffered solutions maintain their pH very well when small amounts of acid or base are added to them or the solution is diluted, they can only withstand the addition of a certain amount of acid or base before becoming overwhelmed.

Buffers are most often made up of a weak acid and its conjugate base, though they can also be made of a weak base and its conjugate acid. A weak acid in aqueous solution will be partially dissociated, the amount of dissociation depending on its pK_a value (the negative logarithm of its acid dissociation constant). The dissociation will be of the form $HA + H_2O \rightarrow H_3O^+ + A^-$, where A^- is the conjugate base of the acid HA. The acidic proton is taken up by a water molecule, forming hydronium. If HA were a strong acid, 100 percent of the acid would become H_3O^+ and A^-, but because it's a weak acid, only a fraction of the HA dissociates, and the rest remains HA.

Solving the earlier expression for K_a for the $[H_3O^+]$ concentration allows you to devise a relationship between the $[H_3O^+]$ and the K_a of a buffer.

$$\left[H_3O^+ \right] = K_a \times \frac{\left[HA \right]}{\left[A^- \right]}$$

Taking the negative logarithm of both sides of the equation and manipulating logarithm rules yields an equation called the *Henderson-Hasselbalch equation* that relates the pH and the pK_a.

$$pH = pK_a + \log\left(\frac{\left[A^- \right]}{\left[HA \right]} \right)$$

This equation can be manipulated using logarithm rules to get $\frac{\left[A^- \right]}{\left[HA \right]} = 10^{pH-pK_a}$, which may be more useful in certain situations.

If you prefer to (or are asked to) calculate the pOH instead, you can use the equation

$$pOH = pK_b + \log\left(\frac{\left[HB^+ \right]}{\left[B \right]} \right)$$

The very best buffers and those best able to withstand the addition of both acid and base are those for which [HA] and [A^-] are approximately equal. Because the [A^-]/[HA] = 1 and the log of 1 is 0, the logarithmic term in the Henderson-Hasselbalch equation disappears, and the equation becomes pH = pK_a. When creating a buffered solution, chemists therefore choose an acid that has a pK_a close to the desired pH.

If you add a strong base such as sodium hydroxide (NaOH) to this mixture of dissociated and undissociated acid, its hydroxide is absorbed by the acidic proton, replacing the exceptionally strong base OH^- with a relatively weak base A^-, and minimizing the change in pH.

$$HA + OH^- \rightarrow H_2O + A^-$$

This causes a slight excess of base in the reaction, but doesn't affect pH significantly. You can think of the undissociated acid as a reservoir of protons that are available to neutralize any strong base that may be introduced to the solution. As we explain in Chapter 15, when a product is added to a reaction, the equilibrium in the reaction changes to favor the reactants or to "undo" the change in conditions. Because this reaction generates A^-, the acid dissociation reaction happens less frequently as a result, further stabilizing the pH.

When a strong acid, such as hydrochloric acid (HCl), is added to the mixture, its acidic proton is taken up by the base A^-, forming HA.

$$H^+ + A^- \rightarrow HA$$

This causes a slight excess of acid in the reaction but doesn't affect pH significantly. It also shifts the balance in the acid dissociation reaction in favor of the products, causing it to happen more frequently and re-creating the base A^-.

The addition of acid and base and their effect on the ratio of products and reactants is summarized in Figure 17-3.

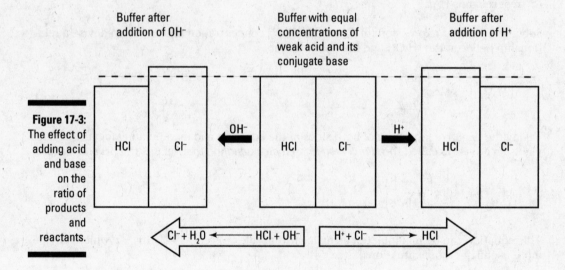

Figure 17-3:
The effect of adding acid and base on the ratio of products and reactants.

Buffer after addition of OH^-

Buffer with equal concentrations of weak acid and its conjugate base

Buffer after addition of H^+

Buffers have their limits, however. The acid's proton reservoir, for example, can only compensate for the addition of a certain amount of base before it runs out of protons that can neutralize free hydroxide. At this point, a buffer has done all it can do, and the titration curve resumes its steep, upward slope.

Chapter 18

Answering Questions on Acids and Bases

- -

In This Chapter

▶ Pointing out what you need to know

▶ Getting in some practice

▶ Going through the answers and explanations

- -

Acids and bases are extremely common participants of chemical reactions, and even have their own special class of reaction (neutralization). Because acidity/basicity is such an important property of aqueous solutions, this topic travels with a whole entourage of concepts and equations. There are different ways to define acids and bases. There are different ways to express acid and base concentrations. There are special quantities for expressing just how acidic or basic a particular compound is. This chapter summarizes key points and skills from Chapter 17, and provides practice questions that test your ability to put acid-base concepts to use.

Soaking Up the Main Points

The acid-base nitty-gritty comes down to two major ideas. First, what are acids and bases, and how can you recognize them? Second, how do acids and bases react with each other? This section organizes highlights of Chapter 17 around these two categories.

Knowing and measuring acids and bases

There are several common ways to define acids and bases, all overlapping to some extent. Having defined them, you can measure their concentrations and express those measurements in different ways, as summarized in the following points.

✔ Three important methods for defining acids and bases are as follows:

- Arrhenius acids are those that dissociate to form hydrogen ions, while bases dissociate to form hydroxide ions in solution.

- Brønsted-Lowry acids donate protons (hydrogen ions), which are accepted by the Brønsted-Lowry base or proton acceptor.

- Lewis bases donate a pair of electrons to form a covalent bond with a Lewis acid, which accepts the pair of electrons to form a coordinate covalent bond.

✔ The acid dissociation constant K_a gives an indication of the acidity of a substance through the expression

$$K_a = \frac{\left[H_3O^+\right] \times \left[A^-\right]}{\left[HA\right]}$$

✔ Similarly, the base dissociation constant K_b indicates the basicity of a substance through the expression

$$K_a = \frac{\left[H^+\right]\left[CH_3COO^-\right]}{\left[CH_3COOH\right]}$$

✔ The pH scale ranges from 0 to 14 and measures acidity, with low numbers indicating acidic solutions, high numbers indicating basic solutions, and 7 indicating a perfectly neutral solution. It is calculated using the equation $pH = -\log[H^+]$.

✔ The pOH scale also ranges from 0 to 14, but measures basicity instead of acidity. Low numbers mean basic solutions and high numbers mean acidic solutions. It is calculated using the equation $pOH = -\log[OH^-]$.

✔ pH and pOH are related through the simple expression $pH + pOH = 14$. In a related way, $[H^+] \times [OH^-] = 10^{-14}$.

Reacting acids and bases

Split H_2O and you can get H^+ and OH^-, acid and base respectively. Add acid to base, and you regain water, neutralizing the acid and base reactants. The idea is simple, but neutralization reactions can be tricky. The following points lay out the core principles.

✔ Neutralization reactions occur when acids and bases are mixed together. The amount of acid needed to neutralize a base or vice versa depends not only on the molarity of the acidic solution, but on the number of hydrogen atoms that the acid can donate to neutralize the hydroxide ions of the base.

✔ Equivalents of acid and base are considered rather than molarities, because different compounds can contribute different numbers of H^+ or OH^-. Equivalents are calculated by multiplying the number of moles of compound by the number of hydrogen or hydroxide ions being contributed by each molecule of compound.

✔ Titrations utilize neutralization reactions in order to determine the concentration of a mystery acid or base. They're accomplished through the following six steps:

1. Measure out a small volume of the mystery acid or base.

2. Add a pH indicator such as phenolphthalein (see Chapter 28 for details about how to choose the best pH indicator for a reaction). Phenolphthalein, for example, is colorless in acidic solutions but pink or purple in basic solutions.

3. Neutralize by dropping the acid or base of known concentration into the solution until the indicator shows that it's neutral (by changing color), keeping careful track of the volume added.

4. Calculate the number of moles added by multiplying the number of liters of acid or base added by the molarity of that acid or base to get the number of moles added.

5. Account for equivalents.

6. Solve for molarity. Divide the number of moles of the mystery acid or base by the number of liters measured out in Step 1, giving you the molarity.

✔ Buffered solutions are made of a weak acid and its conjugate base, or a weak base and its conjugate acid, and resist changes in pH as long as only a small amount of acid or base is added. The best buffers are those that have a pK_a value equal or nearly equal to the desired pH of the solution. Note that pK_a = –logK_a and pK_b = –logK_b.

✔ pH and pK_a are related through the Henderson-Hasselbach equation as follows:

$$0.080L \times \frac{0.315\text{mol CH}_3\text{COOH}}{L} \times \frac{1\text{mol NaOH}}{1\text{mol CH}_3\text{COOH}} \times \frac{1L}{0.250\text{mol NaOH}} = 0.10L\text{ NaOH}$$

Practice Questions

With so many concepts and equations swirling about, its clear that acids and bases just aren't the kind of chemistry you simply read about and leave be. You are guaranteed to encounter them on the AP exam, and you are equally guaranteed to be confused by them — unless you practice, that is. Here are some targeted questions to help give you that practice.

Multiple choice

1. A solution prepared by mixing 15mL of 1.0M H_2SO_4 and 15mL of 2.0M KOH has a pH of

   (A) 6.

   (B) 7.

   (C) 8.

   (D) 9.

   (E) 10.

2. H_2O < NH_3 < $Ca(OH)_2$ < LiOH < KOH < OH–

   Six bases are listed above in order of increasing base strength. Which of the following reactions must have an equilibrium constant with a value less than 1?

   (A) $Ca(OH)_2$ + K_2CO_3 → $CaCO_3$ + 2KOH

   (B) NH_3 + H_2O → NH_4 + OH^-

   (C) 2LiOH + CO_2 → Li_2CO_3 + $3H_2O$

   (D) LiOH + H_2O → LiH + KOH

   (E) KOH + H_2O → KH + OH–

3. Which of the following equations represents the reaction between solid magnesium hydroxide and aqueous hydrochloric acid?

   (A) $Mg(OH)_2(s)$ + 2HCl(l) → $MgCl_2(aq)$ + $H_2O(l)$

   (B) MgOH(s) + HCl(aq) → MgCl(aq) + $2H_2O(l)$

   (C) $Mg(OH)_2(s)$ + 2HCl(aq) → $MgCl_2(aq)$ + $2H_2O(l)$

   (D) $Mg(OH)_2(s)$ + HCl(aq) → $MgCl_2(aq)$ + $H_2O(l)$

   (E) MgOH(s) + 2HCl(aq) → MgCl(aq) + $2H_2O$ (l)

4. What is the ideal pKa for an indicator in a titration when the pOH at the equivalence point is 9.8?

    (A) 2.1

    (B) 4.2

    (C) 4.9

    (D) 9.8

    (E) 10

Free response

5. $CH_3COOH(aq) \rightarrow H^+(aq) + CH_3COO^-(aq)$

Acetic acid, CH_3COOH, is a weak acid that dissociates in water according to the equation above.

    (a) Calculate the pH of the solution if the H^+ concentration is $3.98 \times 10^{-4}M$.

    (b) A solution of NaOH is titrated into a solution of CH_3COOH.

    (i) Calculate the volume of 0.250M NaOH needed to reach the equivalence point when titrated into an 80mL sample of 0.315M CH_3COOH.

    (c) Calculate the number of moles of $CH_3COO^-Na^+$ that would have to be added to 150mL of 0.025M CH_3COOH to produce a buffered solution with $[H^+] = 4.25 \times 10^{-7}M$ if the K_a of acetic acid is 1.8×10^{-5}. Assume that the volume change is negligible.

6. $C_5H_5N(aq) + H_2O(l) \rightarrow C_5H_5NH^+(aq) + OH^-(aq)$

Pyridine (C_5H_5N), a weak base, reacts with water according to the reaction above.

    (a) Write the equilibrium constant expression, K_b, for the reaction above.

    (b) A sample of pyridine is dissolved in water to produce 50mL of a 0.20M solution. The pH of the solution is 8.50. Calculate the equilibrium constant, K_b, for this reaction.

Answers with Explanations

Have any doubts about your answers to the practice questions? Neutralize those doubts by inspecting the answers given here. Even if you find that you've answered all the questions correctly, it's a good idea to read the explanations to solidify your acid-base genius and because, well, you might have been simply lucky on one or two.

1. (B). Problems like this are really just limiting reagent problems. The acid and base will react with one another until one or the other is used up and the amount of remaining reactant will determine the pH. In this case, however, the 1.0M diprotic sulfuric acid and the 2.0M potassium hydroxide exactly balance one another:

$$\frac{0.015L \ H_2SO_4}{1} \times \frac{1.0mol \ H_2SO_4}{1L \ H_2SO_4} \times \frac{2 \ H^+}{1 \ H_2SO_4} = 0.030mol \ H^+$$

$$\frac{0.015L \ KOH}{1} \times \frac{2.0mol \ KOH}{1L \ KOH} \times \frac{2 \ OH^-}{1 \ KOH} = 0.030mol \ OH^-$$

The acid has two protons to donate per molecule and the base is twice as concentrated. They're mixed in equal amounts, so there is no leftover acid or base. The resulting pH is perfectly neutral.

2. (C). If the equilibrium constant is less than 1, then at equilibrium the concentration of reactants must be greater than the concentration of products. Because the information given is regarding base strength, the problem must be dealing with the tendency of dissociation (and therefore concentration) to increase with the strength of the base. You're looking for a reaction where the stronger base appears on the lefthand side of the reaction arrow. The only reaction where this is the case is C.

3. (C). You're looking for a balanced reaction, which shows the proper phases for all of the products and reactants. C is the only reaction that fulfills both of those criteria and contains molecules with atoms combining in the proper ratios for charge balance.

4. (B). The ideal pKa is equal to the pH at the equivalence point. Use the given pOH to determine this value using the equation pH + pOH = 14. This equation gives you 14 − 9.8 = 4.2.

5. (a) Plug the given H$^+$ concentration into the equation pH = −log[H$^+$], which gives you pH = 3.40.

   (b) $0.080L \times \dfrac{0.315mol\ CH_3COOH}{L} \times \dfrac{1mol\ NaOH}{1mol\ CH_3COOH} \times \dfrac{1L}{0.250mol\ NaOH} = 0.10L\ NaOH$

   (c) 159mL. You're asked to solve for the number of moles of $CH_3COO^-Na^+$ to be added. The easiest way to do this is to go through the CH_3COO^- concentration, which is equivalent to the $CH_3COO^-Na^+$ concentration because it dissociates in water. Begin by writing out the expression for the acid dissociation constant.

   Then, solve for the concentration of CH_3COO^-:

   $0.080L \times \dfrac{0.315\ mol\ CH_3COOH}{L} \times \dfrac{1\ mol\ NaOH}{1\ mol\ CH_3COOH} \times \dfrac{1L}{0.250\ mol\ NaOH} = 0.10L\ NaOH$

   Finally, multiply this molarity by the volume to get the number of moles added.

6. (a) The equilibrium constant is the concentration of the products over the concentration of the reactants, excluding water. This results in the expression

   $$K_b = \dfrac{\left[C_5H_5NH^+\right] \times \left[OH^-\right]}{C_5H_5N}$$

(b) 2.3×10^{-9}. Use the given pH to find the pOH and from there the OH$^-$ concentration. Because pH = 9.33, the pOH = 14 − 9.33 = 4.67. Use this to solve for [OH$^-$].

   $$\left[OH^-\right] = 10^{-4.67} = 2.14 \times 10^{-5}$$

This concentration should be equivalent to the [$C_5H_5NH^+$] concentration. Plug these and the given C_5H_5N concentration into the equation derived in part (a) to get your K_b, giving you

   $$K_b = \dfrac{\left(2.14 \times 10^{-5}\right)^2}{0.20} = 2.3 \times 10^{-9}$$

Chapter 19

Transferring Charge: Oxidation-Reduction

*I*n chemistry, electrons get all the action. Among other things, electrons might transfer between reactants during a reaction. Reactions like these are called *oxidation-reduction* reactions, or *redox* reactions for brevity. Redox reactions are as important as they are common. But they're not always obvious. In this chapter, you discover how to recognize redox reactions and how to balance the equations that describe them. All combustion, combination, decomposition, and single replacement reactions are redox. You'll also learn how to figure out what goes on inside electrochemical cells, chemical systems that use redox reactions to generate electrical current.

Using Oxidation Numbers to Recognize Redox Reactions

If electrons move between reactants during redox reactions, then it should be easy to recognize those reactions just by noticing changes in charge, right?

Sometimes.

The following two reactions are both redox reactions:

$$Mg(s) + 2H^+(aq) \rightarrow Mg^{2+}(aq) + H_2(g)$$

$$2H_2(g) + O_2(g) \rightarrow 2H_2O(g)$$

In the first reaction, it's obvious that electrons are transferred from magnesium to hydrogen. In the second reaction, it's not so obvious that electrons are transferred from hydrogen to oxygen. We need a way to keep tabs on electrons as they transfer between reactants. In short, we need *oxidation numbers*.

Oxidation numbers are tools to keep track of electrons. Sometimes an oxidation number describes the actual charge on an atom. Other times an oxidation number describes an imaginary sort of charge — the charge an atom would have if all its bonding partners left town, taking their own electrons with them. In ionic compounds, oxidation numbers come closest to describing actual atomic charge. The description is less apt in covalent compounds, in which electrons are less clearly "owned" by one atom or another. The point is this: Oxidation numbers are useful accounting tools, but they are not direct descriptions of physical reality. And remember that oxidation numbers apply to individual atoms, so it is possible that different atoms of the same element can have different oxidation numbers within the same compound.

Here are some basic rules **that are never broken** for figuring out an atom's oxidation number:

- ✔ Atoms in elemental form have an oxidation number of zero. So, the oxidation number of both $Mg(s)$ and $O_2(g)$ is zero.

- ✔ Single-atom (monatomic) ions have an oxidation number equal to their charge. So, the oxidation number of $Mg^{2+}(aq)$ is +2 and the oxidation number of $Cl^-(aq)$ is –1.

- ✔ In a neutral compound, oxidation numbers add up to zero. In a charged compound, oxidation numbers add up to the compound's charge.

Here are some basic rules that **that have exceptions** for figuring out an atom's oxidation number:

- ✔ In compounds, oxygen usually has an oxidation number of –2. An annoying exception is the peroxides, like H_2O_2, in which oxygen has an oxidation number of –1.

- ✔ In compounds, hydrogen has an oxidation number of +1 when it bonds to nonmetals (as in H_2O), and an oxidation number of –1 when it bonds to metals (as in NaH).

- ✔ In compounds

  - Group IA atoms (alkali metals) have oxidation number +1.

  - Group IIA atoms (alkaline earth metals) have oxidation number +2.

  - Group IIIA atoms have oxidation number +3.

  - Group VIIA atoms (halogens) usually have oxidation number –1.

By applying these rules to chemical reactions, you can discern who supplies electrons to whom. The chemical species that loses electrons is *oxidized,* and acts as the *reducing agent* (or *reductant*). The species that gains electrons is *reduced,* and acts as the *oxidizing agent* (or *oxidant*). All redox reactions have both an oxidizing agent and a reducing agent.

Oxidation may or may not involve bonding with oxygen, breaking bonds with hydrogen or losing electrons — but oxidation always means an increase in oxidation number.

Reduction may or may not involve bonding with hydrogen, breaking bonds with oxygen or gaining electrons — but reduction always means a decrease in oxidation number.

For example, consider the following reaction:

$$Cr_2O_3(s) + 2Al(s) \rightarrow 2Cr(s) + Al_2O_3(s)$$

Which are the oxidizing and reducing agents? To answer this question, you must assign oxidation numbers to each atom and then see how those numbers change between the reactant side and the product side. Recall that atoms in elemental form (like solid Al and solid Cr) have oxidation number zero, and that oxygen typically has oxidation number –2 in compounds. The oxidation number breakdown (shown in Figure 19-1) reveals that $Al(s)$ is oxidized to $Al_2O_3(s)$, and $Cr_2O_3(s)$ is reduced to $Cr(s)$. So, $Al(s)$ is the reducing agent and $Cr_2O_3(s)$ is the oxidizing agent.

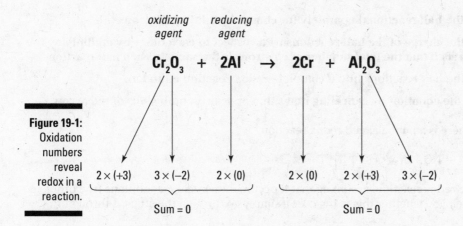

Figure 19-1:
Oxidation numbers reveal redox in a reaction.

Respecting the Hyphen: Oxidation-Reduction Half-Reactions

When you balance a chemical reaction equation, the primary concern is to obey the principle of Conservation of Mass — the total mass of the reactants must equal the total mass of the products. This is done by conserving and accounting for atoms. See Chapter 9 if you need to review this process. In redox reactions, you must obey a second principle as well: the Conservation of Charge. The total number of electrons lost must equal the total number of electrons gained. In other words, you can't just leave electrons lying around. The universe is finicky about that type of thing. People hire accountants to keep the universe straight, so AP chemistry is just Accounting 101!

Sometimes simply balancing a redox reaction with an eye to mass results in a charge-balanced equation as well. In other words, tracking atoms also tracks all the electrons. Like a string of green lights, that's a lovely thing when it occurs, but you can't count on it. So, it's best to have a fail-safe system for balancing redox reactions. The details of that system depend on whether the reaction occurs in acidic or basic condtions — in the presence of excess H^+ or excess OH^-. Both variations use *half-reactions*, incomplete parts of the total reaction that reflect either oxidation or reduction alone.

Balancing under acidic conditions

To properly conserve mass and charge as you balance a redox reaction under acidic conditions, you often need to add water or hydrogen ion to the reactants or products. You are allowed to add water because, as the solvent, water is always available to participate in the reaction. You are allowed to add hydrogen ion because that species is present at higher concentration in acidic solutions.

Here is a summary of the method for balancing a redox reaction equation for a reaction under **acidic conditions:**

1. **Separate the reaction equation into the oxidation half-reaction and the reduction half-reaction.** Use oxidation numbers to identify these component half-reactions.

2. **Balance the half-reactions separately, temporarily ignoring O and H atoms.**

3. **Balance the half-reactions separately, using H_2O to add O atoms and using H^+ to add H atoms.**

4. **Balance the half-reactions separately for charge by adding electrons (e⁻).**

5. **Balance the charge of the half-reactions with respect to each other by multiplying the reactions such that the total number of electrons is the same in each half-reaction.**

6. **Reunite the half-reactions into a complete redox reaction equation.**

7. **Simplify the equation by canceling items that appear on both sides of the arrow.**

For example, here is an unbalanced redox reaction:

$$NO_2^- + Br^- \rightarrow NO + Br_2$$

To balance the redox reaction, assuming that it occurs under acidic conditions (the problem will typically tell you whether this is the case), simply go through the steps, 1 through 7:

1. **Divide the equation into half-reactions for oxidation and reduction:**

   $Br^- \rightarrow Br_2$ (oxidation)

   $NO_2^- \rightarrow NO$ (reduction)

2. **Balance half-reactions, temporarily ignoring O and H:**

   $2Br^- \rightarrow Br_2$

   $NO_2^- \rightarrow NO$

3. **Balance half-reactions for O and H by adding H⁺ and H₂O, respectively:**

   $2Br^- \rightarrow Br_2$

   $2H^+ + NO_2^- \rightarrow NO + H_2O$

4. **Balance the charge within each half-reaction by adding electrons (e⁻):**

   $2Br^- \rightarrow Br_2 + 2e^-$

   $1e^- + 2H^+ + NO_2^- \rightarrow NO + H_2O$

5. **Balance the charge of the half-reactions with respect to each other:**

   $2Br^- \rightarrow Br_2 + 2e^-$

   $2e^- + 4H^+ + 2NO_2^- \rightarrow 2NO + 2H_2O$

6. **Add the half-reactions, reuniting them within the total reaction equation:**

   $2e^- + 4H^+ + 2Br^- + 2NO_2^- \rightarrow Br_2 + 2NO + 2H_2O + 2e^-$

7. **Simplify by canceling items that appear on both sides of the equation:**

   $4H^+ + 2Br^- + 2NO_2^- \rightarrow Br_2 + 2NO + 2H_2O$

Balancing under basic conditions

Fine, you say, but what if the reaction occurs under basic conditions? Do I have to learn a whole different set of rules? No — the process for balancing redox equations under basic conditions is 90 percent identical to the one used for balancing under acidic conditions. In other words, master one and you've mastered both. Instead of using water and hydrogen ion as balancing tools, you are allowed to use water and hydroxide ion, the ion present in greater concentration in basic solutions.

Here's how easy it is to adapt your balancing method for basic conditions:

✔ Perform Steps 1 through 7 as described for balancing under acidic conditions.

✔ Observe where H^+ is present in the resulting equation. Add an identical amount of OH^- to both sides of the equation such that all the H^+ is "neutralized," becoming water.

✔ Cancel any amounts of H_2O that appear on both sides of the equation.

That's it. Really.

For example, here is another unbalanced redox reaction:

$$Cr^{2+} + Hg \rightarrow Cr + HgO$$

Assuming the reaction takes place under basic conditions (again, problems typically tell you whether acidic or basic conditions apply), begin balancing as if the reaction occurs under acidic conditions, and then neutralize any H^+ ions by adding OH^- equally to both sides. Finally, cancel any excess H_2O molecules.

Follow Steps 1 through 7 as under acidic conditions:

$$H_2O + Cr^{2+} + Hg \rightarrow Cr + HgO + 2H^+$$

Neutralize H^+ by adding sufficient and equal amounts of OH^- to both sides:

$$2OH^- + H_2O + Cr^{2+} + Hg \rightarrow Cr + HgO + 2H_2O$$

Simplify by canceling H_2O as possible from both sides:

$$2OH^- + Cr^{2+} + Hg \rightarrow Cr + HgO + H_2O$$

Redox reactions have a bad reputation among chemistry students due the the perceptions that they're difficult to understand and even more difficult to balance. But the perception is only a perception.

The whole process boils down to the following principles:

✔ Balancing redox reaction equations is exactly like balancing other equations; you simply have one more component to balance — the electron.

✔ Use oxidation numbers to discern the oxidation and reduction half-reactions.

✔ Under acidic conditions, balance O and H atoms by adding H_2O and H^+.

✔ Under basic condtions, balance as you do with acidic conditions, but then neutralize any H^+ by adding OH^-.

✔ There are other methods for balancing redox reaction equations, so feel free to use another method if you're more comfortable with it!

Keeping Current: Electrochemical Cells

Redox reactions are responsible for some not-so-useful things, like rust. But they're also responsible for some very useful things, like *electrochemical cells*, devices that convert the chemical energy of a redox reaction into electric energy of flowing current, or vice versa. *Voltaic cells* are electrochemical cells that convert chemical energy into electrical energy.

Batteries are examples of voltaic cells. A device called an *electrolytic cell* converts electrical energy into chemical energy. Keeping track of the relationships between different reaction mixtures and the direction and magnitude of current flow can be tricky. In this section, we describe the fundamentals of electrochemical and electrolytic cells. In addition, we give you the basic math you need to relate redox reactions to current flow in these kinds of cells.

Navigating voltaic cells

The energy provided by batteries is created in a unit called a *voltaic cell,* the kind of electro-chemical cell that converts chemical energy to electrical energy. Many batteries use a number of voltaic cells wired in series, and others use a single cell. Voltaic cells harness the energy released in a redox reaction and transform it into electrical work. A voltaic cell is created by connecting two metals called electrodes in solution with an external circuit. In this way, the reactants are not in direct contact, but are able to transfer electrons to one another through an external pathway, allowing the redox reaction to occur. Obviously, to be useful, a battery involves reactions that are far from equilibrium and are just waiting to power your Mp3 player by sending electrons through it so the reaction can reach equilibrium. At equilibrium, the battery is "dead." The electrode that undergoes oxidation is called the *anode.*

This is easily remembered if you can burn the phrase "an ox" (for *an*ode *ox*idation) into your memory. The phrase "red cat" is equally useful for remembering what happens at the other electrode, called the *cathode* where the reduction reaction occurs.

Electrons created in the oxidation reaction at the anode of a voltaic cell flow along an external circuit to the cathode, where they fuel the reduction reaction taking place there. We will use the spontaneous reaction between zinc and copper as an example of a voltaic cell here, but it is important to realize that there are many powerful redox reactions that power many types of batteries, so they are not limited to reactions between copper and zinc.

Zinc metal will react spontaneously with an aqueous solution of copper sulfate when they are placed in direct contact with one another. Zinc, being a more reactive metal than copper, displaces the copper ions in solution. The displaced copper deposits itself as pure copper metal on the surface of the dissolving zinc strip. It may at first appear to be a simple single replacement reaction, but it is also a redox reaction. The oxidation of zinc proceeds as follows:

$$Zn(s) \rightarrow Zn^{2+}(aq) + 2e^-$$

The two electrons created in this oxidation of zinc are consumed by the copper in the reduction half of the reaction:

$$Cu^{2+}(aq) + 2e^- \rightarrow Cu(s)$$

These two half-reactions make up the total reaction:

$$Cu^{2+}(aq) + 2e^- + Zn(s) \rightarrow Cu(s) + Zn^{2+}(aq) + 2e^-$$

The electron duo appears on both sides of this combined reaction and therefore cancels, leaving

$$Cu^{2+}(aq) + Zn(s) \rightarrow Cu(s) + Zn^{2+}(aq)$$

This is the reaction that happens when the two reactants are in direct contact, but recall that a voltaic cell is created by connecting the two reactants by an external pathway. The electrons created at the anode in the oxidation reaction travel to the reduction half of the reaction along this external pathway. This flow of traveling electrons is electrical current. A voltaic cell utilizing this same oxidation-reduction reaction between copper and zinc is shown in Figure 19-2, which we will examine piece by piece.

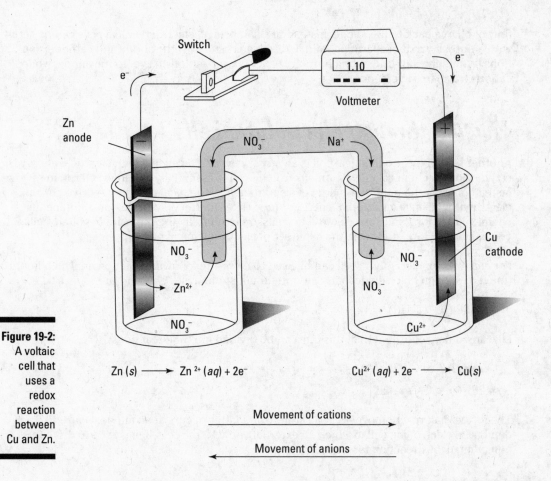

Figure 19-2:
A voltaic cell that uses a redox reaction between Cu and Zn.

$$Zn\,(s) \longrightarrow Zn^{2+}\,(aq) + 2e^-$$

$$Cu^{2+}\,(aq) + 2e^- \longrightarrow Cu(s)$$

Movement of cations →

← Movement of anions

First, note that zinc is being oxidized at the anode, which is labeled with a negative sign. It is important to note that this does not mean that the anode is negatively charged. Rather, it is meant to indicate that electrons are being released there. The oxidation of zinc releases Zn^{2+} cations into the solution as well as two electrons that flow along the circuit to the cathode. This oxidation thus results in an increase of Zn^{2+} ions into the solution and a decrease in the mass of the zinc metal anode.

The electrons released by the oxidation of zinc at the anode carry out the reduction of Cu at the cathode. This pulls Cu^{2+} from solution and deposits more Cu metal on the cathode. The result is the exact opposite of the effect occurring at the anode: The solution becomes less concentrated as Cu^{2+} ions are used up in the reduction reaction, and the electrode gains mass as Cu metal is deposited.

No doubt you have noticed that the diagram also contains a U-shaped tube connecting the two solutions. This is called the *salt bridge* and it permits the system to adjust the charge imbalance created as the anode releases more and more cations into its solution (resulting in a net positive charge) and the cathode uses up more and more of the cations in its solution (leaving it with a net negative charge). The salt bridge contains an electrolytic salt (in this case $NaNO_3$). A good salt bridge is created with an electrolyte whose component ions will not react with the ions already in solution. The salt bridge functions by sucking up the extra NO_3^- ions in the cathode solution and depositing NO_3^- ions from the other end of the bridge into the anode solution. It also sucks up the excess positive charge at the anode by absorbing Zn^{2+} ions and depositing its own cation, Na^+, into the cathode solution. These actions of the salt bridge are necessary because the solutions must be neutral in order for the redox reaction to continue. The bridge also completes the cell's electrical circuit by allowing for the flow of charge back to the anode. Instead of a salt bridge, electrochemical cells can be set up with a porous barrier separating (and in direct contact with) the anode and cathode solutions. Whether a salt bridge or a porous barrier is used, anions travel to the anode, while cations travel to the cathode.

Flowing through electrolytic cells

Another kind of electrochemical cell is an *electrolytic cell,* a cell in which electrical energy is converted into chemical energy. An electrolytic cell is essentially a voltaic cell run in reverse. Instead of using the potential energy of a spontaneous chemical reaction to drive current, electrolytic cells use an external energy source (like a battery) to drive nonspontaneous chemical reactions. Because electrolytic cells are essentially the opposite of voltaic cells, you can understand both with the same set of electrochemical principles.

For example, an electrolytic cell can be used to drive the following reaction, in which liquid (melted) sodium chloride decomposes into liquid sodium and chlorine gas:

$$2NaCl(l) \rightarrow 2Na(l) + Cl_2(g)$$

Like any other redox reaction, this one can be divided into half-reactions:

$$2Na^+(l) + 2e^- \rightarrow 2Na(l)$$
$$2Cl^-(l) \rightarrow Cl_2(g) + 2e^-$$

The former half-reaction occurs at a cathode and latter occurs at an anode. Figure 19-3 depicts an electrolytic cell in which an external voltage source is being used to drive this nonspontaneous reaction forward.

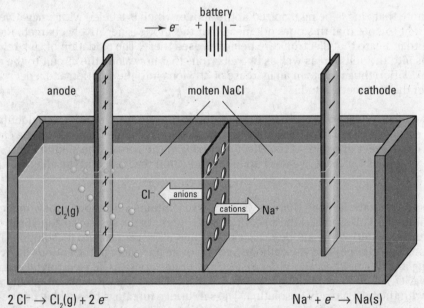

Figure 19-3:
An elec-
trolytic cell.

$2\,Cl^- \rightarrow Cl_2(g) + 2\,e^-$ $Na^+ + e^- \rightarrow Na(s)$

As in a voltaic cell, oxidation occurs at the anode and reduction occurs at the cathode. Thus our "an ox" and "red cat" mnemonics are still valid.

Calculating EMF and standard cell potentials

Voltaic cells connect two halves of a redox reaction with a current-carrying conductor. But what exactly is it that causes the flow of charge across the conductor? To answer this question, you need to understand the concept of electromotive force. To be able to calculate the diriving force behind the flow of current, you need to be able to use standard cell potentials. In this section we describe the connection between electromotive force and standard cell potential, and show you how to use them in the context of a specific redox reaction.

Electromotive force and standard cell potentials have to do with potential energy. When a difference in potential energy is established between two locations, an object has a natural tendency to move from an area of higher potential energy to an area of lower potential energy. When Wile E. Coyote places a large circular boulder at the top of a hill overlooking a road on which he knows his nemesis the Roadrunner will soon be traveling, he takes it for granted that when he releases the boulder it will simply roll down the hill. This is due to the difference in gravitational potential energy between the top and bottom of the hill, which have high and low potential energies respectively. In a similar, though less diabolical, manner, the electrons produced at the anode of a voltaic cell have a natural tendency to flow along the circuit to a location with lower potential: the cathode.

It is this potential difference between the two electrodes that causes the *electromotive force*, or EMF, of the cell. EMF is also often referred to as the *cell potential* and is denoted E_{cell}. The cell potential varies with temperature and concentration of products and reactants and is measured in Volts (V). A Volt is a derived unit — one Volt equals one Joule per Coulomb (V = J C^{-1}); in other words, a Volt is an amount of energy held by an amount of charge. As in many other parts of chemistry, chemists have found it much easier to track relative changes rather than absolute ones. So it has been found to be convenient to have standards by which cells can be compared. The E_{cell} that occurs when concentrations of solutions are all at 1mol L^{-1} and the cell is at STP (1atm pressure, 0°C) is given the special name of *standard cell potential*, or E^0_{cell}.

You can calculate electrochemical (voltaic or electrolytic) cell potentials by taking the difference between the standard potentials of the oxidation and reduction half-reactions. Once upon a time, a group of chemists sitting in a small, dark room decided that all half-reaction potentials should be listed as the potential for reduction (not oxidation). As such, these potentials are often referred to as *standard reduction potentials*. They are calculated using the formula

$$E^\circ_{cell} = E^\circ_{red}(\text{cathode}) - E^\circ_{red}(\text{anode})$$

Note that this equation will not be provided to you on the AP exam — you must memorize it! Standard reduction potentials are reported relative to a reference reduction potential, one that sets the "zero" for the spectrum of other reduction potentials. The reference potential is that for the reduction of 2 moles of hydrogen ion in solution, specifically at 1 molar concentration to 1 mole of hydrogen gas at 1 atmosphere of pressure:

$$2H^+(aq) + 2e^- \rightarrow H_2(g) \qquad E^\circ_{red} = 0 \text{ V}$$

Here's a happy consequence of the fact that E°_{red} values are reported in Volts: Because Volts are already standardized per unit of charge (per Coulomb, specifically), the value of E°_{red} stays the same, even if the stoichiometry of the half-reaction changes. So, you get the same potential for H$^+$(aq) + e$^-$ → ½ H$_2$(g).

Table 19-1 lists some standard reduction potentials along with the reduction half-reactions associated with them. The table is ordered from the most negative E°_{red} (least-favorable reduction) to the most positive E°_{red} (most-favorable reduction). Note that different tables may list items in reverse order, so be sure to orient yourself to the table you're using. The reactions with negative $E°_{red}$ are therefore reactions that happen at anodes, while those with a positive $E°_{red}$ occur at cathodes.

Put another way, the reactants in half-reactions with the most negative $E°_{red}$ are the strongest reducing agents (like Li^+). The reactants in half-reactions with the most positive $E°_{red}$ are the strongest oxidizing agents (like F_2).

Table 19-1

Reduction Half-Reaction	$E°_{red}$ (in V)
$Li^+(aq) + e^- \rightarrow Li(s)$	−3.04
$K^+(aq) + e^- \rightarrow K(s)$	−2.92
$Ca^{2+}(aq) + 2e^- \rightarrow Ca(s)$	−2.76
$Na^+(aq) + e^- \rightarrow Na(s)$	−2.71
$Zn^{2+}(aq) + 2e^- \rightarrow Zn(s)$	−0.76
$Fe^{2+}(aq) + 2e^- \rightarrow Fe(s)$	−0.44
$2H^+(aq) + 2e^- \rightarrow H_2(g)$	0
$Cu^{2+}(aq) + 2e^- \rightarrow Cu(s)$	+0.34
$I_2(s) + 2e^- \rightarrow 2I^-(aq)$	+0.54
$Ag^+(aq) + e^- \rightarrow Ag(s)$	+0.80
$Br_2(l) + 2e^- \rightarrow 2Br^-(aq)$	+1.06
$O_3(g) + 2H^+(aq) + 2e^- \rightarrow O_2(g) + H_2O(l)$	+2.07
$F_2(g) + 2e^- \rightarrow 2F^-(aq)$	+2.87

Note that not all of the reactions in Table 19-1 show the oxidation or reduction of solid metals as in our examples so far. Note that there are liquids and gases in the mix. Not every voltaic cell is fueled by a reaction taking place between the metals of the electrodes. Although the cathode itself must be made of a metal to allow for the flow of electrons, those electrons could be passed into a gas or a liquid in order to complete the reduction half-reaction. Don't panic! You do not need to memorize this table. The folks at the College Board do give you a table of reduction potentials on the AP exam!

Using Concentration and Standard Cell Potentials to Predict Results

Electrochemistry occurs in situations other than in beakers connected by salt bridges. Electrochemistry occurs anywhere in the universe where you find a suitable set of half-reactions. By connecting the lessons you've learned about electrochemical cells with other chemical wisdom about the concentrations of reactants and products, you get a powerful set of tools to predict whether redox reactions will occur in a given situation, or to explain why they did occur. In this section, we show you how to use standard cell potentials and concentrations to make predictions about redox reactions. We also introduce you to the Faraday constant, a key equation that ties electricity to chemistry by relating charge and the mole. To allow you to make predictions about redox reactions under nonstandard, real-life conditions, we add another tool to your mathematical toolbelt, the Nernst equation.

Predicting spontaneity with standard cell potentials

The painstaking measurement of myriad standard reduction potentials serves a greater purpose than simply making Impressive Tables of Official Data even more impressive. These potentials can be used to predict whether a redox reaction is *spontaneous* or *nonspontaneous* — whether it occurs on its own or whether it requires an input of energy to go forward. Remember though, that spontaneity is defined relative to equilibrium. No obvious reaction will occur if a redox system is at equilibrium.

Recall that you can calculate the standard cell potential in a voltaic or electrolytic cell by taking the difference in the standard reduction potentials of the half-reactions occurring at the cathode and anode:

$$E°_{cell} = E°_{red}(\text{cathode}) - E°_{red}(\text{anode})$$

In the same way, you can calculate the electromotive force of any particular redox reaction occurring outside of an electrochemical cell:

$$E° = E°_{red}(\text{reduction half-reaction}) - E°_{red}(\text{oxidation half-reaction})$$

Here's the kicker: If $E°$ is positive, the reaction is spontaneous; if $E°$ is negative, the reaction is nonspontaneous.

Note that this sign convention is different from that typically used in discussions of energy and spontaneity. The *Gibbs free energy change* for a reaction (ΔG), for example, is negative for a spontaneous process. If you pay close attention to the equation for the relationship between Gibbs free energy and cell potential given below, you have a strong reminder about them being opposite in sign.

You can use calculated electromotive force values to predict whether or not a displacement reaction will occur. For example, here is a candidate displacement reaction:

$$Cu(s) + CaCl_2(aq) \rightarrow CuCl_2(aq) + Ca(s)$$

Will this reaction occur spontaneously? Because it is a redox reaction as well as a displacement reaction, we can decide by examining the redox half-reactions:

$$Cu(s) \rightarrow Cu^{2+}(aq) + 2e^- \qquad \text{(oxidation half-reaction)}$$

$$Ca^{2+}(aq) + 2e^- \rightarrow Ca(s) \qquad \text{(reduction half-reaction)}$$

The relevant calculation of electromotive force is

$$E° = E°_{red}(Ca^{2+} \rightarrow Ca) - E°_{red}(Cu^{2+} \rightarrow Cu)$$

Plugging in values from Table 19-1, we get

$$E° = (-2.76\ V) - (+0.34\ V) = -3.10\ V$$

Because the calculated $E°$ is negative, we know that the reaction is nonspontaneous — copper will not displace calcium from calcium chloride without the input of additional energy.

Tying electricity to chemistry: The Faraday constant

Using standard reduction potentials, you can calculate the spontaneity of a redox reaction under standard conditions. In other words, you have to assume that the reaction takes place with the reactants at 1mol L^{-1} concentration. Most real-life reactions don't take place under standard conditions. Wouldn't it be nice if we could use standard reduction potentials to make predictions about real-life reactions? We can, once we have a solid mathematical connection between the flow of current and the individual atoms involved in redox reactions. That connection is provided by Faraday's constant.

As noted briefly above (and discussed at greater length in Chapter 21), spontaneous processes are those that occur with a decrease in Gibbs free energy; the products have lower free energy than the reactants, so the overall change in free energy (ΔG) is negative. What we really need, then, is a way to relate electromotive force for a real-life reaction, E, to the free energy change for that reaction, ΔG. Here is where the work of Michael Faraday helps us out.

Faraday was a high-powered English chemist, physicist, and by most accounts a generally swell guy. Among other things he spent quite a bit of time pondering the relationship between chemistry and electricity. Studying the properties of electrolytic cells, he noticed two important things:

- The mass of substance produced at an electrode (an anode or cathode) is directly proportional to the number of electrons transferred through the cell.

- The amount of the substances produced by electrolysis depends on the atomic masses of the substances as well as the "excess charge" of the substances (in other words, the ionic charge of the atoms).

The upshot of these observations was that Faraday had discovered how to relate current to chemistry at the level of individual atoms. To make this connection explicit and useful for predictions, we now use a number called the *Faraday constant*, F. The Faraday constant is simply the product of Avogadro's number, N_A (that is, one mole) and the charge of a single electron, e:

$$F = N_A \cdot e = (6.022 \times 10^{23} \text{ mol}^{-1})(1.602 \times 10^{-19} \text{ C}) = 9.649 \times 10^4 \text{ C mol}^{-1}$$

The Faraday constant serves as a bridge to allow us to relate the free energy change for a redox reaction to the electromotive force for that reaction:

$$\Delta G = -n \, F \, E$$

In this simple but powerful equation, n is the number of molar equivalents of electrons that are transferred during the reaction. In other words, n is the number of electrons written in a balanced redox reaction equation. The negative sign on the right side of the equation reflects the fact that spontaneous redox reactions have opposite signs in ΔG and E.

Bringing it all together with the Nernst equation

To make predictions about behavior in real-life redox reactions, where concentrations aren't at the standard of 1mol L^{-1}, we need to involve those real-life, nonstandard concentrations in our calculations. This section introduces the right tool for that job: the Nernst equation. This useful equation combines nonstandard concentration information in the form of the reaction quotient, Q, with the electricity-chemistry bridge of the Faraday constant, F.

Nernst was a German chemist and Nobel Prize-winner who was born shortly before Faraday died. So, we can imagine Faraday as sort of handing off the electrochemical baton to Nernst. Nernst ran with it. Nernst knew that the free energy change for a real-life reaction, ΔG, could be related to the standard free energy change for that reaction (ΔG°, when reactants and products are all at 1mol L^{-1}) in the following way:

$$\Delta G = \Delta G^{\circ} + RT \ln Q$$

In this equation, R is the gas constant, T is temperature, and Q is the *reaction quotient*. The reaction quotient is the key component for our purposes because it describes the actual concentrations of reactants and products in a reaction. So, for the general, reversible reaction

$$a\text{A} + b\text{B} \leftrightarrow c\text{C} + d\text{D}$$

the reaction quotient is

$$Q = \frac{\left[C\right]^{c}\left[D\right]^{d}}{\left[A\right]^{a}\left[B\right]^{b}}$$

By combining this relationship with Faraday's observation that $\Delta G = -n\,F\,E$, we obtain the very useful *Nernst equation*:

$$E = E^{\circ} - \frac{RT}{nF} \ln Q$$

Armed with the Nernst equation and a table of standard reduction potentials, you can predict the electromotive force of real life reactions, ones that occur at odd temperatures and with odd concentrations of reactants and products. Or, given an observed electromotive force, you can calculate unknown concentrations or temperatures. Thanks to Faraday and Nernst, you have a quantitative, scientifically deep explanation for why your car battery fails you on a cold winter morning.

Chapter 20

Answering Questions on Oxidation-Reduction

. . . .

In This Chapter

▶ Making sure you remember it all

▶ Getting in some practice

▶ Checking your answers

. . . .

Chapter 19 was charged with must-remember lists and tidbits. Oxidation-reduction reaction problems are certain to be a part of your AP exam experience, and they are the favorite villains of many chemistry students. Arm yourself for what you know is coming, and be sure to master the highlights of Chapter 19, listed here, and try your hand at the practice problems.

Remembering the Rules of Engagement

Conquering redox problems involves two major skills. First, you must be able to identify oxidizing and reducing agents so you can follow the flow of charge in a redox reaction. Second, you need to be able to balance redox reaction equations. The following two sections review the key points and processes of these skills.

Following the flow of charge in redox

✔ During a redox reaction, the oxidizing *agent* becomes reduced and the reducing *agent* becomes oxidized.

✔ Oxidation may or may not involve bonding with oxygen, breaking bonds with hydrogen or losing electrons — but oxidation always means an increase in oxidation number.

✔ Reduction may or may not involve bonding with hydrogen, breaking bonds with oxygen or gaining electrons — but reduction always means a decrease in oxidation number.

✔ Rules for figuring out oxidation numbers

- Atoms in elemental form have an oxidation number of zero.

- Single-atom (monatomic) ions have an oxidation number equal to their charge.

- In a neutral compound, oxidation numbers add up to zero. In a charged compound, oxidation numbers add up to the compound's charge.

- In compounds, oxygen usually has an oxidation number of –2, in which its oxidation number is –1.

- In compounds, hydrogen has an oxidation number of +1 when it bonds to non-metals, and an oxidation number of −1 when it bonds to metals.

- In compounds

    Group IA atoms (alkali metals) have oxidation number +1.

    Group IIA atoms (alkaline earth metals) have oxidation number +2.

    Group IIIA atoms have oxidation number +3.

    Group VIIA atoms (halogens) usually have oxidation number −1.

✔ Oxidation occurs at an anode, which releases electrons that flow to a cathode, where reduction occurs.

✔ $E^\circ_{cell} = E^\circ_{red}(\text{cathode}) - E^\circ_{red}(\text{anode})$ and $E^\circ = E^\circ_{red}$ (reduction half-reaction) $- E^\circ_{red}$ (oxidation half-reaction)

✔ $\Delta G = -nFE$ and the related equation

$$\log K_{eq} = \frac{nE^\circ}{0.0592}$$

✔ $\Delta G = \Delta^\circ G + RT \ln Q$

✔ Nernst equation

$$E = E^\circ - \frac{RT}{nF} \ln Q$$

Balancing redox reaction equations

To balance a redox reaction equation for a reaction under **acidic conditions**

1. Separate the reaction equation into the oxidation half-reaction and the reduction half-reaction. Use oxidation numbers to identify these component half-reactions.

2. Balance the half-reactions separately, temporarily ignoring O and H atoms.

3. Balance the half-reactions separately, using H_2O to add O atoms and using H^+ to add H atoms.

4. Balance the half-reactions separately for charge by adding electrons (e^-).

5. Balance the charge of the half-reactions with respect to each other by multiplying the reactions such that the total number of electrons is the same in each half-reaction.

6. Reunite the half-reactions into a complete redox reaction equation.

7. Simplify the equation by canceling items that appear on both sides of the arrow.

To balance a redox reaction equation for a reaction under **basic conditions,** perform steps 1 through 7 above, and then add the following steps:

1. Observe where H^+ is present in the resulting equation. Add an identical amount of OH^- to both sides of the equation such that all the H^+ is "neutralized," becoming water.

2. Cancel any amounts of H_2O that appear on both sides of the equation.

Testing Your Knowledge

Don't let your hard work in reading Chapter 19 go to waste — seal redox into your brain by attempting these questions.

Questions 1 through 4 refer to the following reaction:

$Cl_2(g) + Sb(s) \rightarrow Cl-(aq) + SbO^+(aq)$

1. Which of the following are the correct oxidation numbers for, in order, Cl in Cl_2, Sb, Cl^- and the components of SbO^+ (ions of Sb and O)?

    (A) $-1, 0, -1, 0, +1$

    (B) $0, 0, -1, +1, 0$

    (C) $0, 0, -1, +3, -2$

    (D) $-1, 0, -1, +3, -2$

    (E) $0, +3, -1, +3, -2$

2. How many electrons are transferred in the reaction?

    (A) 8

    (B) 6

    (C) 4

    (D) 2

    (E) 0

3. What is the balanced reaction equation under acidic conditions?

    (A) $2Sb + 3Cl_2 + 2H_2O \rightarrow 2SbO^+ + 6Cl^- + 4OH^-$

    (B) $2Sb + 3Cl_2 + O_2 \rightarrow 2SbO^+ + 6Cl^- + 4H^+$

    (C) $Sb + 3Cl_2 + 2H_2O \rightarrow SbO^+ + 6Cl^- + 4H^+$

    (D) $2Sb + 3Cl_2 + 2H_2O \rightarrow 2SbO^+ + 6Cl^- + 4H^+$

    (C) $3Sb + 4Cl_2 + 3H_2O \rightarrow 3SbO^+ + 9Cl^- + 6H^+$

4. What is the balanced reaction equation under basic conditions?

    (A) $2Sb + 3Cl_2 + 4OH^- \rightarrow 2SbO^+ + 6Cl^- + 2H_2O$

    (B) $2Sb + 3Cl_2 + 2H_2O \rightarrow 2SbO^+ + 6Cl^- + 4H^+$

    (C) $2Sb + 3Cl_2 \rightarrow 2SbO^+ + 6Cl^-$

    (D) $2Sb + 3Cl_2 + 2H_2O \rightarrow 2SbO^+ + 6Cl^- + 4OH^-$

    (E) $Sb + Cl_2 + 2OH^- \rightarrow SbO^+ + 2Cl^- + 2H_2O$

5. Solid iron (II) sulfide reacts with atmospheric oxygen to form iron (II) oxide and sulfur dioxide. Which of the following statements are true about the reaction?

 I. Sulfur is the reducing agent, oxygen is the oxidizing agent.

 II. Sulfur is reduced, oxygen is oxidized.

 III. Sulfur transfers electrons to iron and oxygen.

 (A) I only

 (B) II only

 (C) III only

 (D) I and II only

 (E) II and III only

Questions 6 and 7 refer to the following galvanic cell (Figure 20-1):

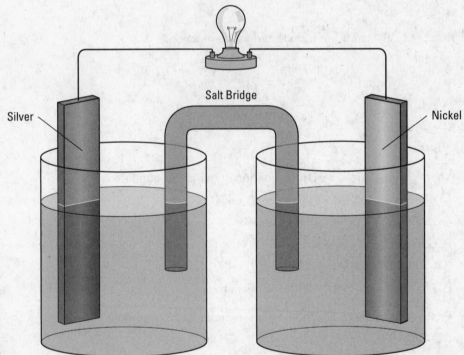

Figure 20-1:
Galvanic
cell.

6. Which of the following statements are true about the spontaneously running electrochemical cell shown? Use the following information and assume standard conditions:

$$Ag^+(aq) + e^- \rightarrow Ag(s) \qquad E^{\circ}_{red} = +0.80 \text{ V}$$

$$Ni^{2+}(aq) + 2e^- \rightarrow Ni(s) \qquad E^{\circ}_{red} = -0.28 \text{ V}$$

   I. Nickel is the cathode.

  II. Silver is the anode.

 III. Electrons flow from nickel to silver.

(A) I only

(B) II only

(C) III only

(D) None are true.

(E) All are true.

7. What is the standard cell potential of the electrochemical cell shown? Use the following information:

(A) +0.52 V

(B) −0.52 V

(C) +1.36 V

(D) −1.08 V

(E) +1.08 V

Questions 8 through 12 refer to the following list of choices:

(A) anode

(B) cathode

(C) electromotive force

(D) Faraday

(E) Nernst equation

8. Occurs when electrodes have a potential difference

9. Mole of electron charges

10. Site of reduction

11. Provides relationship between EMF and nonequilibrium concentrations

12. Site of oxidation

13. Which of the following is true of a nonspontaneous reaction?

   I. $\Delta G < 0$

   II. $E < 0$

   III. $K_{eq} < 1$

   (A) I only

   (B) II only

   (C) III only

   (D) I and III only

   (E) II and III only

Questions 14 through 17 refer to the following reaction that takes place within an electrochemical cell:

Solid copper reacts with oxygen gas under acidic conditions to produce copper (II) cation and water.

$Cu^{2+}(aq) + 2e^- \rightarrow Cu(s)$ $\qquad\qquad E°_{red} = +0.34$ V

$O_2(g) + 4H^+(aq) + 4e^- \rightarrow 2H_2O(l)$ $\qquad E°_{red} = +1.23$ V

14. What is the balanced reaction equation?

   (A) $2Cu + O_2 \rightarrow 2Cu^{2+} + H_2O$

   (B) $Cu + O_2 + H^+ \rightarrow Cu^{2+} + H_2O$

   (C) $2Cu + O_2 + 4H^+ \rightarrow 2Cu^{2+} + 2H_2O$

   (D) $2Cu + O_2 + 2H^+ \rightarrow 2Cu^{2+} + H_2O$

   (E) $Cu + O + 2H^+ \rightarrow Cu^{2+} + H_2O$

15. What is the standard cell potential for the reaction?

   (A) -0.89 V

   (B) $+0.89$ V

   (C) -1.57 V

   (D) $+1.57$ V

   (E) $+1.23$ V

16. What is the standard free energy change for the reaction? (Note: 1 J = 1 C·V, so 1 F = 96,500 J $V^{-1}mol^{-1}$)

   (A) $+610$ kJmol^{-1}

   (B) -610 kJmol^{-1}

   (C) $+150$ kJmol^{-1}

   (D) -340 kJmol^{-1}

   (E) $+340$ kJmol^{-1}

17. What effect would increasing the acid concentration beyond $1M$ have on the otherwise standard reaction?

   I. It would make the reaction less favorable.

   II. It would increase the cell voltage.

   III. It would decrease the value of the reaction quotient, Q.

   (A) I only

   (B) II only

   (C) III only

   (D) II and III only

   (E) I, II and III

Checking Your Work

You made it. Have a snack. Check your answers.

1. (C.) Both Cl_2 and Sb are in elemental form, so their atoms have an oxidation number of zero. The Cl^- oxidation number is the same as its charge. SbO^+ is a bit tricky to unravel, but remember that oxygen has a -2 oxidation number. So, in that compound, Sb must have a $+3$ oxidation number.

2. (B.) Find out that six electrons are transferred by balancing the reaction. Doing the first six steps of balancing gives you:

   Step 1 $Sb \rightarrow SbO^+$ (oxidation)

   $Cl_2 \rightarrow Cl^-$ (reduction)

   Step 2 $Sb \rightarrow SbO^+$

   $Cl_2 \rightarrow 2Cl^-$

   Step 3 $Sb + H_2O \rightarrow SbO^+ + 2H^+$

   $Cl_2 \rightarrow 2Cl^-$

   Step 4 $Sb + H_2O \rightarrow SbO^+ + 2H^+ + 3e^-$

   $Cl_2 + 2e^- \rightarrow 2Cl^-$

   Step 5 $2Sb + 2H_2O \rightarrow 2SbO^+ + 4H^+ + 6e^-$

   $3Cl_2 + 6e^- \rightarrow 6Cl^-$

   Step 6 $2Sb + 3Cl_2 + 2H_2O + 6e^- \rightarrow 2SbO^+ + 6Cl^- + 4H^+ + 6e^-$

   So, six electrons are transferred from Sb to Cl_2.

3. (D.) Completing the final step of balancing under acidic conditions gives you

   Step 7 $2Sb + 3Cl_2 + 2H_2O \rightarrow 2SbO^+ + 6Cl^- + 4H^+$

4. (A.) Adding steps 8 and 9 to account for basic conditions gives you

   Step 8 $2Sb + 3Cl_2 + 2H_2O + 4OH^- \rightarrow 2SbO^+ + 6Cl^- + 4H_2O$

   Step 9 $2Sb + 3Cl_2 + 4OH^- \rightarrow 2SbO^+ + 6Cl^- + 2H_2O$

5. (A.) To figure out this one, just assign oxidation numbers to all the reactants and products:

   ✔ In iron (II) sulfide, FeS, iron has oxidation number +2, sulfur has oxidation number –2.

   ✔ In O_2, which is in elemental form, oxygen has oxidation number 0.

   ✔ In iron (II) oxide, FeO, iron has oxidation number +2, oxygen has oxidation number –2.

   ✔ In sulfur dioxide, SO_2, sulfur has oxidation number +4, oxygen has oxidation number –2.

   So, the oxidation number of oxygen decreased from reactants to products, while the oxidation number of sulfur increased. This means that oxygen was reduced, and therefore is the oxidizing agent. Sulfur was oxidized, and therefore is the reducing agent. So, I is correct and II is not correct. Choice III is wrong because, although sulfur transfers electrons to oxygen, it does not do so to iron.

6. (C.) Electrons flow from the anode to the cathode in an electrochemical cell. This flow of electrons occurs spontaneously from the stronger reducing agent to the weaker reducing agent. Based on the standard reduction potentials given for nickel and silver, it is clear that nickel is the stronger reducing agent — reducing Ni^{2+} is unfavorable whereas reducing Ag^+ is favorable. So, nickel is the anode and silver is the cathode, with electrons flowing from nickel to silver.

7. (E.) Calculate the standard cell potential by subtracting the standard reduction potential of the anode from the standard reduction potential of the cathode:

   $E°_{cell} = E°_{red}(\text{cathode}) - E°_{red}(\text{anode})$

   $E°_{cell} = +0.80 \text{ V} - (-0.28 \text{ V}) = +1.08 \text{ V}$

8. (C.) Electromotive force, or EMF, is the potential difference across the anode and cathode of an electrochemical cell. Electrons flow from high potential to low potential, just as a boulder rolls down a hill from a position of high gravitational potential to one of low potential.

9. (D.) One Faraday (F) is the charge associated with one mole (Avogadro's number, N_A) of electron charges (e):

   $F = N_A \cdot e = (6.022 \times 10^{23} \text{ mol}^{-1})(1.602 \times 10^{-19} \text{ C}) = 9.649 \times 10^4 \text{ C mol}^{-1}$

   This constant is key to relating the number of electrons transferred during a redox reaction to the free energy change for that reaction.

10. (B.) Reduction occurs at cathodes. Think "red cat."

11. (E.) The Nernst equation combines Faraday's constant, F, with the reaction quotient, Q, to produce a relationship between the EMF of an electrochemical cell and the nonstandard concentrations of the reactants and products within that cell.

12. (A.) Oxidation occurs at anodes. Think "an ox."

13. (E.) Nonspontaneous reactions do not occur on their own, Zbut require an input of energy. Spontaneous reactions occur on their own, releasing energy. Remember that the free energy change varies with conditions, so all conditions must be known for you to compute the sign

of ΔG. Chemical species (that is, reactants and products) move spontaneously from high energy states to low energy states. So, in a spontaneous reaction, the final free energy is lower than the initial free energy. Because $\Delta G = G_{final} - G_{initial}$, $\Delta G > 0$ for a nonspontaneous reaction, in which the final free energy would be greater than the initial free energy. The equilibrium constant, K_{eq}, for a reaction is the quotient of the product concentration(s) divided by the reactant concentration(s). Nonspontaneous reactions favor reactants, and therefore have $K_{eq} < 1$. Electromotive force is different from ΔG in that favorable redox reactions generate a positive electromotive force; this force is what produces current across the cell. So, a nonspontaneous reaction is associated with a negative electromotive force, where $E < 0$.

14. (C.) Balance the reaction as follows:

Step 1 $Cu \rightarrow Cu^{2+}$ (oxidation)

$O_2 \rightarrow H_2O$ (reduction)

Step 2 (already done)

Step 3 $Cu \rightarrow Cu^{2+}$

$O_2 + 4H^+ \rightarrow 2H_2O$

Step 4 $Cu \rightarrow Cu^{2+} + 2e^-$

$O_2 + 4H^+ + 4e^- \rightarrow 2H_2O$

Step 5 $2Cu \rightarrow 2Cu^{2+} + 4e^-$

$O_2 + 4H^+ + 4e^- \rightarrow 2H_2O$

Step 6 $2Cu + O_2 + 4H^+ + 4e^- \rightarrow 2Cu^{2+} + 2H_2O + 4e^-$

Step 7 $2Cu + O_2 + 4H^+ \rightarrow 2Cu^{2+} + 2H_2O$

15. (B.) First, figure out which is the reduction half-reaction and which is the oxidation half-reaction. This has already been done in Steps 1 through 4 of the answer to the previous question:

$Cu \rightarrow Cu^{2+} + 2e^-$ (oxidation)

$O_2 + 4H^+ + 4e^- \rightarrow 2H_2O$ (reduction)

Next, recall the formula for calculating the standard, EMF, $E°$, for a redox reaction from the standard reduction potentials of the half-reactions:

$E° = E°_{red}$(reduction half-reaction) $- E°_{red}$(oxidation half-reaction)

$E° = +1.23\text{ V} - (+0.34\text{ V}) = +0.89\text{ V}$

16. (D.) The relationship between standard free energy change and standard electromotive force is given by

$\Delta G° = -nFE°$

We know from the balanced reaction equation that four electrons are transferred during the reaction, so $n = 4$. Faraday's constant is 96,500 $Cmol^{-1}$ (or $J\ V^{-1}mol^{-1}$). The standard electromotive force or cell potential (calculated in question 15) is +0.89 V. So,

$\Delta G° = -4\ (\ (96{,}500\ J\ V^{-1}mol^{-1}) \times (0.89\ V) = -340{,}000\ Jmol^{-1} = -340\ kJmol^{-1}$

17. (D.) The concentration of acid corresponds to the concentration of H⁺, a species that sits on the reactants side of the reaction equation. Chemical systems respond to an increase in reactant concentrations by shifting mass toward the products; in other words, increasing the concentration of H⁺ makes the reaction more favorable. Increasing H⁺ decreases the reaction quotient, Q, which is calculated as the concentration of products divided by the concentration of reactants. If Q decreases, then the electromotive force increases. This is made clear by the Nernst equation.

$$E = E° - \frac{RT}{nF}\ln Q$$

If Q decreases, then the entire term $(RT\ln Q) / (nF)$ decreases. This in turn means that a smaller quantity is subtracted from $E°$ to compute E. The electromotive force, E, is therefore larger.

Chapter 21

Knowing Speed from Spontaneity: Kinetics and Thermodynamics

In This Chapter

▶ Getting comfy with the difference between kinetics and thermodynamics

▶ Using rate laws to describe reaction kinetics

▶ Using state functions and heat to describe reaction thermodynamics

▶ Keeping tabs on heat flow by using calorimetry and Hess's Law

Most people don't like waiting. And nobody likes waiting for nothing. Research has tentatively concluded that chemists are people, too. It follows that chemists don't like to wait, and if they must wait, they'd prefer to get something for their trouble.

To address these concerns, chemists study things such as

✔ **Kinetics** is concerned with the rates of things, such as chemical reactions. Kinetics answers the question, "How much product will I get in the next minute?"

✔ **Thermodynamics** is concerned with the relationship between energy and chemical systems. Thermodynamics answers questions like, "How far will my reaction go?" and "How hot will my beaker get?"

Kinetics and thermodynamics are separate. There's no simple connection between how long it takes for a reaction to proceed and how productive a reaction can be. In other words, chemists have good days and bad days, like everyone else. At least they have a little bit of science to help them make sense of these things. In this chapter, you get an overview of this science. Don't wait. Read on.

Getting There Rapidly or Slowly: Kinetics and Reaction Rates

So, you've got this beaker, and a reaction is going on inside of it. Is the reaction a fast one or a slow one? How fast or how slow? How can you tell? These are questions about rates. You can measure a reaction rate by measuring how fast a reactant disappears or by measuring how fast a product appears. If the reaction occurs in solution, the molar concentration of reactant or product changes over time, so rates are often expressed in units of molarity per second ($M\,s^{-1}$).

For the following reaction,

$$A + B \rightarrow C$$

you can measure the reaction rate by measuring the decrease in the concentration of either reactant A (or B) or the increase in the concentration of product C over time. So, by measuring changes in concentration over a defined change in time, you can measure *average rate,* the rate of the reation averaged over a specific time period:

$$\text{Average rate} = -\Delta[A]/\Delta t = \Delta[C]/\Delta t$$

Note that Δ means "average change in." Here, the expression for average rate is given in terms of reactant A or in terms of product C. The expression in terms of A includes a minus sign because that reactant disappears in the course of the reaction. No sign in the C expression means "+" as [C] increases with time. Note that if the reaction were reversed so that [C] decreased and [A] increased over time, the signs of [A] and [C] would still be opposite, but reversed.

As the period of time gets smaller (that is, as Δt approaches zero), the average rate approaches the *instantaneous rate,* the rate of the reaction at a given instant:

$$\text{Instantaneous rate} = -d[A]/dt = d[C]/dt$$

In these kinds of equations, if you have done calculus you'll smile down upon those less-fortunate mortals, but "*d*" is only math-speak for a change in the amount of something at any given moment. If you plot the concentration of product against the reaction time, for example, you might get a curve like the one shown in Figure 21-1. Each tiny point on the graph corresponds to a slope that is an instantaneous value of $d[\text{product}]/dt$.

Reactions usually occur most quickly at the beginning of a reaction, when the concentration of products is the lowest and the concentration of reactants is highest. Chemists usually measure "initial rates" under these conditions and use these rates to characterize the reactions.

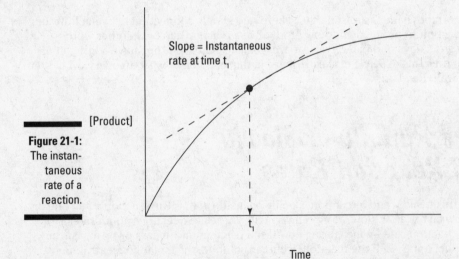

Figure 21-1:
The instan-
taneous
rate of a
reaction.

The stoichiometry of a reaction helps determine the relative rates at which reactant and product concentrations change. For example, in the reaction

$$A + 2B \rightarrow C$$

the following is true:

$$\text{Instantaneous rate} = -d[A]/dt = -1/2\,d[B]/dt = d[C]/dt$$

Because two moles of reactant B are consumed for every mole of reactant A, moles of B disappear at twice the rate of moles of A.

Measuring and Calculating with Rates

How can chemists follow the concentration of reactants and products as a reaction proceeds, anyway? One common method is by using *absorption spectroscopy,* which measures the concentration of substances in a sample by their ability to absorb electromagnetic radiation. If the substance you want to measure absorbs green light, you shine green light through the sample and measure how much green light passes through to the other side — less transmitted light means more absorbed light, which means a greater concentration of sample. *Beer's Law* is a useful equation that expresses this idea:

$$A = a \cdot b \cdot c$$

where A is the light absorbed by the sample, a is the molar absorptivity of the substance (a property that varies from one substance to another), and b is the length of the path traveled by the radiation through the sample. To make this easy in practice, solutions are often put into flat sided cells, so the length of the path is easily measured. The variable c is the concentration of the substance.

Equations that relate the rate of a reaction to the concentration of some species (reactant or product) in solution are called *rate laws.* The exact form a rate law assumes depends on the reaction involved. Countless research studies have described the intricacies of rate laws in chemical reactions. Here, we focus on rate laws for only the simplest reactions.

In general, rate laws take the form

$$\text{Reaction rate} = k\,[\text{reactant A}]^{m}[\text{reactant B}]^{n}$$

The rate law shown describes a reaction whose rate depends on the concentration of two reactants, A and B. Other rate laws for other reactions may include factors for greater or fewer reactants. In this equation, k is the *rate constant,* a number that must be experimentally measured for different reactions and reaction conditions (that is, different temperatures, solvents, etc.). The exponents m and n are called *reaction orders,* and must also be measured for different reactions. A reaction order reflects the impact of a change in concentration in overall rate. If $m > n$, then a change in the concentration of A affects the rate more than does changing the concentration of B. The sum, $m + n$, is the overall reaction order.

Some simple kinds of reaction rate laws crop up frequently, so they're worth your notice:

- **Zero-order reactions:** Rates for these reactions don't depend on the concentration of any species, but simply proceed at a characteristic rate:

    $$\text{Rate} = k$$

- **First-order reactions:** Rates for these reactions depend linearly on the concentration of a single species:

    $$\text{Rate} = k[A]$$

✔ **Second-order reactions:** Rates for these reactions may depend on the concentration of one or two species (or some intermediate combination):

Rate = $k[A]^2$

or

Rate = $k[A][B]$

Figure 21-2 illustrates how changing the reaction order changes the progress of a reaction; the more the rate of a reaction depends on the concentration of a reactant, the more severely the reaction slows down as that reactant is used up.

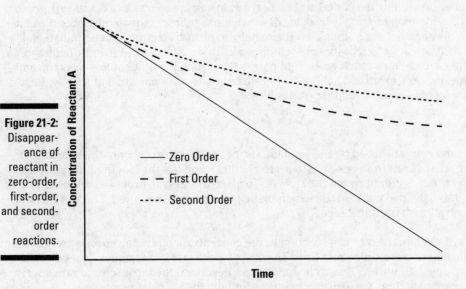

Figure 21-2: Disappearance of reactant in zero-order, first-order, and second-order reactions.

Because the initial concentration of reactants may vary from one reaction to another, chemists often find it useful to employ *integrated rate equations,* which describe the reactant concentration at time t ($[A]_t$) with respect to the initial concentration of a reactant, usually designated $[A]_0$.

Another useful mathematical tool is *half-life,* the time it takes for the concentration of a reactant to drop to one half of its initital value. This value is usually given the symbol $t_{1/2}$.

Figure 21-3 summarizes the rate laws, integrated rate laws, rate constant units, and half-life equations for the three most common reaction types. Other types are not expected in AP.

Figure 21-3: Summary of expressions for the kinetics of zero-, first-, and second-order reactions.

	Zero Order	First Order	Second Order
Rate Law	$r = k$	$r = k[A]$	$r = k[A]^2$
Integrated Rate Law	$[A] = [A]_0 - kt$	$[A] = [A]_0 e^{-kt}$	$\dfrac{1}{[A]} = \dfrac{1}{[A]_0} + kt$
Units of Rate Constant	$\dfrac{M}{s}$	$\dfrac{1}{s}$	$\dfrac{1}{M \cdot s}$
Half-life	$t_{1/2} = \dfrac{[A]_0}{2k}$	$t_{1/2} = \dfrac{\ln(2)}{k}$	$t_{1/2} = \dfrac{1}{[A]_0 k}$

Changing Rates: Factors That Alter Kinetics

Despite the impression given by their choice in clothing, chemists are finicky, tinkering types. They often want to change reaction rates to suit their own needs. What can affect rates, and why? Temperature, concentration, and catalysts influence rate. In this section, we describe how these factors can change reaction rates, and we show how those changes are expressed mathematically in the Arrhenius equation, an important equation that relates reaction rate to the activation energy of a chemical reaction. Finally, we give a brief summary of how reaction rates are used to help figure out chemical reaction mechanisms, the nitty-gritty details about how exactly chemical reactions occur.

Seeing rate changes in terms of collisions

The effects of these changes can be understood in light of the *collision model,* in which chemical reactions occur when reactatnts slam into one another with sufficient energy and in just the right way, As shown in Figure 21-4, if the collision isn't "just right" the reactants will bounce away and no new products are formed.

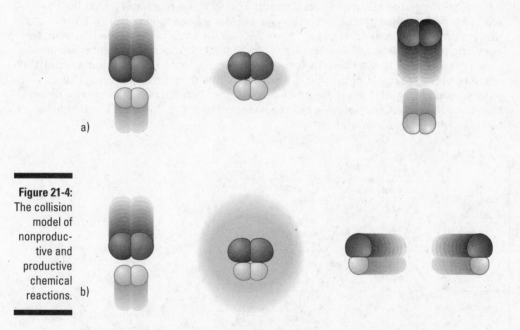

a)

Figure 21-4:
The collision
model of
nonproduc-
tive and
productive
chemical
reactions. b)

Here is a summary of the major factors that affect reaction rates:

✔ **Reaction rates tend to increase with temperature.** This trend results from the fact that reactants must collide with one another to have the chance to react. If reactants collide with the right orientation and with enough energy, the reaction can occur. So, the greater the number of collisions, and the greater the energy of those collisions, the more actual reacting takes place. An increase in temperature corresponds to an increase in the average kinetic energy of the particles in a reacting mixture — the particles move faster, colliding more frequently and with greater energy. So an increase in temperature provides a "double whammy" effect — increased number of colissions (per second) and increased average energy per collision. Make sure you include both of these in your AP exam answer.

✔ **Increasing concentration or surface area tends to increase reaction rate.** The reason for this trend also has to do with collisions. Higher concentrations mean there are more reactant particles to undergo more collisions per second and have a greater chance of reacting. Increasing the concentration of reactants may mean dissolving more of those reactants in solution. Some reactants aren't completely dissolved, but come in larger, undissolved particles. In these cases, smaller particles lead to faster reaction. Smaller particles have a greater fraction of the reactant molecules (like Compound A) on the surface where they can more easily react with molecules of Compound B, making a greater portion of the particles available for reaction.

✔ **Catalysts increase reaction rates.** Catalysts don't themselves undergo overall chemical change, and they don't alter the amount of product a reaction can eventually produce (the *yield*). However, they do enhance the probablility of a reaction occurring. Most catalysts form temporary complexes with one or both reactant molecules, holding them in more appropriate orientation for reaction. Then the product is released, leaving the catalyst to "fight another day" or engage more reactants over and over. Thus catalysts are often only needed in very small amounts to increase the rate of a reaction. Chemists (who like to define things, if you didn't realize that by now), separate catalysts into two types. *Homogenous catalysts* are of the same phase (solid, liquid, gas) as the reacting molecules, and *heterogeneous catalysts* are in a different phase. Catalysts operate mostly in the way described, but all of those ways have to do with decreasing *activation energy,* the hill in chemical potential energy that reactants must climb to reach a *transition state,* the highest-energy state along a reaction pathway. Lower activation energies mean faster reactions. Figure 21-5 shows a *reaction progress diagram,* the energetic pathway reactants must traverse to become products. The figure also shows how a catalyst affects reactions, by lowering the activation energy without altering the energies of the reactants or products. If you think this diagram looks much like a roller coaster, you are right! In that case, the energy is gravitational potential energy, rather than chemical potential energy — but the effect is the same. Too little kinetic energy or too big a hill — no ride or no reaction.

Figure 21-5: A reaction progress diagram highlighting the effect of a catalyst on activation energy.

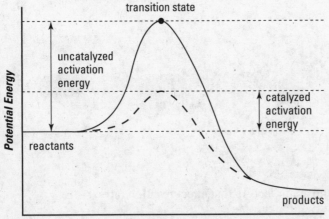

Relating rate to activation energy

Concentration, temperature, and the the activation energy, E_a, help determine the rate of a reaction. The influence of concentration is clearly evident in the rate laws shown in Figure 21-3. But where is the influence of temperature and activation energy? These two factors help determine the rate constant, k, as shown in the *Arrhenius equation*:

$$k = Ae^{-E_a/RT}$$

In the Arrhenius (yes the same Arrhenius who worked out what acids are doing) equation, k is calculated by raising the natural logarithm (base number e, approximately 2.718) to a rational exponent. The exponent, $-E_a / RT$, is the negative of the activation energy divided by the product of the gas constant, R, and the temperature, T. This whole power is multiplied by a prefactor (sometimes called the frequency factor), A. This factor has the same units as k and depends on the reaction order. It is hard to calculate theoretically, though it relates to the effectiveness of the collisions in "going over the hill," and is often an experimental measurement.

The Arrhenius equation can be rearranged into different forms that are more or less convenient for different purposes. If you know a variety of values for k and T, and would like to determine E_a, then you may want to use this form:

$$\ln k = -\frac{E_a}{RT} + \ln A$$

The previous equation corresponds to a linear plot of $\ln k$ versus $1/T$. The slope of the line is equal to $-E_a / RT$ and the y-intercept equals $\ln A$.

If you know the rate constant at one temperature and would like to predict it at another temperature, then you may want to use this form:

$$\ln \frac{k_1}{k_2} = \frac{E_a}{R}\left(\frac{1}{T_2} - \frac{1}{T_1}\right)$$

Using rates to determine mechanisms

Measuring reaction rates can help you figure out *mechanism,* the molecular details by which a reaction takes place. Chemical reactions can be broken down into a series of *elementary steps,* simple events involving only one or two molecules that occur with characteristic rates. Each elementary step has a certain *molecularity,* the number of molecular bits that must collide within that step. Unimolecular steps require one molecule, bimolecular steps require two molecules, and termolecular steps require three molecules. Termolecular steps are rare, as are collisions at one intersection of three cars (except in the movies).

Most reaction mechanisms consist of several elementary steps. Only a few of these steps consume or produce chemical players that you see in the formal reaction equation. The other players, the ones on the path between the formal reactants and products, are called *intermediates*. Much chemical research on reaction mechanisms is concerned with establishing the important elementary steps and detecting and defining intermediates.

Each elementary step in a multireaction has its own rate law. The slowest step acts as a bottleneck and determines the overall rate of the reaction. For this reason, the slow step is called the *rate determining step*. If an intermediate step is the slow step, chemists often assume that the previous, faster steps come to equilibrium during all the time afforded to them by the bottleneck.

Whatever the details, two things are always true:

✔ For a mechanism to be true, the rate law proposed by that mechanism must agree with the observed rate law.

✔ Mechanisms can never be proved correct directly by kinetics, but incorrect mechanisms can be disproved.

Getting There at All: Thermodynamics and State Functions

Figure 21-6 shows a reaction progress diagram like the one shown in Figure 21-5, but highlights the difference in energy between reactants and products. This difference is completely independent of activation energy, which is discussed in the previous section. Although activation energy controls the rate of a reaction, the difference in energy between reactants and products determines the extent of a reaction — how much reactant will have converted to product when the reaction is complete.

Figure 21-6:
A reaction progress diagram highlighting the difference in energy between reactants and products.

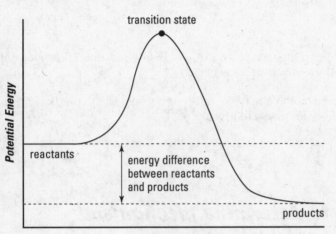

A reaction that has produced as much product as it ever will is said to be at *equilibrium*. We discuss equilibrium at length in Chapter 15. You may want to refer to that chapter for a quick refresher on that important concept, particularly on how equilibrium is quantified by the equilibrium constant, K_{eq}. Here, we are primarily concerned with the *thermodynamics* of equilibrium — how the equilibrium position of a reaction relates to energy changes as a reaction proceeds.

The difference in energy between reactants and products is built into the rules that determine the K_{eq} for a reaction. The particular form of energy important in this relationship is called *free energy, G*. You may sometimes see this called the Gibbs free energy (hence the *G*), in honor of one of the earliest American chemists, Josiah Willard Gibbs. According to Gibbs every chemical possesses a quantity of "free energy" *G* depending on what it is and what the conditions (phase, temperature, pressure) are. While it is extremely difficult to find out what this is, changes in this value can easily be tracked by smart chemists. The relationship between free energy and equilibrium is

$$\Delta G = -RT \ln K_{eq}$$

or

$$K_{eq} = e^{-\Delta G / RT}$$

In these equations, *R* is the gas constant and *T* is the temperature in Kelvin degrees. Favorable (spontaneous) reactions possess negative values for ΔG and unfavorable (nonspontaneous) reactions possess positive values for ΔG. Energy must be added to drive an unfavorable reaction forward. If the ΔG for a set of reaction conditions is 0, the reaction is at equilibrium. But be careful! These values only apply at a specific set of conditions. A reaction that is nonspontaneous under one set of conditions may be spontaneous if those conditions change. For example, most people think of hydrogen and oxygen reacting pretty spontaneously to form water, but heat water to thousands of degrees and it will split into hydrogen and oxygen!

Free energy is an example of a state function, a property of a system that depends only on its characteristics in the given moment. State functions don't care about how you got here, only about who you are. But they change as conditions change! Your state function is different in bed at home as compared to sitting at your desk in school. In addition to free energy, other state functions include *enthalpy, H,* and *entropy, S.* At constant pressure, enthalpy equals the the thermal energy content of a system. Entropy is a measure of the energy of disorder in a system.

Energy is a pretty abstract concept. Before looking further into how energy governs chemical reactions, it might be useful to remind yourself of three things that are always true — the laws of thermodynamics:

> **First Law: In any process, the total energy of the universe remains constant.** Remember, chemists divide the universe into two parts, the system and the surroundings. Universe = system + surroundings. The system and the surroundings can exchange energy; the universe cannot.

> **Second Law: In any spontaneous process, the overall entropy of the universe increases.** Things that happen on their own do so with an increase in disorder. If you focus only on the system, the increased disorder may not be apparent, as it may occur in the surroundings. In fact, complex chemical reactions (such as the biochemical ones in our bodies) often increase order at the expense of the disorder in the surroundings.

> **Third Law: As temperature approaches absolute zero, the entropy of a system approaches a constant minimum.** If the system is a pure, perfectly ordered crystalline solid, then that minimum of entropy is also zero.

Chemical reactions are processes that occur in the universe. So, chemical reactions obey the laws of thermodynamics. If you encounter a chemical reaction that fails to obey these laws, patent it. You'll probably find some investors willing to lose money on it.

Two other concepts follow closely on the heels of the thermodynamic laws: *reversibility* and *irreversibility.* In a reversible process, a system can go back to its original state by reversing its steps *along the same path* it originally took. When a reaction is at equilibrium, mass (in the form of atoms) moves between reactant and product states reversibly.

In an irreversible process, the system may be able to return to its original state, but must do so along a different path. Spontaneous processes are irreversible, as they irrevocably increase the entropy of the universe. Never fear, irreversible processes and their properties do not show up on AP chemistry exams.

Adding Up the Energies: Free Energy, Enthalpy, and Entropy

Three different kinds of energy are important for determining whether or not a particular change occurs, such as whether or not a chemical reaction goes forward. These three kinds of energy are

✔ **Gibbs free energy, *G*,** the portion of a system's energy that is available to do work

✔ **Enthalpy, *H*,** the thermal energy content of a system at constant pressure

✔ **Entropy, *S*,** a measure of the portion of a system's energy that is unavailable to do work; entropy is related to disorder

By tracking changes in these three energies between different states (like the reactant and product states of a chemical reaction), you can determine whether a change will occur on its own, or whether that change needs to be driven forward by the addition of energy. The change in Gibbs free energy, ΔG, is negative for spontaneous processes, ones that occur on their own. The change in Gibbs free energy can be seen in terms of changes in enthalpy and entropy in the *Gibbs equation*:

$$\Delta G = \Delta H - T\Delta S \text{ (where } T \text{ is temperature in Kelvin)}$$

The Gibbs equation reveals that spontaneity arises from the interplay of enthalpy and entropy, two state functions that underlie free energy. A lot of AP questions revolve around which is the winning term on the righthand side, so get all your ducks (sorry, arguments) in a row here. Negative changes in enthalpy are spontaneous, as are positive changes in entropy. Note that the effect of entropy changes on the free energy (which are generally 1,000 times smaller than enthalpy changes) is magnified by the temperature. The upshot of this equation is that chemical reactions can be driven forward by spontaneous changes in enthalpy, entropy, or both. If the changes in neither enthalpy nor entropy are spontaneous, then the reaction will not occur spontaneously. These patterns are summarized in Figure 21-7.

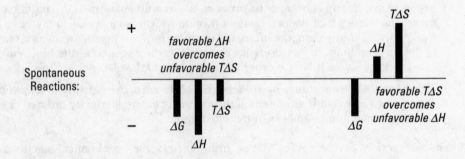

Figure 21-7:
The interplay of enthalpy and entropy components of free energy.

But wait! The second law of thermodynamics states that all spontaneous processes occur with a positive change in entropy — how can a chemical reaction be driven forward by enthalpy alone? Good question. Answer: The second law refers to the entropy of the *universe*. The Gibbs equation refers to the enthalpy and entropy of the *system*. A decrease in the enthalpy of a system means an increase in the entropy of the surroundings, which means an increase in the entropy of the universe. The second law holds true. What a relief.

Under standard conditions (typically, 1atm, pure substances at $1M$ concentration), the state functions are said to be in standard state and the state functions of the Gibbs equation are annotated with the ° symbol:

$$\Delta G° = \Delta H° - T\Delta S°$$

However, while most free energy changes are quoted at 298K, technically this is not defined by the "standard conditions." Texts will provide tables where the temperature is clearly indicated, as in ΔG°_{298}. *Some texts may omit this, in which case 298K temperature must be assumed.*

Changes in standard free energy, standard enthalpy, and standard entropy for a reaction can be calculated directly from the stoichiometry of the reaction along with tabulated values. The tabulated values are those for the *standard free energies, enthalpies, or entropies of formation* of the reactants and products, ΔG_f°, ΔH_f° and ΔS_f°, respectively:

$$\Delta G^\circ = \Sigma\, a\Delta G_f^\circ(\text{products}) - \Sigma\, b\Delta G_f^\circ(\text{reactants})$$

$$\Delta H^\circ = \Sigma\, a\Delta H_f^\circ(\text{products}) - \Sigma\, b\Delta H_f^\circ(\text{reactants})$$

$$\Delta S^\circ = \Sigma\, a\Delta S_f^\circ(\text{products}) - \Sigma\, b\Delta S_f^\circ(\text{reactants})$$

where Σ refers to the sum of products or reactants and where *a* and *b* refer to the stoichiometric coffiecients of each product or reactant. Standard enthalpies and entropies of formation are the enthalpies and entropies associated with forming a compound under standard conditions (1atm, 298K, 1mol L^{-1}) from its component elements, in their elemental forms under standard conditions.

The relationship between the standard free energy and the free energy under nonstandard conditions is:

$$\Delta G = \Delta G^\circ + RT \ln Q$$

where Q is the reaction quotient (see Chapter 15). Compare this equation to $K_{eq} = e^{-\Delta G/RT}$ and be sure you understand the differences.

Measuring Heat: Thermochemistry and Calorimetry

Energy shifts between many forms. It may be tricky to detect, but energy is always conserved. Sometimes energy reveals itself as heat. Thermodynamics explores how energy moves from one form to another. *Thermochemistry* investigates changes in thermal energy that accompany chemical reactions. To understand how thermochemistry is done, you need to first understand how the particular form of energy called thermal energy and the elusive term "heat" fit into the overall dance of energy and matter. In this section, we discuss how thermal energy fits into the overall energy picture, and we describe how you can measure heat transfer during chemical reactions by using a concept called heat capacity and a technique called calorimetry.

Seeing heat within the bigger energy picture

You're probably familiar with the idea that objects can have greater or lesser amounts of energy associated with them. One of the kinds of energy an object can possess is thermal energy. In this section, we show how thermal energy relates to the overall energy of a system (like a chemical system), and how thermal energy relates to heat.

Energy itself can be divided into

- ✓ *Potential energy* (*PE*) is energy due to position. *Chemical energy* is a kind of potential energy, arising from the positions of particles within systems.

- ✓ *Kinetic energy* (*KE*) is the energy of motion. *Thermal energy* is a kind of kinetic energy, arising from the movement of particles within systems.

The total *internal energy* of a system (*E*) is the sum of its potential and kinetic energies. When a system moves between two states (as it does in a chemical reaction), the internal energy may change as the system exchanges energy with the surroundings. The difference in energy (Δ*E*) between the initial and final states derives from heat (*q*) added to or lost from the system, and from work (*w*) done by the system or on the system. "Heat" in physics and chemistry is strictly defined as a quantity of thermal energy being transferred between one system and another. As a result nothing can "contain" heat!

We can summarize these energy explanations with the help of a couple of handy formulas:

$$E_{total} = KE + PE$$

$$\Delta E = E_{final} - E_{initial} = q + w$$

What kind of "work" can atoms and molecules do in a chemical reaction? Remember (maybe) from your physics class that work is defined in terms of motion against a force. No motion, no work! One kind of work that is easy to understand is *pressure-volume work*. Consider the following reaction:

$$CaCO_3(s) \rightarrow CaO(s) + CO_2(g)$$

Solid calcium carbonate decomposes into solid calcium oxide and carbon dioxide gas. At constant pressure (*P*), this reaction proceeds with a change in volume (*V*). The added volume comes from the production of carbon dioxide gas. As gas is made, it expands, pushing against the surroundings. The carbon dioxide gas molecules do work as they push into a greater volume:

$$w = -P\Delta V$$

The negative sign in this equation means that the system loses internal energy due to the work it does on the surroundings. If the surroundings did work on the system, thereby decreasing the system's volume, then the system would gain internal energy. Note that if there were no pressure (resistive force), then $P = 0$, so $w = 0$ . . . so there is no work! So, expanding into a vacuum a gas does no work!

So, pressure-volume work can partly account for changes in internal energy during a reaction. When pressure-volume work is the only kind of work involved, any remaining changes come from heat. As mentioned in the previous section, *enthalpy* (*H*) corresponds to the thermal energy content of a system at constant pressure. An *enthalpy change* (Δ*H*) in such a system corresponds to heat. The enthalpy change equals the change in internal energy minus the energy used to perform pressure-volume work:

$$\Delta H = \Delta E - (-P\Delta V) = \Delta E + P\Delta V$$

Although *E*, *P*, *V*, and *H* are state functions, heat (*q*) is *not* a state function, but is simply a quantity of thermal energy that transfers from a warmer object to a cooler object.

Now breathe. The practical consequences of all this theory are the following:

- Chemical reactions involve an amount of heat, q.
- Chemists monitor changes in thermal energy by measuring changes in temperature.
- At constant pressure, the change in thermal energy content equals the change in enthalpy, ΔH.
- Knowing ΔH values helps to explain and predict chemical behavior.

Using heat capacity in calorimetry

Heat is a quantity of thermal energy that transfers from warmer objects to cooler objects. But how much thermal energy can an object hold? If objects have the same thermal energy content, does that mean they are the same temperature? You can measure temperature changes, but how do these temperatures relate to heat? These kinds of questions revolve around the concept of *heat capacity*. Heat capacity is the amount of heat required to raise the temperature of a system by 1K (or 1°C). In this section, we describe how the concept of heat capacity is used in an experimental technique called calorimetry.

It takes longer to boil a large pot of water than a small pot of water. With the burner set on high, the same amount of heat transfers into each pot, but the larger pot of water has a higher heat capacity. So, it takes more heat transfer to increase the temperature of the larger pot.

You'll encounter heat capacity in different forms, each of which is useful in different scenarios. Any system has a heat capacity. To best compare heat capacities between chemical systems you use one of the following:

- *Molar heat capacity,* which is the heat capacity of 1 mole of a substance
- *Specific heat capacity,* or just *specific heat,* which is the heat capacity of 1 gram of a substance

How do you know whether you're dealing with heat capacity, molar heat capacity, or specific heat capacity? Look at the units:

- **Heat capacity:** Energy / K
- **Molar heat capacity:** Energy / (mol·K)
- **Specific heat capacity:** Energy / (g°K)

Fine, but what are the units of energy? Well, that depends. The SI unit of energy is the *joule (J)* (see Chapter 3 for more about the International System of units), but the units of *calorie (cal)* and *liter-atmosphere (L·atm)* may also be used. Here's how the joule, the calorie, and the liter-atmosphere are related (AP will stick to joules!):

1 J = 0.2390 cal

101.3 J = 1 L·atm

Note that in everyday language about nutrition, a "calorie" actually refers to a "Calorie" or a kilocalorie — a calorie of cheesecake is 1,000 times larger than you think it is. Why do you think the American food industry wants to keep it that way? In Europe and Canada, candy bars are rated in kilojoules, so a 100 calorie bar is 418 kJ of energy.

Calorimetry is a family of techniques that puts all this thermochemical theory to use. When chemists do calorimetry, they initiate a reaction within a defined system, and then measure any temperature change that occurs as the reaction progresses. There are a few variations on the theme of calorimetry, each measuring heat transfer under different conditions:

- *Constant-pressure calorimetry* directly measures an enthalpy change (ΔH) for a reaction because it monitors heat transferred at constant pressure: $\Delta H = q_P$.

  Typically, heat is observed through changes in the temperature of a reaction solution. If a reaction warms a solution, then that reaction must have released thermal energy into the solution. In other words, the change in thermal energy content of the reaction ($q_{reaction}$) has the same magnitude as the change in thermal energy for the solution ($q_{solution}$) so the heat into one is the same as the heat out of the other, but has opposite sign: $q_{solution} = -q_{reaction}$.

  So, measuring $q_{solution}$ allows you to calculate $q_{reaction}$, but how can you measure $q_{solution}$? You do so by measuring the difference in temperature (ΔT) before and after the reaction:

  $$q_{solution} = (\text{mass of solution}) \times (\text{specific heat of solution}) \times \Delta T$$

  In other words,

  $$q = mC_P\Delta T$$

  Here, "m" is the mass of the solution and C_P is the specific heat capacity of the solution at constant pressure. ΔT is equal to $T_{final} - T_{initial}$.

  When you use this equation, be sure that all your units match. For example, if your C_P has units of J g^{-1} K^{-1}, don't expect to calculate heat flow in kilocalories.

- *Constant-volume calorimetry* directly measures a change in internal energy, ΔE (not ΔH) for a reaction because it monitors heat flow at constant volume. Often, ΔE and ΔH are very similar values, especially if no gases are involved to do pressure-volume work.

  A common variety of constant-volume calorimetry is *bomb calorimetry*, a technique in which a reaction (often, a combustion reaction) is triggered within a sealed vessel called a bomb. The vessel is immersed in a water bath of known volume. The vessel and water together are considered the "calorimeter" and are enclosed by an insulated container. The temperature of the water is measured before and after the reaction. Because the heat capacity of the calorimeter (C_{cal}) is known, you can calculate heat from the change in temperature:

  $$q = -C_{cal} \times \Delta T$$

Dealing with Heat by Using Hess's Law

So, as you find out in the previous section, you can monitor heat by measuring changes in temperature. But what does any of this have to do with chemistry? Chemical reactions transform both matter and energy. Though reaction equations usually list only the matter components of a reaction, you can consider thermal energy as a reactant or product as well. When chemists are interested in heat appearing during a reaction (and when the reaction is run at constant pressure), they may list an enthalpy change (ΔH) to the right of the reaction equation. As we explain in the previous section, at constant pressure, heat equals ΔH:

$$q_P = \Delta H = H_{final} - H_{initial}$$

If the ΔH listed for a reaction is negative, then that reaction releases heat as it proceeds — the reaction is *exothermic*. If the ΔH listed for the reaction is positive, then that reaction absorbs heat as it proceeds — the reaction is *endothermic*. In other words, exothermic reactions release heat as a product, and endothermic reactions consume heat as a reactant.

The sign of the ΔH tells us the direction of heat, but what about the magnitude? The coefficients of a chemical reaction represent molar equivalents. So, the value listed for the ΔH refers to the enthalpy change for one molar equivalent of the reaction. Here's an example:

$$CH_4(g) + 2O_2(g) \rightarrow CO_2(g) + 2H_2O(g) \qquad \Delta H = -802 \text{ kJ}$$

The reaction equation shown describes the combustion of methane, a reaction you might expect to release heat. The enthalpy change listed for the reaction confirms this expectation: For each mole of methane that combusts, 802 kJ of heat are released. The reaction is highly exothermic. Based on the stoichiometry of the equation, you can also say that 802 kJ of heat are released for every 2 moles of water produced.

So, reaction enthalpy changes (or reaction "heats") are a useful way to measure or predict chemical change. But they're just as useful in dealing with physical changes, like freezing and melting, evaporating and condensing, and others. For example, water (like most substances) absorbs heat as it melts (or *fuses*) and as it evaporates:

Molar enthalpy of fusion: $\Delta H_{fus} = 6.01$ kJ

Molar enthalpy of vaporization: $\Delta H_{vap} = 40.68$ kJ

The same sorts of rules apply to enthalpy changes listed for chemical changes and physical changes. Here's a summary of the rules that apply to both:

✔ **The heat absorbed or released by a process is proportional to the moles of substance that undergo that process.** Two moles of combusting methane release twice as much heat as does one mole of combusting methane.

✔ **Running a process in reverse produces heat flow of the same magnitude but of opposite sign as running the forward process.** Freezing one mole of water releases the same amount of heat that is absorbed when one mole of water melts.

Yikes. Earlier in the chapter we described that chemical reactions can be complicated, multi-step sorts of things. Now we're telling you that both chemical and physical processes can be associated with heat. How can you possibly keep track of all these heats? The answer is simple: Hess's Law.

Imagine that the product of one reaction serves as the reactant for another reaction. Now imagine that the product of the second reaction serves as the reactant for a third reaction. What you have is a set of coupled reactions, connected in series like the cars of a train:

$$A \rightarrow B \quad \text{and} \quad B \rightarrow C \quad \text{and} \quad C \rightarrow D$$

So,

$$A \rightarrow B \rightarrow C \rightarrow D$$

You can think of these three reactions adding up to one big reaction, $A \rightarrow D$. What is the overall enthalpy change associated with this reaction, ΔH_{AD}? Here's the good news:

$$\Delta H_{AD} = \Delta H_{AB} + \Delta H_{BC} + \Delta H_{CD}$$

Enthalpy changes are additive. Hess's Law declares that the total enthalpy change for the overall reaction is the sum of the enthalpy changes for each of the steps, as summarized in Figure 21-8.

But the good news gets even better. Imagine that you're trying to figure out the total enthalpy change for the following multistep reaction:

$$X \rightarrow Y \rightarrow Z$$

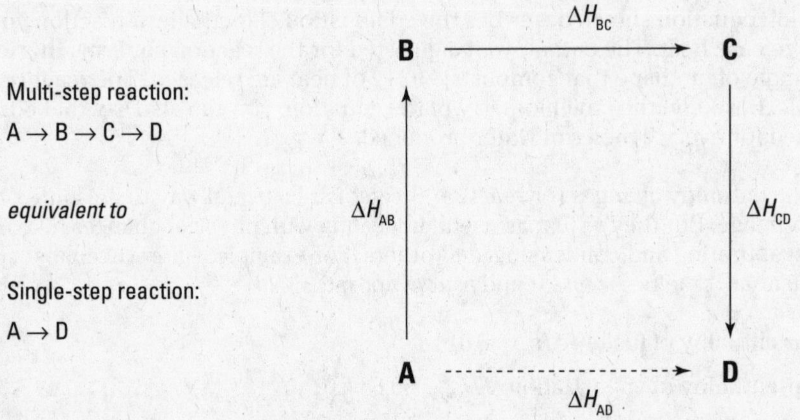

Multi-step reaction:

$A \rightarrow B \rightarrow C \rightarrow D$

equivalent to

Figure 21-8:
Schematic illustration of Hess's Law.

Single-step reaction:

$A \rightarrow D$

Here's a wrinkle: For technical reasons, you can't measure this enthalpy change, ΔH_{XZ}, directly, but must calculate it from tabulated values for ΔH_{XY} and ΔH_{YZ}. No problem, right? You simply look up the tabulated values and add them. But here's another wrinkle: When you look up the tabulated values, you find the following:

$\Delta H_{XY} = -37.5 \ \text{kJmol}^{-1}$

$\Delta H_{ZY} = -10.2 \ \text{kJmol}^{-1}$

Gasp! You need ΔH_{YZ}, but you're provided only ΔH_{ZY}! Relax: Reactions are reversible. The enthalpy change for a reaction has the same magnitude and opposite sign as the enthalpy change for the reverse reaction. So, if $\Delta H_{ZY} = -10.2 \ \text{kJmol}^{-1}$, then $\Delta H_{YZ} = 10.2 \ \text{kJmol}^{-1}$. It really is that simple. So,

$$\Delta H_{XZ} = \Delta H_{XY} + (-\Delta H_{ZY}) = -37.5 \ \text{kJmol}^{-1} + 10.2 \ \text{kJmol}^{-1} = -27.3 \ \text{kJmol}^{-1}$$

Thanks be to Hess.

Chapter 22

Answering Questions on Kinetics and Thermodynamics

In This Chapter

- Going through what you can't forget
- Trying out some sample questions
- Checking out your answers

Chapter 21 gave a broad overview of the relationships between differences in energy and the rates and equilibria observed for chemical reactions. These kinds of relationships are fundamental to chemistry and are ripe pickings for the AP chemistry exam. Do yourself a favor: Be sure you know the principles at work behind kinetics and thermodynamics and can apply those principles to problems. The best way to ensure that you know what you need to know about kinetics and thermodynamics is to review the tips and suggestions in this chapter and then practice the questions included.

Picking through the Important Points

For many students, the most difficult thing about kinetics and thermodynamics is telling the two apart. Both subjects describe things about chemical reactions. Both use a wealth of mathematical equations. Both have something to do with energy. Before launching into this summary of the highlights of Chapter 21, make sure you understand that kinetics and thermodynamics, though interrelated, deal with entirely separate issues:

- **Kinetics** deals with the rate at which reactants convert to products and how that rate depends on concentrations during a reaction.
- **Thermodynamics** deals with differences **in different types of energy (enthalpy and entropy)** between reactants and products and how those differences relate to concentrations at equilibrium.

Important points about rates

The rules for measuring and describing the speed of a reaction are expressed in equations. Luckily, the equations are pretty simple and easy to interpret. Seeing all of them together helps make clear how all the equations relate to each other. Instead of spending time trying to memorize all these equations, spend time trying to understand how they all relate to one another.

In summary, here are the important points about rate made in Chapter 21:

✔ The rate of a reaction does *not* determine the final extent of a reaction, *nor* does the final extent of a reaction determine the rate.

✔ For the reaction: A + B → C

- Average rate = $-\Delta[A]/\Delta t = -\Delta[B]/\Delta t = \Delta[C]/\Delta t$.

- Instantaneous rate = $-d[A]/dt = -d[B]/dt = d[C]/dt$.

✔ For the reaction: A + 2B → C

- Instantaneous rate = $-d[A]/dt = -1/2\,d[B]/dt = d[C]/dt$.

✔ You can extract rates from the shapes of reaction progress curves, which plot the concentration of a reactant or product versus time.

✔ Concentrations of reactants or products can be measured spectrophotometrically, using **Beer's law:** $A = abc$

✔ For a reaction with two reactants in its rate-determining step:
Rate = k [reactant A]^m[reactant B]^n.

- k is the **rate constant** and depends on reaction conditions.

- Exponents m and n are **reaction orders** and depend on reaction conditions.

- The sum, $m + n$, is the **overall reaction order.**

✔ **Zero-order reactions:** Rates don't depend on any concentration.

- Rate = k.

✔ **First-order reactions:** Rates depend on the concentration of a single species.

- Rate = $k[A]$.

✔ **Second-order reactions:** Rates may depend on the concentration of one or two species:

- Rate = $k[A]^2$ or   Rate = $k[A][B]$.

✔ **Integrated rate equations** describe the rate of a reaction with respect to the initial concentration of a reactant.

For a first-order reaction: $[A] = [A]_0 \cdot e^{-kt}$.

or equivalently $\ln[A] = -kt + \ln[A]_0$.

✔ **Half-life** describes the time it takes for the concentration of a reactant to drop to one half of its initital value.

- For a first-order reaction: $t_{1/2} = (\ln 2) / k$.

✔ Reaction rates tend to increase with temperature.

✔ Increasing concentration tends to increase reaction rate.

✔ Catalysts increase reaction rates, but do not alter concentrations or final amounts produced at equilibrium.

- Catalyts decrease **activation energy,** the energetic hill reactants must climb to reach a **transition state.**

✔ The Arrhenius equation relates the rate constant, k, to activation energy, E_a, and temperature, T, and a constant, A (the "frequency factor"), related to reactant orientation and frequency of collisions:

$$k = Ae^{-E_a/RT}$$

✔ The slowest elementary step in a reaction mechanism is the **rate-determining step.**

Important points about thermodynamics

As described in Chapter 21, thermodynamics is about big ideas and (usually) small equations. Thermodynamic concepts can seem a bit abstract, so as you peruse the highlights in this summary, anchor your brain to the idea that thermodynamics has nothing whatever to do with how fast a reaction proceeds — only how far it goes. Thermodynamics cares only about beginning and end states.

Re-energize your grasp of thermodynamics by perusing these points:

- **Laws of Thermodynamics:**

  - **First Law:** In any process, the total energy of the universe remains constant. Universe = system + surroundings.

  - **Second Law:** In any spontaneous process, the overall entropy of the universe increases.

  - **Third Law:** As temperature approaches absolute zero, the entropy of a system approaches a constant minimum.

- The equilibrium constant, K_{eq}, is related to the difference in free energy, ΔG, between products and reactants:

  $$\Delta G = -RT \ln K_{eq}$$

  This equation can be rearranged to the following form:

  $$K_{eq} = e^{-\Delta G/RT}$$

- Favorable (spontaneous) reactions possess negative values for ΔG and unfavorable (nonspontaneous) reactions possess positive values for ΔG.

- The **Gibbs equation** relates the difference in free energy (ΔG) between products and reactants to the differences in enthalpy (ΔH) and entropy (ΔS) and the temperature of the system in Kelvin:

  - $\Delta G = \Delta H - T\Delta S$

- Under standard conditions (usually 1atm, pure substances at $1M$ concentration), the Gibbs equation reflects "standard state" quantities:

  - $\Delta G° = \Delta H° - T\Delta S°$

- Nonstandard ΔG relates to standard $\Delta G°$ via corrections concentration (as reflected in Q, the reaction quotient):

  - $\Delta G = \Delta G° + RT \ln Q$

- **Heat capacity** is the amount of heat required to raise the temperature of a system by 1K.

  - **Molar heat capacity** is the heat capacity of 1 mole of a substance.

  - **Specific heat capacity** is the heat capacity of 1 gram of a substance.

- **Constant-pressure calorimetry** directly measures an enthalpy change (ΔH) for a reaction because it monitors heat flow at constant pressure: $\Delta H = q_P$.

  - $q = mC_P\Delta T$

- **Constant-volume calorimetry** directly measures a change in internal energy, ΔE (not ΔH) for a reaction because it monitors heat flow at constant volume.

- $q = -C_{cal} \times \Delta T$

- **Exothermic reactions** release heat as a product, $\Delta H < 0$. **Endothermic reactions** consume heat as a reactant, $\Delta H > 0$.

✔ **Hess's Law (enthalpies are additive):**

- For the set of coupled reactions,

  $$A \rightarrow B \rightarrow C \rightarrow D$$

  the following is true:

  $$\Delta H_{AD} = \Delta H_{AB} + \Delta H_{BC} + \Delta H_{CD}$$

- Before adding reaction equations, you may need to manipulate those equations, and doing so has consequences on the associated ΔH values:

  Reaction: $A \rightarrow B$ Enthalpy change: ΔH_{AB}

  compare that to:

  Reaction: $2A \rightarrow 2B$ Enthalpy change: $2 \cdot \Delta H_{AB}$

  compare that to:

  Reaction: $B \rightarrow A$ Enthalpy change: $-\Delta H_{AB}$

Testing Your Knowledge

Concentration, rate, and equilibrium. Free energy, enthalpy, and entropy. Heat, mass, and temperature. Time spent on these kinds of problems is never wasted.

Multiple choice

Questions 1 through 4 refer to the following reactions:

Reaction 1: $A + B \rightarrow C$ Rate $= k[A]^2$

Reaction 2: $D + E \rightarrow F + 2G$ Rate $= k[D][E]$

1. What are the overall reaction orders for Reaction 1 and Reaction 2, respectively?

   (A) First-order, Second-order

   (B) Second-order, First-order

   (C) First-order, First-order

   (D) Second-order, Second-order

   (E) Not enough information

2. For Reaction 1, how will the initial rate change if the concentration of A is doubled and the concentration of B is halved?

   (A) Twofold increase

   (B) Threefold increase

   (C) Fourfold increase

   (D) Twofold decrease

   (E) Depends on the concentration of C

3. For Reaction 2, how will the initial rate change if the concentration of D is doubled and the concentration of E is tripled?

   (A) Twofold increase

   (B) Threefold increase

   (C) Fivefold increase

   (D) Sixfold increase

   (E) Depends on the concentrations of F and G

4. For Reaction 2, what is the relationship between the rates of change in [D], [E], [F], and [G]?

   (A) Rate = $d[D]/dt = d[E]/dt = -d[F]/dt = -2d[G]/dt$

   (B) Rate = $d[D]/dt = d[E]/dt = -d[F]/dt = 2d[G]/dt$

   (C) Rate = $d[D]/dt = d[E]/dt = -d[F]/dt = -0.5d[G]/dt$

   (D) Rate = $d[D]/dt = d[E]/dt = -d[F]/dt = 0.5d[G]/dt$

   (E) Rate = $2d[D]/dt = 2d[E]/dt = -d[F]/dt = -0.5d[G]/dt$

5. Methane combusts with oxygen to yield carbon dioxide and water vapor:

   $CH_4 + 2O_2 \rightarrow CO_2 + 2H_2O$

   If methane is consumed at 2.79mol s^{-1}, what is the rate of change in the concentrations of carbon dioxide and oxygen?

   (A) +2.79mol s^{-1} CO_2 and +5.58mol s^{-1} O_2

   (B) −2.79mol s^{-1} CO_2 and +5.58mol s^{-1} O_2

   (C) +5.58mol s^{-1} CO_2 and −5.58mol s^{-1} O_2

   (D) +2.79mol s^{-1} CO_2 and −2.79mol s^{-1} O_2

   (E) +2.79mol s^{-1} CO_2 and −5.58mol s^{-1} O_2

6. You study the following reaction: $D + E \rightarrow F + 2G$

   You vary the concentration of reactants D and E, and observe the resulting rates:

	[D], M	[E], M	Rate, M s^{-1}
Trial 1:	2.7×10^{-2}	2.7×10^{-2}	4.8×10^{6}
Trial 2:	2.7×10^{-2}	5.4×10^{-2}	9.6×10^{6}
Trial 3:	5.4×10^{-2}	2.7×10^{-2}	9.6×10^{6}

   At what rate will the reaction occur in the presence of 1.3×10^{-2} M reactant D and 9.2×10^{-3} M reactant E?

   (A) 7.9×10^{5} M s^{-1}

   (B) 1.2×10^{-4} M s^{-1}

   (C) 6.6×10^{9} M s^{-1}

   (D) 8.6×10^{7} M s^{-1}

   (E) 6.1×10^{7} M s^{-1}

7. For the reaction: $A + 2B \leftrightarrow 2C$

   If the $K_{eq} = 1.37 \times 10^3$, what is the free energy change for the reaction at 25 °C?

   (Note: $R = 8.314 \ J \ K^{-1} mol^{-1}$)

   (A) $-1.50 \times 10^3 \ Jmol^{-1}$

   (B) $+1.50 \times 10^3 \ Jmol^{-1}$

   (C) $-1.79 \times 10^4 \ Jmol^{-1}$

   (D) $+1.79 \times 10^4 \ Jmol^{-1}$

   (E) Not enough information

8. Which of the following statements is correct?

   (I) The entropy of the system can decrease during a spontaneous reaction.

   (II) The free energy of the system can increase during a spontaneous reaction.

   (III) The enthalpy of the system can increase during a spontaneous reaction.

   (A) I only

   (B) II only

   (C) III only

   (D) I and II only

   (E) I and III only

9. Which of the following statements is incorrect?

   (I) Rates typically vary with time during a reaction.

   (II) Rate constants typically vary with time during a reaction.

   (III) Activation energies typically vary with time during a reaction.

   (A) I only

   (B) II only

   (C) III only

   (D) I and II only

   (E) II and III only

Questions 10 through 14 refer to the following choices:

   (A) Total energy

   (B) Change in total energy

   (C) Potential energy

   (D) Kinetic energy

   (E) Work

10. Chemical bonds store this due to the relative positions of electrons and nuclei

11. Sum of work and heat flow

12. Equal and opposite to pressure multiplied by a change in volume

13. Translation, vibration, and rotation give evidence of this

14. Remains constant in the universe

15. A 375g plug of lead is heated and placed into an insulated container filled with 0.500L water. Prior to the immersion of the lead, the water is at 293K. After a time, the lead and the water reach the same maximum temperature, 297K. The specific heat capacity of lead is 0.127 J g^{-1} K^{-1} and the specific heat capacity of water is 4.18 J g^{-1} K^{-1}. How hot was the lead before it entered the water? (Note: Density of water = 1.00 kg / L)

    (A) 473K

    (B) 121K

    (C) 590K

    (D) 295K

    (E) 237K

16. What is the molar reaction enthalpy for the following reaction?

    $C(s)$ + $H_2O(g)$ → $CO(g)$ + $H_2(g)$

    Use the following data:

    Reaction 1: $C(s)$ + $O_2(g)$ → $CO_2(g)$; ΔH = –605 kJ

    Reaction 2: $2CO(g)$ + $O_2(g)$ → $2CO_2(g)$; ΔH = –966 kJ

    Reaction 3: $2H_2(g)$ + $O_2(g)$ → $2H_2O(g)$; ΔH = –638 kJ

    (A) 999 kJ

    (B) 197 kJ

    (C) –1407 kJ

    (D) 394 kJ

    (E) 500 kJ

Answers with Explanations

Having worked through all those problems on rates, energy, work, and heat, you may be feeling a bit depleted of energy yourself. Hang in there long enough to check your answers.

1. **(D).** Rate laws include a concentration factor for each reactant whose concentration affects the rate. Each concentration factor has an exponent, even if that exponent is simply 1. The exponent on each factor is the individual reaction order for that component. The overall reaction order is the sum of the individual reaction orders.

2. **(C).** The rate observed for Reaction 1 depends on the rate law for the reaction: Rate = $k[A]^2$. Component B doesn't even show up in the rate law, so cutting in half the concentration of B has no effect on the rate. However, doubling the concentration of A has a major effect on the rate because the reaction is second order in A (meaning that the concentration factor for A has an exponent of two). You can calculate the rate change associated with doubling A by substituting in simple numbers for A within the rate law. If the original concentration of A is 1 and the doubled concentration is 2, then the original and final rates are k and $4k$, respectively. So, doubling the concentration of A increases the rate fourfold.

3. (D). As described in the answer to question 2, the rate observed for Reaction 2 depends on the rate law for the reaction: Rate = k[D][E]. The reaction is first order in D and first order in E (and second order overall). You can calculate the effect of changing the concentrations of D and E by substituting simple numbers. If the original concentrations of D and E are 1 and 1, respectively, then the new concentrations are 2 and 3, respectively. So, the original rate is $k(1)(1) = k$, and the new rate is $k(2)(3) = 6k$. The changes in concentration result in a sixfold increase in rate.

4. (C). You can figure out the relationship between the rates of change of all the reactants and products by inspecting the balanced reaction equation. D + E → F + 2G. First, because D and E are reactants and F and G are products, it's clear that the rates of change in D and E have opposite sign to those of F and G — as D and E decrease in concentration, F and G increase in concentration. Second, you must account for the stoichiometric coefficients in the reaction equation. Two moles of product G are made for every one mole made of product F and for every one mole consumed of reactants D and E. So, the rate of change in D, E, and F is one half the rate of change in G, as indicated by the coefficient of 0.5. Note that in answer C, whether each term is positive or negative has nothing whatever to do with whether a reactant or product is gained or lost during the reaction; positive versus negative values simply indicate that the rates occur in opposite directions.

5. (E). The reaction equation makes clear that for each mole of methane consumed, 2 moles of oxygen gas are consumed and 1 mole of carbon dioxide is produced. So

   $d[CH_4]/dt = 0.5d[O_2]/dt = -d[CO_2]/dt$

   This means that the disappearance of 2.79mol s^{-1} of methane corresponds to the appearance (positive change) of 2.79mol s^{-1} of CO_2 and the disappearance (negative change) of 2 – 2.79 = 5.58mol s^{-1} of O_2. The concentration of methane changes at half the rate of oxygen, so the concentration oxygen changes at twice the rate of methane.

6. (A). Doubling the concentration of either reactant (D or E) doubles the rate. So, the data are consistent with the rate law, Rate = k[D][E]. Solve for k by substituting known values of rate [D] and [E] from any of the trial reactions:

   k = Rate / ([D][E]) = $4.8 \times 10^6\, M\,s^{-1}$ / ($2.7 \times 10^{-2}\, M \times 2.7 \times 10^{-2}\, M$) = $6.6 \times 10^9\, M^{-1}\, s^{-1}$.

   Use this calculated value of k to determine the rate in the presence of $1.3 \times 10^{-2}\, M$ reactant D and $9.2 \times 10^{-3}\, M$ reactant E:

   Rate = ($6.6 \times 10^9\, M^{-1}\, s^{-1}$)($1.3 \times 10^{-2}\, M$)($9.2 \times 10^{-3}\, M$) = $7.9 \times 10^5\, M\,s^{-1}$.

7. (C). You can calculate the free energy change directly from the equilibrium constant and the temperature by using the equation: $\Delta G = -RT\ln K_{eq}$. But to use the equation properly, you must make sure that all your units match. Specifically, notice that the temperature given to you in the problem has units of Celsius degrees, while the gas constant given has temperature units of Kelvin. Convert the temperature to Kelvin first: 25 °C = (25 + 273)K = 298K. Then, substitute into $\Delta G = -RT\ln K_{eq}$. You are told that $K_{eq} = 1.37 \times 10^3$, so:

   ΔG = (–1)(8.314 J K^{-1}mol^{-1})(298K)ln(1.37 × 10^3) = –1.79 × 10^4 J mol^{-1}

8. (E). Although the free energy of a system must decrease during a spontaneous process (otherwise the process would not be spontaneous), both etropy and enthalpy can either decrease or increase during that process. The reason for this flexibility in entropy and enthalpy is made clear by the Gibbs equation:

   $\Delta G = \Delta H - T\Delta S$

Decreases in enthalpy are favorable and increases in entropy are favorable, it is true. However, an unfavorable change in enthalpy can be compensated by a sufficiently favorable change in entropy and vice versa, such that the overall free energy change for the process is favorable, and the process is spontaneous. Note that, although entropy may decrease within a system, entropy always increases for the universe as a whole (system + surroundings = universe).

9. (E). Rates can certainly change during the course of a reaction because rates are frequently dependent on concentrations, as expressed by rate laws. Because concentrations change during the reaction, so do the rates. Rate constants and activation energies are a different matter, however. For a given reaction at a given temperature, the rate constant and activation energy are related by the Arrhenius equation:

$$k = Ae^{-E_a/RT}$$

The Arrhenius equation contains no terms for concentration of reactants or products, so the rate constant and activation energy typically remain constant for a given reaction.

10. (C). Potential energy is the portion of total energy due to the position of particles. Chemical energy is essentially potential energy because energy is stored within chemical bonds as a result of the favorable position of electrons between adjacent nuclei.

11. (B). The change in total energy of a system is the sum of heat flow and work:

$$\Delta E = E_{final} - E_{initial} = q + w$$

Positive heat flow ($+q$) is heat flow into the system, which increases the total energy of the system. Positive work ($+w$) is work done on the system (*not by* the system), and also increases the total energy of the system.

12. (E). Pressure-volume work is one kind of work that can be done on a system or by a system, especially by expanding or contracting gases: $w = -P\Delta V$. If gas within a system expands against the surroundings (increasing the system's volume), then the system has done work on the surroundings and the sign of work is negative. If the surroundings expand against the system (decreasing the system's volume), then the surroundings have done work on the system and the sign of work is positive.

13. (D). Kinetic energy is the portion of total energy due to the motion of particles. In an ideal gas, kinetic energy consists entirely of translating particles. In real substances, other forms of motion contribute to the kinetic energy, like vibration and rotation. The average kinetic energy of the particles of a system is proportional to the temperature of that system.

14. (A). Transform as it may between different forms and slip as it may between system and surroundings, the total energy of the universe remains constant.

15. (A). The key to setting up this problem is to realize that whatever heat flows out of the lead flows into the water, so: $q_{lead} = -q_{water}$. Calculate each quantity of heat by using $q = mC_P\Delta T$. Recall that $\Delta T = T_{final} - T_{initial}$. The unknown in the problem is the initial temperature of lead.

$$q_{lead} = (375g)(0.127J\ g^{-1}\ K^{-1})(297K - T_{initial})$$

To calculate q_{water}, you must first calculate the mass of 0.500L water by using the density of water: $0.500L \times (1.00kg\ L^{-1}) = 0.500kg = 500g$:

$$q_{water} = (500g)(4.18J\ g^{-1}\ K^{-1})(297K - 293K) = 8.36 \times 10^3 J$$

Setting q_{lead} equal to $-q_{water}$ and solving for $T_{initial}$ yields 473K:

$$(375g)(0.127J\ g^{-1}\ K^{-1})(297K - T_{initial}) = -8.36 \times 10^3 J$$

$$297K - T_{initial} = -176K$$

$$T_{initial} = 473K$$

16. (B). Reverse Reaction 2 and divide it by 2 to get Reaction 2' (ΔH = +483kJ). Reverse Reaction 3 and divide it by 2 to get Reaction 3' (ΔH = +319kJ). Add Reactions 1, 2', and 3', yielding ΔH = +197kJ. Reactions 2 and 3 need reversing to put the right compounds on the reactant and product sides of the equation. Reaction 1 already has the right orientation. Reactions 2 and 3 must be divided by 2 so that the stoichiometry of the final equation matches the stoichiometry of the target equation (namely, one mole each of H_2O, CO and H_2). Even though this division temporarily creates noninteger coefficients for O_2 in these equations, the noninteger coefficients cancel out when you add reactions 1, 2', and 3'.

Part V

Describing Patterns and Predicting Properties

The 5th Wave   By Rich Tennant

"Your formula for a carbohydrate is close, but not entirely accurate. I'm pretty sure carbohydrates consist of carbon, hydrogen, oxygen, sour cream, and bacon bits."

In this part . . .

In earlier times, chemists relied extensively on high-tech instruments like "eyes" and "noses" and so forth. Chemistry used to be largely a descriptive science, without the extensive calculations and precise measurements that characterize it today. Qualitative observations are still a useful part of chemistry, and understanding qualitative patterns in things like color and solubility is a big part of developing an elusive and prized skill called "chemical intuition." This part gives an overview of descriptive chemistry as it applies to the AP Chemistry exam. Organic chemistry is another branch of chemical science with deep roots and time-honored, qualitative traditions. But beyond its history, organic chemistry is a huge modern enterprise with massive application in medicine and industry, among other things. We close this part by giving you a glimpse into organic chemistry, the chemistry of carbon-based compounds, preparing you for its inevitable appearance on the exam.

Chapter 23

Knowing Chemistry When You See It: Descriptive Chemistry

There are a bunch of elements. These elements assemble into a ridiculously large variety of compounds. Then, just to make things more complicated, the compounds react with one another in different ways. To reduce the headache-inducing complexity of chemistry, it helps to know the lay of the land, to be familiar with a few details and themes that pop up over and over. Knowing these patterns helps you to develop a sense for whether certain compounds will react and what products they might generate. Also, sometimes things happen during a chemical reaction that are visible to the naked eye — solid particles or bubbles appear as if from nowhere — and these events give important clues as to the microscopic details of the reaction. Getting a feel for these kinds of things is what some people call "chemical intuition." In this chapter, you gain some chemical intuition so you can make predictions about how reactions will turn out.

Using Your Senses: Qualitative Aspects of Common Reaction Types

Seen from the outside, chemistry can seem weird. Mix two clear solutions and you never know what might happen. Maybe a solid (called a *precipitate*) will form, settling to the bottom of the beaker. Maybe gas will bubble out of the mixture. Maybe the final solution will be green. Maybe nothing will happen. How can you predict? Knowing which compounds you're mixing is a good start. Armed with that information, you might be able to identify a likely reaction. The sections that follow list some common themes and patterns that might prove helpful on the AP chemistry exam.

Precipitation reactions

Precipitation occurs when an insoluble compound forms during the course of a reaction. Usually, the insoluble product is an ionic compound, formed between metal and nonmetal components. Being able to predict when precipitates form requires you to know which compounds are soluble and which are insoluble. Here are some helpful guidelines to help you sniff out which compounds are likely to precipitate and which are likely to stay in solution.

Precipitation reactions can occur when ionic compounds react, typically in *double displacement* reactions. In double displacement reactions, the cation components of ionic compounds swap anionic partners. So, if you are presented with two ionic reactant compounds,

$$AB(aq) + CD(aq) \rightarrow ?$$

perform the following steps to complete the reaction and predict the products.

1. **Swap the anion and cation partners to predict the reaction equation:**

   $$AB(aq) + CD(aq) \rightarrow CB(?) + AD(?)$$

2. Next, **inspect the products to see if either or both are insoluble.** Here are some basic rules for judging whether an ionic compound is soluble.

   Compounds that contain the following ions are **soluble:**

   - NH_4^+
   - Group IA (column 1) cations (that is, alkali metal ions)
   - NO_3^-
   - ClO_3^-, ClO_2^-
   - CH_3COO^- (acetate ion, which can also be written as $C_2H_3O_2^-$)

   Compounds that contain the following ions are **usually soluble:**

   - Cl^-, Br^- and I^- (**except** when paired with Ag^+, Pb^{2+}, and Hg_2^{2+})
   - SO_4^{2-} (**except** when paired with Ag^+, Pb^{2+}, Hg_2^{2+}, Sr^{2+}, Ca^{2+}, and Ba^{2+})

   Compounds that contain the following ions are **usually insoluble:**

   - OH^- (that is, hydroxides, **except** when paired with Group IA cations, Sr^{2+}, Ca^{2+}, and Ba^{2+})
   - CO_3^{2-}, PO_4^{3-}, SO_3^{2-}, and S^{2-} (**except** when paired with Group I cations or NH_4^+)

3. After you determine whether you have a product that is insoluble, **rewrite the completed reaction equation as a *net ionic equation.*** Net ionic equations list only those components of the reaction that change state during the reaction. Ions in aqueous solution are listed separately, as ions, and not as part of a compound. Ions that appear on both the reactant and product sides (these are called *spectator ions*) are omitted. Thus,

   $$Na_2CO_3(aq) + Ba(NO_3)_2(aq) \rightarrow ?$$

   undergoes a double displacement reaction, yielding:

4. $Na_2CO_3(aq) + Ba(NO_3)_2(aq) \rightarrow 2NaNO_3(?) + BaCO_3(?)$ Once you have the completed reaction and have assigned phases, **write a *total* ionic equation.** The solubility rules state that the sodium nitrate pair is soluble, but when barium ion and carbonate ion combine, a solid precipitate forms. The resulting *total* ionic equation is

   $$2Na^+(aq) + CO_3^{2-}(aq) + Ba^{2+}(aq) + 2NO_3^-(aq) \rightarrow 2Na^+(aq) + 2NO_3^-(aq) + BaCO_3(s)$$

5. Then, remove the spectator ions (Na^+ and NO_3^-), leaving the *net* ionic equation:

   $$CO_3^{2-}(aq) + Ba^{2+}(aq) \rightarrow BaCO_3(s)$$

Aside from precipitation reactions, you should be familiar with the basic patterns and common players in other reaction types, including gas-releasing reactions, reactions of hydrides and oxides, combustion reactions, reactions involving colored compounds, and acid-base reactions.

Gas-releasing reactions

Gas-releasing reactions often include decomposition or the reaction of metals/metal hydrides with water.

Gas-releasing decomposition can result from several processes:

- **Heating:** Adding heat to some compounds can cause them to decompose (break apart) into several products, some of which are gases. For example
  - $CaCO_3(s) \rightarrow CaO(s) + CO_2(g)$
  - $2KNO_3(s) \rightarrow 2KNO_2(s) + O_2(g)$
  - $2KClO_3(s) \rightarrow 2KCl(s) + 3O_2(g)$
  - $2H_2O_2(l) \rightarrow 2H_2O(l) + O_2(g)$

- **Electrolysis:.** Molten ionic compounds can be decomposed when an electric current passes through them, releasing gas in the process:

  $2Na^+(l) + 2Cl^-(l) \rightarrow 2Na(l) + Cl_2(g)$

  Or, shown as two redox half-reactions

  $2Na^+(l) + 2e^- \rightarrow 2Na(l)$ (reduction)

  $2Cl^-(l) \rightarrow Cl_2(g) + 2e^-$ (oxidation)

- **Unstable acids:** Some weak acids are unstable and can decompose, especially if they aren't in solution while heat is added.

  $H_2SO_3(s) \rightarrow H_2O(l) + SO_2(g)$

  $H_2CO_3(s) \rightarrow H_2O(l) + CO_2(g)$

Metals and metal hydride reactions

Metals and metal hydrides can release hydrogen gas when they react with water. These reactions result in a basic solution as hydrogen is taken from water leaving excess hydroxide ions:

- $LiH(s) + H_2O(l) \rightarrow Li^+(aq) + OH^-(aq) + H_2(g)$
- $CaH_2(s) + 2H_2O(l) \rightarrow Ca^{2+}(aq) + 2OH^-(aq) + 2H_2(g)$
- $2Na(s) + 2H_2O(l) \rightarrow 2Na^+(aq) + 2OH^-(aq) + H_2(g)$

Combustion reactions

Complete combustion occurs when hydrocarbons react with oxygen, forming carbon dioxide and water.

- $2C_2H_6(g) + 7O_2(g) \rightarrow 4CO_2(g) + 6H_2O(g)$ (ethane combusts)
- $C_2H_4(g) + 3O_2(g) \rightarrow 2CO_2(g) + 2H_2O(g)$ (ethene combusts)
- $2C_2H_2(g) + 5O_2(g) \rightarrow 4CO_2(g) + 2H_2O(g)$ (ethyne combusts)

Oxide reactions

The pattern of an oxide reaction depends on whether it is a metal oxide or nonmetal oxide that reacts.

Metal oxides often react with water to produce basic solutions.

$$Li_2O(s) + H_2O(l) \rightarrow 2Li^+(aq) + 2OH^-(aq)$$

Nonmetal oxides often react with water to produce acidic solutions.

$$N_2O_5(s) + H_2O(l) \rightarrow 2NO_3^-(aq) + 2H^+(aq)$$

Colored compounds

Knowing something about which compounds are colored and what colors those compounds are can help you identify an unknown compound.

Salts and oxides that contain transition metals are often colored.

- ✔ Ni-containing compounds are often green.

- ✔ Mn-containing compounds are often pink/purple.

- ✔ Colored soluble salts

    - • Fe-containing soluble salts are red and/or brown.

    - • Cu-containing soluble salts are blue and/or green.

    - • Co-containing soluble salts are blue.

- ✔ Colored insoluble salts

    - • Precipitates containing CrO_4^{2-} (chrmate) are yellow.

    - • Precipitates containing $Cr_2O_7^{2-}$ (dichromate) are orange.

    - • Precipitates containing OH^- (hydroxide) are white.

    - • Precipitate of silver chloride (AgCl) is white.

    - • Precipitate of silver bromide (AgBr) is pale yellow.

Acids and bases

Acid-base reactions are very common and confuse many students. Key to understanding these reactions is being able to identify the acid and the base, and understanding how strong acids and bases behave differently from weak acids and bases. Acids and bases have predictable properties and reactions:

- ✔ A strong acid is a weak base. A strong base is a weak acid.

- ✔ The conjugate acid of a weak base is a weak acid. The conjugate base of a weak acid is a weak base.

- ✔ Strong acids react with strong bases to form neutral salts and water.

- ✔ Strong acids react with weak bases to form acidic solutions.

- ✔ Strong bases react with weak acids to form basic solutions.

Surveying the Table: Predicting Properties and Reactivity

There are two basic ways to look at the periodic table. You can pick a property and observe the trends in that property as you move left and right or up and down the table. Or, you can pick a row or column of the table and observe changes or the lack of changes in various properties as you move along the members of the column or row.

Reading trends across the table

The key periodic property is electronegativity, the tendency of an atomic nucleus to draw electrons toward itself within a bond. Electronegativity tends to increase to the right and upward within the main body of the periodic table, as shown in Figure 23-1. Electronegativity represents the result of a balance between an unbound atom's ionization energy (tendency to lose an electron) and its electron affinity, the ability of an unbound atom to attract an electron. Ionization energy is the energy it takes to remove an electron from a gas-phase atom.

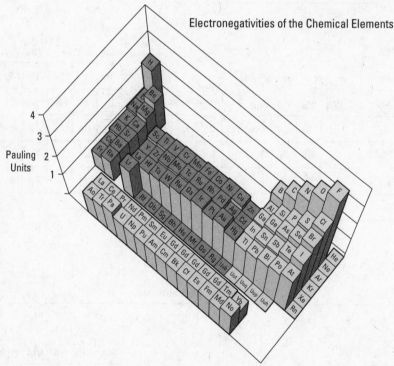

Electronegativities of the Chemical Elements

Figure 23-1: Electro-negativity increases upward and to the right within the periodic table.

Trends in electronegativity, ionization energy (Figure 23-2), and electron affinity underlie other periodic trends, such as atomic radius (Figure 23-3), and metallic character (Figure 23-4). Atomic radius and metallic character are also largely determined by electronegativity, but both follow a pattern that is the mirror image of electronegativity, as each increases downward and to the left within the periodic table.

Increasing Ionization Energy → High Energy

Increasing Ionization Energy ↑

1	IA																VIIIA
1	H	IIA															He
2	Li	Be										B	C	N	O	F	Ne
3	Na	Mg	IIIB	IVB	VB	VIB	VIIB	VIIIB		IB	IIB	Al	Si	P	S	Cl	Ar
4	K	Ca	Sc	Ti	V	Cr	Mn	Fe	Co	Ni	Cu	Zn	Ga	Ge	As	Se	Br	Kr
5	Rb	Sr	Y	Zr	Nb	Mo	Tc	Ru	Rh	Pd	Ag	Cd	In	Sn	Sb	Te	I	Xe
6	Cs	Ba	La	Hf	Ta	W	Re	Os	Ir	Pt	Au	Hg	Tl	Pb	Bi	Po	At	Rn
7	Fr	Ra	Ac

Figure 23-2: Ionization energy increases upward and to the right within the periodic table.

Low Energy

Increasing Size ← Small Radii

Increasing Size ↓

	IA																VIIIA
1	H	IIA															He
2	Li	Be										B	C	N	O	F	Ne
3	Na	Mg	IIIB	IVB	VB	VIB	VIIB	VIIIB		IB	IIB	Al	Si	P	S	Cl	Ar
4	K	Ca	Sc	Ti	V	Cr	Mn	Fe	Co	Ni	Cu	Zn	Ga	Ge	As	Se	Br	Kr
5	Rb	Sr	Y	Zr	Nb	Mo	Tc	Ru	Rh	Pd	Ag	Cd	In	Sn	Sb	Te	I	Xe
6	Cs	Ba	La	Hf	Ta	W	Re	Os	Ir	Pt	Au	Hg	Tl	Pb	Bi	Po	At	Rn
7	Fr	Ra	Ac

Figure 23-3: Atomic radius increases downward and to the left within the periodic table.

Large Radii

Increasing Metallic Character ← Least Metallic

Increasing Metallic Character ↓

	IA																VIIIA
1	H	IIA															He
2	Li	Be										B	C	N	O	F	Ne
3	Na	Mg	IIIB	IVB	VB	VIB	VIIB	VIIIB		IB	IIB	Al	Si	P	S	Cl	Ar
4	K	Ca	Sc	Ti	V	Cr	Mn	Fe	Co	Ni	Cu	Zn	Ga	Ge	As	Se	Br	Kr
5	Rb	Sr	Y	Zr	Nb	Mo	Tc	Ru	Rh	Pd	Ag	Cd	In	Sn	Sb	Te	I	Xe
6	Cs	Ba	La	Hf	Ta	W	Re	Os	Ir	Pt	Au	Hg	Tl	Pb	Bi	Po	At	Rn
7	Fr	Ra	Ac

Figure 23-4: Metallic character increases downward and to the left within the periodic table.

Most Metallic

Finding patterns in columns and rows

The trends described in the previous section help explain why the elements within a vertical group of the periodic table tend to have properties similar to one another, whereas elements across a horizontal period display incremental changes in their properties, with a consistent trend.

Some groups hang together more closely than others. Within the periodic table, the two leftmost and two rightmost groups show the greatest kinship, keeping tightly to the characteristic properties of their respective vertical families. So, you can feel most confident making predictions based on the group similarity of leftmost columns (columns 1 and 2) and two columns on the right (columns 16 and 17). In addition, the elements in group VIIIA (column 18), the noble gases, have closely identifiable properties to one another. Note that two methods are used to identify columns in the periodic table. The older one uses A and B columns, the newer one has been adopted by the International Union of Pure and Applied Chemistry (IUPAC) and the American Chemical Society. Both will be used here.

✔ Alkali metals (Group IA or 1)

- Elemental alkali metals are extremely reactive, and their cations are extremely unreactive.

- Alkali metals are the least electronegative, easily giving up electrons to form cations.

- In compounds, alkali metals have an oxidation number of +1.

- Though listed with Group IA, **hydrogen is not a metal** and has some very different properties:

  - In compounds with nonmetals, hydrogen assumes an oxidation number of +1.

  - In compounds with metals, hydrogen assumes "hydride" form, with an oxidation number of –1.

✔ Alkaline earth metals (Group IIA or 2)

- Alkaline earth metals are very reactive, and their cations are very unreactive.

- Alkaline earth metals have very low electronegativity.

- In compounds, alkaline earth metals have an oxidation number of +2.

✔ VIA or 16

- These elements are reactive because they have six electrons in their valence shell, not the stable, eight-electron valence shell of a noble gas.

- Ions of this group tend to have –2 charge.

✔ Halogens (Group VIIA or 17)

- Halogens are very electronegative, with the most electronegative at the top.

- In compounds, halogens usually have an oxidation number of –1.

- Halogens range in reactivity (though none are *un*reactive), with the most reactive at the top.

- Halogens form ionic compounds with metals and molecular compounds with nonmetals.

✔ Noble gases (Group VIIIA or 18)

- Noble gases are the least reactive of the elements, due to their filled valence shells.

Chapter 24

Answering Questions on Descriptive Chemistry

....

In This Chapter

▶ Making sure you remember what's important

▶ Practicing with questions

▶ Getting your questions answered and explained

....

Descriptive chemistry, as you may have gathered from Chapter 23, is perhaps the most memorization-intensive area covered by the AP chemistry exam. But fight the impulse to sweep it under the rug — having an intuitive feel for periodic trends and the broad patterns of common chemical reactions is an invaluable asset. Help gel your chemical intuition by reviewing the highlights of Chapter 23 and applying your insights to the practice questions provided here.

Keeping the Important Stuff Straight

Because descriptive chemistry includes a lot of "just-have-to-know" details, this summary of ideas from Chapter 23 is one that you'll probably want to review well and often, even more than the summaries of many other chapters. To make the task easier, the bits and pieces are grouped into categories that reflect specific reaction types or specific perioidic patterns. Focus on one category at a time, attempting to master each one before moving on to the others. That way, you won't lose your way in the swirl of details.

Precipitation reactions

Much of the previously mentioned memorization has to do with knowing which ionic compounds are soluble in water and which are not. You need this knowledge to predict the phases of products of ionic double displacement reactions, so you can properly write net ionic equations.

Here is a summary of solubility rules and the process of formulating net ionic equations:

✔ Precipitation reactions can occur when ionic compounds react, typically in *double displacement* reactions.

✔ Compounds that contain the following ions are soluble: NH_4^+, Group IA cations, NO_3^-, ClO_3^-, ClO_2^-, CH_3COO^-.

✔ Compounds that contain the following ions are usually soluble: Cl^-, Br^-, and I^- (except with Ag^+, Pb^{2+}, and Hg_2^{2+}); SO_4^{2-} (except with Ag^+, Pb^{2+}, Hg_2^{2+}, Sr^{2+}, Ca^{2+}, and Ba^{2+}).

✔ Compounds that contain the following ions are usually insoluble: OH^- (except with Group IA cations, Sr^{2+}, Ca^{2+}, and Ba^{2+}); CO_3^{2-}, PO_4^{3-}, SO_3^{2-}, and S^{2-} (except with Group IA cations or NH_4^+).

✔ Net ionic equations list only those components of the reaction that change state during the reaction. Spectator ions, those ions that appear in the same fashion on both the reactant and product sides, are omitted.

Patterns of common reaction types

Beyond precipitation, it pays to know the patterns of common reaction types that you'll likely encounter on the AP exam, including gas-releasing reactions, complete combustion, reactions of oxides, and acid-base reactions.

Here is a concise summary of the patterns you can expect from the most commonly tested reaction types:

✔ Gas-releasing decomposition can result from heating, electrolysis, and unstable acids (especially H_2CO_3 and H_2SO_4).

✔ Metals and metal hydrides can release hydrogen gas when they react with water, producing a basic solution (excess hydroxide ion).

✔ Combustion occurs when hydrocarbons and other compounds containing carbon-hydrogen bonds, such as alcohols, react with oxygen, forming carbon dioxide and water.

✔ Metal oxides often react with water to produce basic solutions.

✔ Nonmetal oxides often react with water to produce acidic solutions.

✔ Salts and oxides that contain transition metals are often colored.

✔ Acids and bases

- A strong acid is a weak base. A strong base is a weak acid.

- The conjugate anion of a strong acid is a *very* weak base. The conjugate cation of a strong base is a *very* weak acid.

- The conjugate acid of a weak base is a weak acid. The conjugate base of a weak acid is a weak base.

- The conjugate acid of a weak base is a *stronger* acid *than the base* and the conjugate base of of a weak acid is a *stronger* base *than the acid*.

- Strong acids react with strong bases to form neutral salts and water.

- Salts of weak acids or bases react in water to form weakly acidic or basic solutions.

Periodic trends and group properties

The periodic table is periodic for a reason — the elements of the table are arranged to reflect recurring patterns in the properties of the elements. Those patterns emerge from trends in the number of valence electrons held by the elements. Although learning these patterns requires a bit of initial memorization, knowing them saves you a load of memorization in the long run.

Here are the key periodic trends and the properties of important groups of elements:

✔ Electronegativity, electron affinity, and ionization energy tend to increase upward and to the right within the main body of the periodic table.

✔ Atomic radius and metallic character tend to increase downward and to the left within the main body of the periodic table.

✔ **Alkali metals (Group IA)** are extremely reactive, and their cations are extremely unreactive; they are the least electronegative elements; they have an oxidation number of +1 within compounds.

> *Hydrogen is not a metal:* In compounds with nonmetals, hydrogen has oxidation number +1; in compounds with metals, hydrogen has oxidation number −1.

✔ **Alkaline earth metals (Group IIA)** are very reactive, and their cations are very unreactive; they have very low electronegativity; they have an oxidation number of +2 within compounds.

✔ **Halogens (Group VIIA)** are very electronegative and range in reactivity, with the most electronegative and most reactive at the top; they form ionic compounds with metals and molecular compounds with nonmetals; they usually have an oxidation number of −1 within compounds.

✔ **Noble gases (Group VIIIA)** are extremely unreactive. The larger atoms (Kr and Xe) *can* form compounds with strongly electronegative atoms such as fluorine and oxygen.

Testing Your Knowledge

You put in the hard work of climbing to the top of Chapter 23. Now peer from the heights, seeing what you can see from the summit. Don't foget to make full use of the periodic table as you work these questions. The more you use it, the quicker you will find your way around it.

1. Which of the following statements are true about the combustion of propane, C_3H_8?

    (I) Oxygen, O_2, is produced.

    (II) Total moles of products outnumber total moles of reactants.

    (III) Products can react to form carbonic acid, H_2CO_3.

    (A) I only

    (B) II only

    (C) III only

    (D) I and II only

    (E) II and III only

2. Which element has the greatest electronegativity?

    (A) Be

    (B) Ba

    (C) Fe

    (D) O

    (E) Po

3. Which element has the most metallic character?

   (A) Si

   (B) Pb

   (C) Fr

   (D) K

   (E) Mg

4. Which element has the lowest ionization energy?

   (A) H

   (B) Li

   (C) Ne

   (D) Cs

   (E) Ra

5. Which of the following pairs of solutions produce a precipitate if mixed?

   (I) $HBr(aq)$ and $Ca(OH)_2(aq)$

   (II) $Pb(NO_3)_2(aq)$ and $LiI(aq)$

   (III) $(NH_4)_3PO_4(aq)$ and $MgCl_2(aq)$

   (A) I only

   (B) II only

   (C) III only

   (D) I and II only

   (E) II and III only

6. Which of the following reactants are least likely to yield a gaseous product?

   (A) $LiClO_3$ and heat

   (B) $Ca(OH)_2$ and heat

   (C) NaH and H_2O

   (D) K and H^+

   (E) Li and H_2O

7. Which mixture(s) will produce a solution with pH < 7?

   (I) P_4O_{10} and H_2O

   (II) MgO and H_2O

   (III) SrH_2 and H_2O

   (A) I only

   (B) II only

   (C) III only

   (D) II and III only

   (E) I, II, and III

8. Which of the following manipulations increase the solubility of calcium carbonate, $CaCO_3(s)$, in aqueous solution?

    (I) Decrease temperature

    (II) Decrease pH

    (III) Add Na_2CO_3

    (A) I only

    (B) II only

    (C) III only

    (D) I and II only

    (E) None

Questions 9 through 13 refer to the following set of choices. Pick the *best* answer in each case.

    (A) Transition metals

    (B) Strong acids conjugates

    (C) Weak acids

    (D) Salts

    (E) Hydrides

9. Are weak bases

10. Often form colored compounds

11. Result from neutralization reactions

12. Release hydrogen gas in the presence of wate

13. Are conjugates of weak base

14. Which of the following represent(s) a net ionic equation?

    (I) $Zn(s) + HCl(aq) \rightarrow ZnCl_2(aq) + H_2(g)$

    (II) $CaCO_3(s) + 2H^+(aq) + 2Cl^-(aq) \rightarrow Ca^{2+}(aq) + 2Cl^-(aq) + H_2O(l) + CO_2(g)$

    (III) $Ag^+(aq) + Cl^-(aq) \rightarrow AgCl(s)$

    (A) I only

    (B) II only

    (C) III only

    (D) I and II only

    (E) II and III only

Questions 15 through 16 refer to the diagram of an electrolytic cell in Figure 24-1:

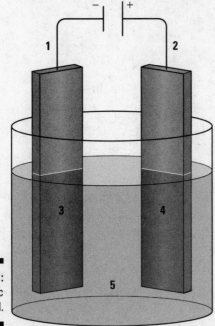

Figure 24-1:
Electolytic
cell.

15. Electrons travel out of the liquid along this wire:

(A) 1

(B) 2

(C) 1 and 2 simultaneously

(D) Neither 1 nor 2

(E) Alternates between 1 and 2

16. If molten LiCl is the liquid indicated by 5, then the following is true:

(I) Solid metal may accumulate at 3.

(II) $Cl_2(g)$ may bubble from 4.

(III) $Li_2(g)$ may bubble from 3.

(A) I only

(B) II only

(C) III only

(D) I and II only

(E) II and III only

Checking Your Work

If you feel like a few of your answers may have been less than perfect, check them against the answers and explanations below. If you're certain that you nailed the questions, check your answers anyway, if only to revel in your perfection.

1. (E). Combustion of a hydrocarbon (like propane) involves a combination reaction with oxygen gas, yielding carbon dioxide and water products. Here's the balanced equation for the combustion of propane:

$$C_3H_8(g) + 5O_2(g) \rightarrow 3CO_2(g) + 4H_2O(l)$$

From the balanced equation, it's clear that oxygen is consumed, not produced. Moreover, the moles of products (3 + 4 = 7) outnumber the moles of reactants (1 + 5 = 6). Finally, the carbon dioxide and water products can undergo a further reaction to form carbonic acid:

$$CO_2(g) + H_2O(l) \rightarrow H_2CO_3(aq)$$

2. (D). Overall, electronegativity increases upward and to the right on the periodic table.

3. (C). Overall, metallic character increases downward and to the left on the periodic table.

4. (D). Overall, ionization energy increases upward and to the right on the periodic table (just like electronegativity).

5. (E). HBr and $Ca(OH)_2$ are a strong acid and a strong base, respectively, that undergo a neutralization reaction to produce water and the soluble salt, $CaBr_2$. Although the nitrate salt of lead is soluble (like virtually all nitrate salts), most other lead salts (like PbI_2) are insoluble. Although ammonium phosphate is soluble (like virtually all ammonium salts), only Group IA and ammonium salts of phosphate are soluble — other phosphate salts (like $Mg_3(PO_4)_2$) are insoluble.

6. (B). Chlorate (ClO_3^-) salts tend to decompose with heat, yielding oxygen gas. Ionic compounds containing hydroxide (OH^-) do not tend to decompose with heat. Metal hydrides (like NaH) tend to produce hydrogen gas when they react with water. Reactive metals (like Group IA and Group IIA metals) tend to react with acid (H^+) to produce hydrogen gas. Reactive metals also tend to react with water to produce hydrogen gas.

7. (A). Nonmetal oxides (like P_4O_{10}) tend to react with water to produce acidic (pH < 7) solutions:

$$P_4O_{10}(g) + 6H_2O(l) \rightarrow 12H^+(aq) + 4PO_4^{3-}(aq)$$

Metal oxides (like MgO) tend to react with water to produce basic (pH > 7) solutions:

$$MgO(s) + H_2O(l) \rightarrow Mg^{2+}(aq) + 2OH^-(aq)$$

Metal hydrides (like SrH_2) tend to react with water, releasing hydrogen gas and producing a basic solution:

$$SrH_2(s) + 2H_2O(l) \rightarrow Sr^{2+}(aq) + 2OH^-(aq) + 2H_2(g)$$

8. (B). Decreasing temperature usually decreases the solubility of solid solutes. Adding CO_3^{2-} anion by means of sodium carbonate decreases the solubility of $CaCO_3$ due to the common ion effect, in which the added carbonate shifts the equilibrium toward undissolved $CaCO_3$.

Decreasing pH increases $CaCO_3$ solubility because the added H^+ undergoes a secondary reaction with CO_3^{2-}, shifting the overall equilibrium toward dissolved $CaCO_3$:

$$CaCO_3(s) + H_2O(l) \rightarrow Ca^{2+}(aq) + CO_3^{2-}(aq)$$

$$H^+(aq) + CO_3^{2-}(aq) \rightarrow HCO_3^-(aq)$$

9. (B). Conjugates of strong acids are _very_ weak bases.

10. (A). Transition metals are notorious for forming intensely colored coordination complexes.

11. (D). In a neutralization reaction, acid and base reactants react to form a salt and water.

12. (E). Hydrides (especially metal hydrides) tend to react with water to produce hydrogen gas, H_2.

13. (C). Weak acids are the conjugate acids _of_ weak bases.

14. (C). To generate a net ionic equation, list dissolved (soluble) compounds as dissociated ions. Then, cancel any spectator ions, which are those ions that appear in identical form on both sides of the reaction equation. What remains is the net ionic equation. For example:

$$AgNO_3(aq) + NaCl(aq) \rightarrow NaNO_3(aq) + AgCl(s)$$

$$Ag^+(aq) + NO_3^-(aq) + Na^+(aq) + Cl^-(aq) \rightarrow Na^+(aq) + NO_3^-(aq) + AgCl(s)$$

$$Ag^+(aq) + Cl^-(aq) \rightarrow AgCl(s)$$

15. (B). The positive and negative signs at the top of the figure indicate the polarity of the voltage source for the electrolytic cell. Because electrons have negative charge, they migrate out of the liquid toward the positive pole along 2, and into the liquid away from the negative pole along 1.

16. (D). If the liquid indicated by 5 is molten LiCl, then the salt is dissociated into Li^+ cations and Cl^- anions. Molten salts can be decomposed by electrolysis in this kind of cell. The electrode at 3 acts as a cathode, delivering electrons to $Na^+(l)$ cations, which deposit there as $Na(s)$ metal. The electrode at 4 acts as an anode, accepting electrons from $Cl^-(l)$ anions, which react with one another to form $Cl_2(g)$ at the electrode.

Chapter 25

Living with Carbon: Organic Chemistry

• •

In This Chapter

▶ Checking out the nomenclature of organic compounds

▶ Taking in testable trends in the properties of organic compounds

▶ Discovering isomerism and chirality

▶ Running through common organic reactions

• •

*O*rganic chemistry is a huge subfield, and the AP chemistry exam merely scratches the surface of it. Still, a good grasp of the concepts and naming rules covered in this chapter could get you those few extra points needed to push your score up one grade on exam day.

Don't spend too much time here if you're still struggling with the big-point-gaining chapters earlier in this book. Your time is better spent polishing your knowledge of the concepts in those earlier chapters, if you're struggling, because so few questions on the AP exam address organic chemistry. But if you feel comfortable with the big-point-gaining chapters, then tackle this chapter with vigor and get ready to file lots of new information away in that chemistry-packed brain of yours — organic chemistry is heavy on memorization.

In addition to the few questions on each AP exam directly addressing organic chemistry, a basic knowledge of the information in this chapter, including basic structure and naming. can aid you in answering many other questions on the exam. Organic compounds are sometimes used in questions testing concepts such as kinetics, stoichiometry, and colligative properties.

Naming Names: Nomenclature and Properties of Hydrocarbons

Any study of organic chemistry begins with the study of *hydrocarbons*. Hydrocarbons are some of the simplest and most important organic compounds and they contain only hydrogen and carbon atoms. *Organic compounds* are based on carbon skeletons. Carbon skeletons can be modified — you can dress them up with chemically interesting atoms like oxygen, nitrogen, halogens, phosphorous, silicon, or sulfur. This cast of atomic characters may seem like a rather small subset of the more than 100 elements in the periodic table. It's true: Organic compounds typically use only a very small number of the naturally occurring elements. Yet these molecules include the most biologically important compounds in existence. In this section you will learn how to recognize and name all of the types of hydrocarbons that appear on the AP exam, including alkenes, alkanes, alkynes, and cyclic hydrocarbons.

Starting with straight-chain alkanes — linear hydrocarbons

The simplest of the hydrocarbons fall into the category of *alkanes.* Alkanes are chains of carbon atoms connected by single covalent bonds. Chapter 5 describes how single covalent bonds result when atoms share pairs of valence electrons. Carbon molecules have four valence electrons. So, carbon atoms are eager to donate their four valence electrons to covalent bonds so that they can receive four donated electrons in turn, filling their valence shell. In other words, carbon really likes to form four bonds. In alkanes, each of these is a single bond with a different carbon or hydrogen partner.

The simplest of the alkanes, called *continuous-* or *straight-chain alkanes,* consist of one straight chain of carbon atoms linked with single bonds. Hydrogen atoms fill all the remaining bonds. Other types of alkanes include closed circles and branched chains.

We begin with straight-chain alkanes because they make clear the basic strategy for naming hydrocarbons. From the standpoint of naming, the hydrogen atoms in a hydrocarbon are more or less "filler atoms." Alkanes' names are based on the largest number of consecutively bonded carbon atoms. So, the name of a hydrocarbon tells you about that molecule's structure. To name a straight-chain alkane, simply match the appropriate chemical prefix with the suffix *-ane.* The prefixes relate to the number of carbons in the continuous chain, which we list in Table 25-1.

Table 25-1		The Carbon Prefixes	
Number of Carbons	*Prefix*	*Chemical Formula*	*Alkane*
1	Meth-	CH_4	Methane
2	Eth-	C_2H_6	Ethane
3	Prop-	C_3H_8	Propane
4	But-	C_4H_{10}	Butane
5	Pent-	C_5H_{12}	Pentane
6	Hex-	C_6H_{14}	Hexane
7	Hept-	C_7H_{16}	Heptane
8	Oct-	C_8H_{18}	Octane
9	Non-	C_9H_{20}	Nonane
10	Dec-	$C_{10}H_{22}$	Decane

The naming method in Table 25-1 tells you how many carbons are in the chain. Because you know that each carbon has four bonds and because you are fiendishly clever, you can deduce the number of hydrogen atoms in the molecule as well. Any carbon at the end of a chain must be bonded to three hydrogens and all interior carbons must be bonded to two. Consider the carbon structure of pentane, for example, shown in Figure 25-1.

Figure 25-1:
Pentane's
carbon
skeleton.

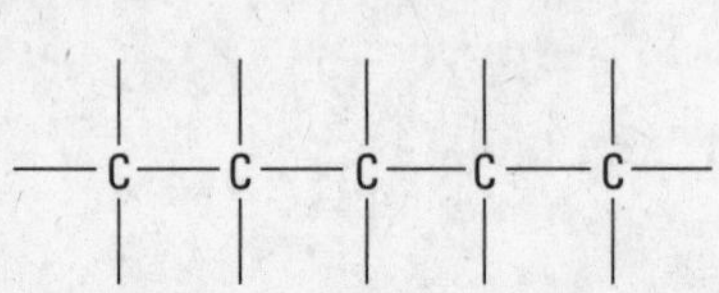

—C—C—C—C—C—

Only four carbon-carbon bonds are required to produce the five-carbon chain of pentane. This leaves many bonds open — two for each interior carbon and three for each of the terminal carbons. These open bonds are satisfied by carbon-hydrogen bonds, thereby forming a hydrocarbon, as shown in Figure 25-2.

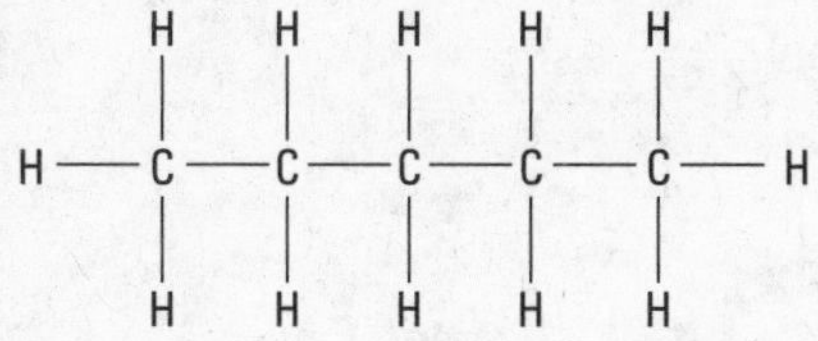

Figure 25-2:
Pentane's
hydro-
carbon
structure.

If you add up the hydrogen atoms in Figure 25-2, you get 12. So, pentane contains 5 carbon atoms and 12 hydrogen atoms.

As the organic molecules you study get more and more complicated, it will become more and more important to draw the molecular structure to visualize the molecule. In the case of straight-chain alkanes, the simplest of all organic molecules, you can remember a convenient formula for calculating the number of hydrogen atoms in the alkane without actually drawing the chain:

Number of hydrogen atoms = (2 × number of carbon atoms) + 2

You can refer to the same molecule in a number of different ways. For example, you can refer to pentane by its name (ahem . . . *pentane*), by its molecular formula, C_5H_{12}, or by the complete structure in Figure 25-2. Clearly, these different names include different levels of structural detail. A *condensed structural formula* is another naming method, one that straddles the divide between a molecular formula and a complete structure. For pentane, the condensed structural formula is $CH_3CH_2CH_2CH_2CH_3$. This kind of formula assumes that you understand how straight-chain alkanes are put together. Carbons on the end of a chain, for example, are only bonded to one other carbon, so they have three additional bonds to be filled by hydrogen and are labeled as CH_3 in a condensed formula. Interior carbons are bonded to two neighboring carbons and have only two hydrogen bonds, and so are labeled CH_2.

Going out on a limb: Making branched alkanes by substitution

Not all alkanes are straight-chain alkanes. That would be too easy. Many alkanes are so-called *branched alkanes*. Branched alkanes differ from continuous-chain alkanes in that carbon chains substitute for a few hydrogen atoms along the chain. Atoms or other groups (like carbon chains) that substitute for hydrogen in an alkane are called *substituents*.

Naming branched alkanes is slightly more complicated, but you need only to follow these simple set of steps to arrive at a proper (and often lengthy) name:

1. **Count the longest continuous chain of carbons.** Tricky chemistry problems often show branched alkanes with the longest chain snaking through a few branches instead of obviously lined up in a row. Consider the two carbon structures shown in Figure 25-3. The two are actually the same structure, drawn differently! Yikes. In either case, the longest continuous chain in this structure has eight carbons. That wasn't hard — and yippee, now you already know the basic name for the compound. It is an octane.

```
                                   C
                                   |
                                   C        C
                                   |        |
        C₁ ─── C₂ ─── C₃ ─── C₄ ─── C₅ ─── C
                                            |
                                            C₆
                                            |
                                            C₇
                                            |
                                            C₈
```

Figure 25-3:
One carbon
structure
drawn two
different
ways.

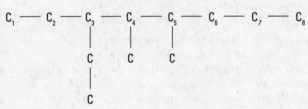

2. **Number the carbons in the chain** *starting with the end which is closest to a branch.*
   You can always check to be sure you have done this step correctly by numbering the carbon chain from the opposite end as well. The correct numbering sequence is the one in which the substituent branches extend from the lowest-numbered carbons. For example, as it is drawn and numbered in Figure 25-3, the alkane has substituent groups branching off of its third, fourth, and fifth carbons. If the carbon chain had been numbered backward, these would be the fourth, fifth, and sixth carbons in the chain. Because the first set of numbers is lower, the chain is numbered properly. The longest chain in a branched alkane is called the *parent chain.*

3. **Count the number of carbons in each branch.** These groups are called *alkyl groups* and are named by adding the suffix *-yl* to the appropriate alkane prefix (Table 25-1 awaits your visit). The three most common alkyl groups are the methyl (one carbon), ethyl (two carbons), and propyl (three carbons) groups. Figure 25-3 has two methyl groups, one ethyl group, and no propyl groups.

   Be careful when you find yourself dealing with alkyl groups made up of more than just a few carbons. A tricky drawing may cause you to misnumber the parent chain!

4. **Attach the number of the carbon from which each substituent branches to the front of the alkyl group name.** For example, if a group of two carbons is attached to the third carbon in a chain, like it is in Figure 25-3, the group is called 3-ethyl.

5. **Check for repeated alkyl groups.** If multiple groups with the same number of carbons branch off the parent chain, do not repeat the name. Rather, include multiple numbers, separated by commas, before the alkyl group name. Also, specify the number of instances of the alkyl group by using the prefixes di-, tri-, tetra-, and so on. For example, if one-carbon groups (in other words, methyl groups) branch off carbons four and five of the parent chain, the two methyl groups appear together as "4,5-dimethyl."

6. **Place the names of the substituent groups in front of the name of the parent chain *in alphabetical order.*** Prefixes like di-, tri-, and tetra- do not figure into the alphabetizing. So, the proper name of the organic molecule in Figure 25-3 is 3-ethyl-4,5-dimethyloctane.

Note that hyphens are used to connect all the naming elements except for the last connection to the parent chain, which includes neither a hyphen nor a space (. . . dimethyl-octane is wrong). We recommend practicing alkanes till you get them in your sleep before moving on to other organic compounds!

Getting unsaturated: Alkenes and alkynes

Carbons can do more than engage in four single bonds. There's more to organic molecules than substituent-for-hydrogen swaps:

- ✔ When carbons in an organic compound fill their valence shells entirely with single bonds, we say the compound is *saturated*.

- ✔ When hydrocarbons contain carbons that bond to each other more than once, creating double or triple covalent bonds, we say these hydrocarbons are *unsaturated* because they have fewer than the maximum possible number of hydrogens or substituents.

For every additional carbon-carbon bond formed in a hydrocarbon, two fewer covalent bonds to hydrogen are formed.

When neighboring carbons share four valence electrons to form a double bond, the resulting hydrocarbon is called an *alkene*. Alkenes are characterized by these chemically interesting double bonds, which are more reactive than single carbon-carbon bonds (see Chapter 7 for a review of sigma and pi bonding). Double bonds also change the shape of a hydrocarbon, because the sp^2 hybridized valence orbitals assume a trigonal planar geometry, as shown by the carbons of ethene in Figure 25-4. Saturated carbon is sp^3 hybridized and has tetrahedral geometry (again, see Chapter 7 to review hybridization).

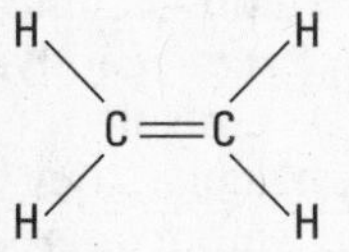

Figure 25-4:
Ethene
carbons.

$$H_2C = CH_2$$

Naming alkenes is slightly more complicated than naming alkanes. In addition to the number of carbons in the main chain and any branching substituents, you must also note the location of the double bonds in an alkene and incorporate that information into the name. Nevertheless, the essential naming strategy for alkenes is quite similar to that for alkanes:

1. **Locate the longest carbon chain that includes the double bond, and number it, *starting at the end closest to the double bond.*** In other words, double bonds trump substituents when it comes to numbering the parent chain. Build the name of the parent chain by using the same prefixes as used for alkanes (refer to Table 25-1), but match the prefix with the suffix *-ene*. A three-carbon chain with a double bond, for example, is called "propene."

2. **Number and name substituents that branch off the alkene *in the same way as done for alkanes* (see the section earlier in this chapter, "Starting with straight-chain alkanes — linear hydrocarbons" for more on naming alkanes).** List the number of the substituted carbon, followed by the name of the substituent. Separate the substituent number and name with a hyphen.

3. **Identify the lowest-numbered carbon that participates in the double bond, and put that number *between the substituent names and the parent chain name* (sandwiched by hyphens), but after all the substituent names.** For example, if the second and third carbons of a five-carbon alkene engage in a double bond, then the molecule is called 2-pentene, not 3-pentene. If that same molecule has a methyl substituent at the fourth carbon, then the molecule is called 4-methyl-2-pentene.

Alternately, and especially when there are substituents present, the position of an unsaturation is indicated between the prefix and suffix of the parent chain name. So, 4-methyl-2-pentene may also be written 4-methylpent-2-ene.

Alkynes are hydrocarbons in which neighboring carbons share six electrons to engage in triple covalent bonds. The naming strategy for alkynes is the same as that for alkenes, except that the alkyne parent chain is named by matching the prefix with the suffix *-yne*.

An important consequence of the presence of multiple bonds in alkenes and alkynes is their marked increase in reactivity over alkanes. Generally speaking, the more multiple bonds a compound has, the more reactive it tends to be. This is because of the propensity of multiple bonded carbons to undergo addition reactions, which are discussed at the end of this chapter.

Rounding 'em Up: Circling Carbons with Cyclics and Aromatics

The compounds we discuss earlier in this chapter are linear or branched. However, hydrocarbons can be circular, or *cyclical*, hydrocarbons There are two important categories of cyclical hydrocarbons:

- *Cyclic aliphatic* hydrocarbons
- *Aromatic* hydrocarbons

Chemists sometimes divide hydrocarbons into aliphatic and aromatic categories to highlight important differences in structure and reactivity. Without going into more technical detail than is useful here, aliphatic molecules and aromatic molecules have significantly different electronic configurations (which electrons go into which orbitals). As a result, the two types of hydrocarbons typically undergo different kinds of reactions. In particular, they tend to undergo different kinds of substitution reactions, ones in which some atom or group substitutes for hydrogen.

Shutting the circle with cyclic aliphatic hydrocarbons

Cyclic aliphatic hydrocarbons are like the hydrocarbons that we discuss earlier in this chapter, except that they form a closed ring. The rules for naming these compounds build on the rules for naming alkanes, alkenes, and alkynes. For example, a cyclical six-carbon alkane includes the name "hexane," but is preceded by the prefix *cyclo-*, making the final name "cyclohexane."

A single substituent or unsaturation on a cyclic aliphatic hydrocarbon does not require a number. So, a single methyl-for-hydrogen substitution on cyclohexane yields a compound with the name methylcyclohexane. Likewise, a lone double bond unsaturation on cyclohexane yields a compound with the name cyclohexene.

Multiple substitutions or unsaturations require numbering. In these cases, the same rules apply for deciding the rank of substituents as applied to alkenes and alkynes. Triple bonds outrank double bonds. Double bonds outrank other substitutions. So, number the carbons in the way that respects these rankings and produces the lowest overall numbers. A cyclohexane molecule with two methyl substituents on neighboring carbons, for example, is called 1,2-dimethylcyclohexane.

Analyzing aromatic hydrocarbons

Aromatic hydrocarbons have special properties because of their electronic structure. Aromatics are both cyclic and conjugated:

- ✔ **Cyclic:** The carbons form a ringlike structure.

- ✔ **Conjugated:** Conjugation results from an alternation of double or triple bonds with single bonds.

Aromatic molecules have clouds of *delocalized pi electrons,* electrons that move freely through a set of overlapping *p* orbitals. The model aromatic compound is benzene. Because of its cyclical, conjugated bonding, pi electrons delocalize evenly into rings above and below the plane of the flat benzene molecule. Aromatic compounds are very stable compared to their aliphatic counterparts.

Numbering substituents on aromatics follows the same basic pattern as followed for cyclic aliphatic compounds. A single substituent requires no numbering, as in methylbenzene. Multiple substituents are numbered by rank, with the highest-ranked substituent placed on carbon number one, and proceeding in a way that results in the lowest overall numbers. A benzene ring with ethyl and methyl substituents situated two carbons away from one another, for example, would be called 1-ethyl-3-methylbenzene.

Decorating Skeletons: Organic Functional Groups

Surely you've noticed that in this chapter we have thus far limited our study of organic compounds to those made entirely of hydrogen and carbon. When studying the complex rules governing hydrocarbons, it is easy to forget that the periodic table contains more than 100 other elements. Although organic chemistry often concentrates on just a few of those elements, you'll need to be familiar with many other compound types, and this time they'll be made of more than just hydrogen and carbon. All of these exotic new compounds contain plain vanilla carbon chains, but each has a distinguishing feature composed of another, more exotic element. These embellishments are called *functional groups*. These groups include noncarbon elements, and give organic compounds a dizzying array of different properties.

Because hydrocarbons are old news, we sometimes refer to them simply using the letter R, representing any hydrocarbon group: branched or unbranched, saturated or unsaturated.

Table 25-2 summarizes the other important organic compounds, their distinguishing features, and their endings. Some molecules contain more than one of these distinguishing features. Locations where one of these groups or a multiple bond appear are prime sites for reactions to occur. These sites are therefore called *functional groups.* Carefully study the functional group column in Table 25-2 so you can recognize them quickly. These are the structures you should keep an eye out for later in this chapter.

Table 25-2		Functional Groups	
Compound Name	*Compound Formula*	*Functional Group*	*Prefix or Suffix*
Alcohol	R-OH	$-\overset{\displaystyle |}{\underset{\displaystyle |}{C}} - \overset{..}{\underset{..}{O}} - H$	-ol
Ether	R-O-R	$-\overset{\displaystyle |}{\underset{\displaystyle |}{C}} - \overset{..}{\underset{..}{O}} - \overset{\displaystyle |}{\underset{\displaystyle |}{C}} -$	ether
Carboxylic acid	R-COOH	$\overset{\displaystyle \overset{..}{O}{\cdot}}{\overset{\displaystyle \|}{-C} - \overset{..}{\underset{..}{O}} - H}$	-oic acid
Ester	R-COOR	$\overset{\displaystyle {\cdot}\overset{..}{O}{\cdot}}{\overset{\displaystyle \|}{-C} - \overset{..}{\underset{..}{O}} - \overset{\displaystyle |}{\underset{\displaystyle |}{C}} -}$	-oate
Aldehyde	R-CHO	$\overset{\displaystyle {\cdot}\overset{..}{O}{\cdot}}{\overset{\displaystyle \|}{-C} - H}$	-al
Ketone	R-COR	$-\overset{\displaystyle |}{\underset{\displaystyle |}{C}} - \overset{\overset{\displaystyle {\cdot}\overset{..}{O}{\cdot}}{\displaystyle \|}}{C} - \overset{\displaystyle |}{\underset{\displaystyle |}{C}} -$	-one
Halocarbon	R-X	$-\overset{\displaystyle |}{\underset{\displaystyle |}{C}} - \overset{..}{\underset{..}{X}}{:}$ (X = halogen)	-oro, omo, or odo
Amine	R-NH$_2$	$-\overset{\displaystyle |}{\underset{\displaystyle |}{C}} - \overset{..}{N} -$	-amine

Alcohols: Hosting a hydroxide

Alcohols are hydrocarbons with a hydroxide group attached to them. Their general form is R-OH. To name an alcohol, you simply count the longest number of consecutive carbon atoms in the chain, find its prefix Table 25-2, and then attach the ending *-ol* to that prefix. For example, a one-carbon chain with an OH group attached is methanol, two is ethanol, three is propanol, and so on. Substituents (groups branching off the main chain) must, as always, be accounted for, and you must specify the number of the carbon atom in the chain to which the hydroxyl (OH) group is attached. Begin your numbering at the end of the chain closest to the OH, and attach the number before the prefix + *-ol*. For example, the compound shown in Figure 25-5 contains a six-carbon chain with a hydroxyl group. Its base name is therefore hexanol. It has methyl groups (with one carbon each) on the second and fourth carbons and its OH group lies on the third carbon. Its proper name is therefore 2,4-dimethyl-3-hexanol.

Some alcohols are *polyhydroxyl,* or contain multiple hydroxyl groups. In order to avoid confusion with multiples of any substituent groups, the prefixes indicating numbers of hydroxyl groups in an alcohol are attached to the suffix *ol,* making them diols, triols, etc. For example, a four carbon alcohol containing hydroxyl groups on the first and third carbons is 1,3-butanediol. The numbers indicating the locations of the hydroxyl groups must come before the base name but after any substituent groups.

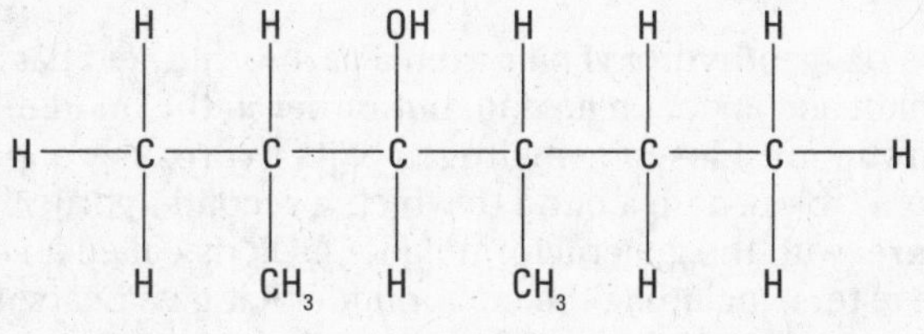

Figure 25-5:
An example of an alcohol.

$$H-\overset{\overset{\displaystyle H}{|}}{\underset{\underset{\displaystyle H}{|}}{C}}-\overset{\overset{\displaystyle H}{|}}{\underset{\underset{\displaystyle CH_3}{|}}{C}}-\overset{\overset{\displaystyle OH}{|}}{\underset{\underset{\displaystyle H}{|}}{C}}-\overset{\overset{\displaystyle H}{|}}{\underset{\underset{\displaystyle CH_3}{|}}{C}}-\overset{\overset{\displaystyle H}{|}}{\underset{\underset{\displaystyle H}{|}}{C}}-\overset{\overset{\displaystyle H}{|}}{\underset{\underset{\displaystyle H}{|}}{C}}-H$$

Ethers: Invaded by oxygen

Perhaps most commonly known for their use as early anesthetics, *ethers* are carbon chains that have been infiltrated by an oxygen atom. These oxygen atoms lie conspicuously in the middle of a carbon chain like badly disguised spies in an enemy camp and have the general formula R-O-R. Ethers are named by naming the alkyl groups on either side of the lone oxygen as substituent groups individually (adding the ending *-yl* to their prefixes), and then attaching the word *ether* to the end. For example, the compound shown in Figure 25-6 is an ether with a methyl group (one carbon) on one side of the oxygen and an ethyl (two carbons) on the other. Placing the substituents in alphabetical order, the compound's proper name is ethyl methyl ether.

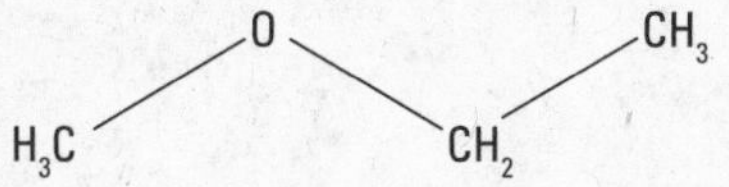

Figure 25-6:
An example of an ether.

$$H_3C \diagdown \overset{\displaystyle O}{} \diagdown \underset{\displaystyle CH_2}{} \diagup \overset{\displaystyle CH_3}{}$$

Oxygen atoms in ethers are often surrounded by two identical alkyl groups, in which case the prefix *di-* must be attached to the name of that substituent. For example, an oxygen surrounded by two propyl groups (with three carbons each) is called dipropyl ether.

Carboxylic acids: R-COOH brings up the rear

Carboxylic acids appear to be ordinary hydrocarbons until you reach the very last carbon in their chain, whose three ordinary hydrogens have been usurped by a double-bonded oxygen and a hydroxyl group (an OH group). This gives a carboxylic acid the general form R-COOH. These compounds are named by attaching the suffix *-oic acid* to the end of the prefix. For example, the compound shown in Figure 25-7 has four carbons, the last in the chain attached to a double-bonded oxygen and a hydroxyl group. The compound is therefore called butanoic acid. The hydrogen of the COOH group can pop off as H^+ leaving an R-COO$^-$ ion, which is why these compounds are called acids. Carboxylic acids are a favorite compound type for AP exam writers because they are both organics and acids. It is important to note, however, that all carboxylic acids are weak acids, so learn to spot them among acid lists and keep their relative weakness in mind.

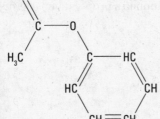

Figure 25-7:
An example of a carboxylic acid.

Esters: Creating two carbon chains

What if the same double-bonded oxygen/hydroxyl pair from a carboxylic acid has infiltrated deeper into the hydrocarbon chain and is not on an end, but rather in the middle? In order to accomplish this, the COOH group must lose the hydrogen of its hydroxyl group (which, as an acidic hydrogen, is no problem) to open up a bond to which a second hydrocarbon can attach. A compound of this nature, with the general formula R-COOR, is called an *ester*. Lots of nice, smelly coumpounds are esters, including the compounds that give pears, apples, and bananas their pleasant aromas.

Because the carbon chain in an ester is also broken by an oxygen, you'll need to choose one carbon chain as a lowly substituent group and the other to bear the proud suffix of *-oate*. This high-priority group is always the carbon chain that includes the carbon double-bonded to one oxygen and single bonded to the other. The group on the other side of the single-bonded oxygen is named as an ordinary substituent. For example, the compound shown in Figure 25-8 is phenyl ethanoate, because the high priority group has two carbons and the low priority group is a benzene ring.

Figure 25-8:
An example of an ester.

That's right! The low-priority group on the other side of the oxygen doesn't have to be a hydrocarbon chain! It can be a ring or a metal as well. In Figure 25-8, this group is a benzene ring.

Aldehydes: Holding tight to one oxygen

Aldehydes are much like carboxylic acids except that they lack the second oxygen in the COOH group. Their final carbon shares a single bond with its neighboring carbon, a double bond with an oxygen, and a single bond with hydrogen. Aldehydes are named with the suffix *-al* (not to be confused with the *-ol* of alcohols) and have the general formula R-CHO. For example, the compound in Figure 25-9 is pentanal, a five-carbon chain with a double-bonded oxygen taking the place of two of the hydrogen bonds on the final carbon.

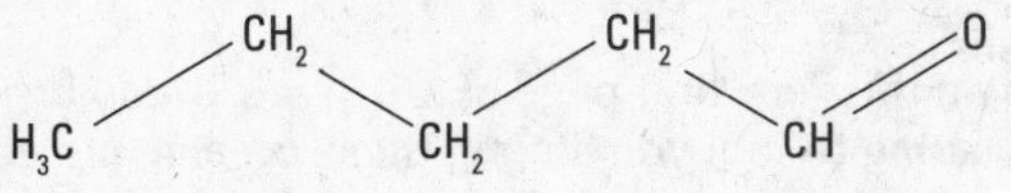

Figure 25-9:
An example
of an
aldehyde.

Ketones: Lone oxygen sneaks up the chain

Much like esters are to carboxylic acids, *ketones* share the same basic structure as aldehydes, except that their double-bonded oxygen can be found hiding out in the midst of the carbon chain, giving them the general form R-COR. Ketones are named by adding the suffix *-one* to the prefix generated by counting up the longest carbon chain. Unlike esters, however, the carbon chain of a ketone is not broken by the double-bonded oxygen group, making naming them much simpler. The compound shown in Figure 25-10, for example, is called simply 2-butanone; the number before the name specifies the number of the carbon to which the oxygen is attached.

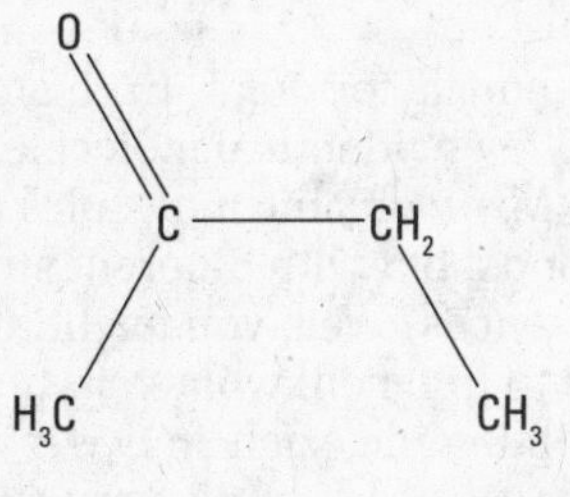

Figure 25-10:
An example
of a ketone.

Halocarbons: Hello, halogen!

Halocarbons are simply hydrocarbons with a halogen or more tacked on to a carbon in place of a hydrogen atom. (A *halogen* is a group VIIA element.) Halogens in a halocarbon are always named as substituent groups — *fluoro*, *chloro*, *bromo*, or *iodo*, one for each of the four halogens that form halocarbons (F, Cl, Br or I). The shorthand for a halocarbon is R-X, where X is the halogen. The compound shown in Figure 25-11 has two bromo groups attached, one to the second carbon and one to the third in a five-carbon alkane. Its official name is, therefore, 2,3-dibromopentane.

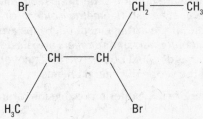

Figure 25-11:
An example
of a
halocarbon.

Amines: Hobnobbing with nitrogen

Amines are as conspicuous as Waldo on the very first page of a *Where's Waldo* book. No tricksters wearing red-striped shirts are waiting to fool you among these organic molecules. Amines and their derivatives are the only compounds that you'll encounter in basic organic chemistry that contain nitrogen atoms. Amines have the basic form $R-NH_2$, and they're named by naming the carbon chain as a substituent (with the ending *-yl*), and then adding the suffix *-amine*. The structure in Figure 25-12, for example, is called ethylamine because it's a two-carbon ethyl chain with an amine group.

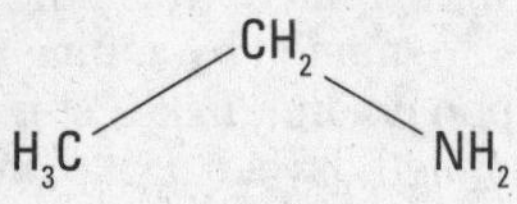

Figure 25-12: An example of an amine.

Rearranging Carbons: Isomerism

How about this: Two organic molecules have identical chemical formulas. Each atom in one molecule is bonded to the same groups as in the other. They're identical molecules, right? Wrong! (Mischievous chemistry gods point and snicker.) Many organic molecules are *isomers,* compounds with the same formula and types of bonds, but with different structural or spatial arrangements. Who cares about such subtle differences? Well, you might. Consider thalidomide, a small organic molecule widely prescribed to pregnant women in the late 1950s and early 1960s as a treatment for morning sickness. Thalidomide exists in two isomeric forms that rapidly switch from one to the other in the body. One isomer is very effective at combating morning sickness. The other isomer causes serious birth defects. Isomers matter.

Isomers can be confusing. They fall into different categories and subcategories. So, before committing your brain to a game of isomeric Twister, peruse the following breakdown: Keep in mind, however that all isomers, no matter what their specific classification, have both the same number and type of atoms.

- ✔ *Structural isomers* have identical molecular formulas but differ in the arrangement of bonds.

- ✔ *Stereoisomers* have identical connectivities — all atoms are bonded to the same types of other atoms — but differ in the arrangement of atoms in space.

  - • *Diastereomers* are stereoisomers that are *not* non-superimposable mirror images of each other. Two types of diastereomers exist: *Geometric isomers* (or *cis-trans isomers*) are diastereomers that differ in the arrangement of groups around a double bond, or the plane of a ring. *Conformers* and *rotamers* are diastereomers that differ because of rotations about individual bonds (we don't cover them in this book because they are beyond the scope of general chemistry).

  - • *Enantiomers* are stereoisomers that *are* non-superimposable mirror images of each other.

You already know how to recognize and name stereoisomers appropriately by carefully studying the structure of the main chain and branching substituents. This section focuses on the trickier category: stereoisomers.

Picking sides with geometric isomers

Geometric isomers or *cis-trans isomers* are a good place to start in the world of stereoisomers because they're the easiest of the stereoisomers to understand. In the following sections, we explain how isomers relate to alkenes, alkanes that aren't straight-chain, and alkynes.

Straight-chain alkanes are immune from geometric isomerism because their carbon-carbon single bonds can rotate freely. Unsaturate (or add another bond to) one of those bonds, however, and you've got a different story. Alkenes have double bonds that resist rotation. Furthermore, the sp^2 hybridization of double-bonded carbons gives them trigonal planar bonding geometry (see Chapter 7 for an introduction to hybridization). The result is that groups attached to these carbons are locked on one side or the other of the double bond. Convince yourself of this by examining Figure 25-13.

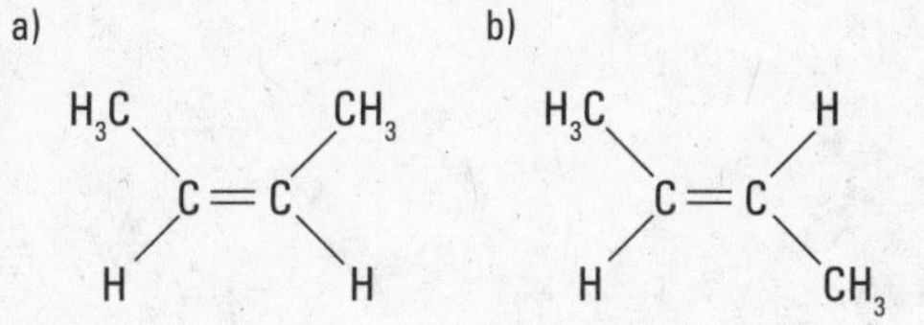

Figure 25-13:
Cis and *trans* isomers of an alkene.

In the left-hand structure of Figure 25-13, the carbon chain continues along the same side of the carbon-carbon double bond. Both methyl (-CH$_3$) groups lie on the same side of the unsaturation. This is called *cis* configuration. In the right-hand structure, the carbon chain swaps sides as it proceeds across the double bond. The methyl groups lie on opposite sides of the unsaturation. This is called a *trans* configuration.

Naming *cis-trans* isomers is simple. Attach the appropriate *cis-* or *trans-* prefix before the number referring to the carbon of the double bond. For example, the left-hand structure in Figure 22-1 is *cis*-2-butene, while the structure on the right is *trans*-2-butene.

Although straight-chain alkanes happily avoid isomerism by rotating merrily about their single bonds, the four bonds of sp^3-hybridized carbons assume tetrahedral geometry. Detailed representations like the one shown for methane in Figure 25-14 reveal this geometry. In the structure of methane, the bonds depicted as straight lines run in the plane of the page. The bond drawn as a filled wedge projects outward from the page. The bond drawn as a dashed wedge projects behind the page. These filled and dashed wedge symbols are known as *stereo bonds* because they are helpful in identifying stereoisomers.

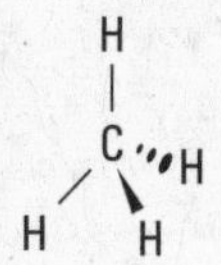

Figure 25-14:
Stereo bonds in methane.

When alkanes close into rings, they can no longer freely rotate about their single bonds, and the tetrahedral geometry of sp^3-hybridized carbons creates *cis-trans* isomers. Groups bonded to ring carbons are locked above or below the plane of the ring, as shown in Figure 25-15.

The figure depicts two different versions of *trans*-1,2-dimethylcyclohexane. In both versions, the adjacent methyl substituents are locked in *trans* positions, on opposite sides of the ring.

✔ The upper set of structures shows the plane of the ring as seen from above, and highlights the *trans-* configuration of the methyl groups with stereo bonds.

✔ The lower set of structures shows the same rings rotated 90 degrees downward and toward you.

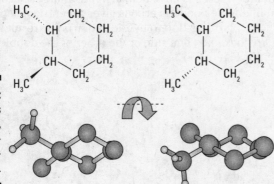

Figure 25-15:
Two isomers of *trans*-1, 2-dimethyl-cyclo-hexane.

The *trans* configuration of the methyl groups is most clear in the lower structures. The front-most methyl group is highlighted with explicit hydrogen atoms to emphasize its position above or below the plane of the ring.

Alkynes also contain carbon-carbon bonds that cannot rotate freely. However, the *sp* hybridization of the carbons in these bonds leads to linear bonding geometry. The two-carbon alkyne, ethyne, is shown in Figure 25-16. Each carbon locks three of its valence electrons into the axis of the triple bond. Each has only one valence electron remaining with which to bond to hydrogen. No *cis-trans* isomerism is possible in this scenario.

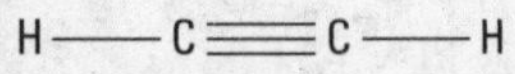

Figure 25-16:
No isomerism is possible at the triple bonds of alkynes, such as ethyne.

$$H \longrightarrow C \equiv C \longrightarrow H$$

Changing Names: Some Common Organic Reactions

The carbon-based organic molecules presented in this chapter can morph into one another through chemical reactions. Such reactions are rather common, and the two most common types are substitution and addition reactions.

Alcohols are particularly prone to a chemical process called *substitution* in which their OH group is replaced by a different atom. For example, the OH group on 2-pentanol can be replaced by the halogen fluorine, turning the alcohol into a halocarbon called 2-fluoro-pentane (see Figure 25-17).

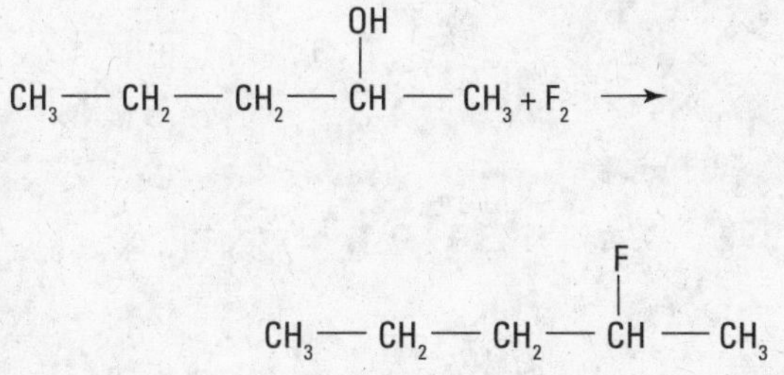

Figure 25-17:
The process of substitution.

$$CH_3 — CH_2 — CH_2 — \overset{\overset{\displaystyle OH}{|}}{CH} — CH_3 + F_2 \longrightarrow$$

$$CH_3 — CH_2 — CH_2 — \overset{\overset{\displaystyle F}{|}}{CH} — CH_3$$

The double bonds of alkenes also make them particularly likely to react with other compounds through a process called *addition*. If the double bond between two carbon molecules is broken, it allows each of those two carbon atoms to form a bond with another atom or molecule. The reaction shown in Figure 25-18, for example, shows water being added across the double bond of 1-hexene. The water molecule itself is split into two pieces — simple hydrogen and a hydroxide — and each of them is added to one of the two carbons that used to share a double bond, forming either 1-hexanol or 2-hexanol.

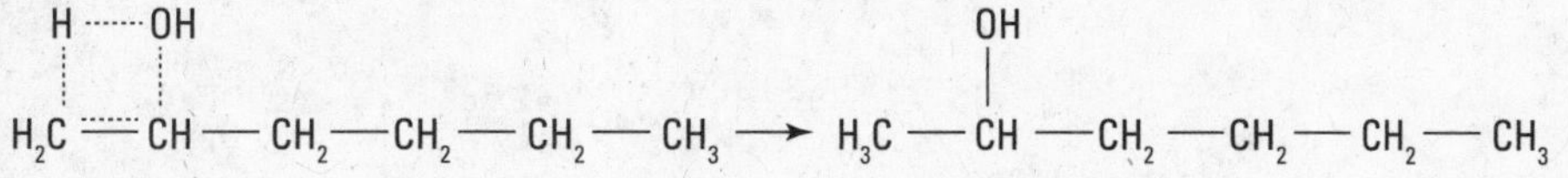

Figure 25-18:
The process of addition.

$$H_2C \overset{\overset{\displaystyle H\cdots\cdots OH}{\vdots \quad \vdots}}{—} CH — CH_2 — CH_2 — CH_2 — CH_3 \longrightarrow H_3C — \overset{\overset{\displaystyle OH}{|}}{CH} — CH_2 — CH_2 — CH_2 — CH_3$$

Chapter 26

Answering Questions on Organic Chemistry

. . . .

In This Chapter
▶ Keeping the facts in mind
▶ Digging into some practice questions
▶ Seeing how you did

. . . .

Chapter 25 is chock-full of juicy little organic chemistry tidbits. Make sure that you have everything filed away in your brain properly by reviewing the concepts in this chapter and attempting the practice problems we include near the end of the chapter. Remember, although there are relatively few questions that directly test organic chemistry on the AP exam, a familiarity with the names and structures of organic compounds can help you to answer many other types of questions.

Making the Main Points Stick

To master the organic chemistry portion of the AP curriculum you need to memorize, memorize, memorize! Specifically, you must remember how to identify the 13 types of organic compounds, and 3 types of isomer listed in the sections below.

Important organic compounds

Listed below are the 13 types of organic compound most likely to appear on the AP exam, together with their naming rules. Make sure that you have a thorough understanding of what makes each unique before you move on.

- **Alkanes** are hydrocarbons containing only single bonds, **alkenes** contain double bonds and **alkynes** contain triple bonds.

- Alkanes, alkenes, and alkynes are named by first numbering the longest carbon chain. Numbering starts at the end closest to the nearest multiple bond for alkenes and alkynes, while alkanes are numbered starting at the end that gives its branches the lowest numbers.

- **Substituents** on all organic compounds are named with the suffix *yl* and are preceded by the number of the main chain carbon atom from which they branch.

- **Cyclic aliphatic hydrocarbons** contain rings of carbon atoms and can contain multiple bonds. **Aromatic hydrocarbons** also contain rings of carbon atoms, but the bonds between carbons alternate between double or triple bonds and single bonds.

- **Alcohols** are hydrocarbon chains that include a hydroxide (OH) group. They are named with the ending -ol.

- **Ethers** contain an oxygen atom in the midst of their carbon chain. They are named by naming the substituents on either side of the oxygen and then adding the ending *ether*, as in ethyl methyl ether.

- The final carbon atom in a **carboxylic acid** is double bonded to an oxygen atom as well as being single bonded to a hydroxide group. In condensed structural formulas, this configuration is written as R-COOH. Carboxylic acids are named by attaching the suffix -oic acid.

- **Esters** are related to carboxylic acids in that they too have a carbon atom that is double bonded to one oxygen. This time, however, rather than also being bonded to a hydroxide group, the final carbon is bonded to a lone oxygen atom. This oxygen is itself bonded to another hydrocarbon chain, giving esters the condensed formula R-COOR. You name an ester by first naming the lower-priority hydrocarbon chain (the one not containing a carbon double bonded to oxygen) as a substituent and then attaching the suffix -oate to the high-priority chain.

- **Aldehydes** are also similar to carboxylic acids in that their final carbon atom is double bonded to an oxygen atom, however in this case the fourth available carbon bond is taken up by a simple hydrogen atom rather than a hydroxide group, giving it the formula R-CHO. Aldehydes are named by attaching the suffix -al.

- **Ketones** have the same structure as aldehydes, except that the double-bonded oxygen has infiltrated its way up to a midchain carbon. These compounds have the condensed formula ROR and are named with the suffix -one. Unlike esters, ketones contain uninterrupted carbon chains, so you need not worry about prefixes.

- **Halocarbons** are hydrocarbon chains with a halogen substituent. The hydrocarbon portion is named normally, with the prefix *fluoro*, *chloro*, *bromo*, or *iodo* attached, depending on which halogen it contains.

- **Amines** are hydrocarbon chains that contain an NH_2 group and have the basic form R-NH_2. They are named by naming the carbon chain as a substituent then adding the ending *amine*.

Isomers

You will also be expected to be able to identify the following three types of isomer on the AP exam. Keep in mind that all isomers contain the same number and type of atoms, but their connectivities may be different.

- **Structural isomers** contain the same number of atoms of carbon, hydrogen, oxygen, etc., but in different configurations.

- **Stereoisomers** must have the same connectivities (in other words, all of their bonds need to be in the same order), and are divided into two categories — compounds that are nonsuperimposable mirror images of one another (enantiomers) and compounds that are superimposable mirror images of one another (diasteriomers).

- **Geometric,** or **cis-trans,** isomers are stereoisomers that differ in the location of substituent groups about a double bond. When the two groups are on the same side of the double bond, you are dealing with the cis isomer, while groups on opposite side of the double bond indicate a trans isomer.

Testing Your Knowledge

Apply your knowledge of the concepts outlined in Chapter 25 with these practice questions. Make sure that you are confident in your ability to distinguish between all of the different types of organic compounds and isomers before you attempt them, though! Remember that organic chemistry in particular involves a lot of memorization.

Questions 1 through 4 refer to the following types of organic compounds

(A) Alcohols

(B) Esters

(C) Ethers

(D) Aldehydes

(E) Halocarbons

1. These organic compounds do not contain any oxygen atoms.

2. You may need to use the prefix *di* in naming one of these organic compounds.

3. These organic compounds are known for having pleasant aromas.

4. These organic compounds are most closely related to ketones.

5. Of the following organic compounds, which is MOST soluble in water at 298K?

(A) Hexane

(B) Hexene

(C) Ethyl Methyl Ether

(D) Octanol

(E) Methanol

6. (a) Write out the condensed structural formula of 2-pentanone.

   (b) Draw the complete structural formula of 2-pentanone, showing all atoms and bonds explicitly.

   (c) Draw a complete structural formula for an isomer of pentanone.

Answers with Explanations

1. (E). The halocarbon is the only one of these organic compounds that doesn't contain an oxygen atom.

2. (C). Ethers are named by naming the substituents on either side of the oxygen atom. If the groups on either side are the same, then the prefix *di-* is used as in CH_3OCH_3 (dimethyl ether).

3. (B). Esters are known for their pleasant aromas.

4. (D). Aldehydes and ketones both share a double C-O bond. In an aldehyde, the carbon double bonded to the oxygen is also bonded to a hydrogen atom, ending the carbon chain. In a ketone, the carbon chain continues beyond the double bond with oxygen.

5. (E). Following the "like dissolves like" rule (review solubility in Chapter 10 for more details), alcohols should be the most soluble in water among the compounds on this list because they have the same intermolecular forces as water. Octanol's long nonpolar hydrocarbon chain, however, balances out the polar end of the molecule to some extent, making methanol the more soluble alcohol.

6. (a) $CH_3CH_2CH_2COCH_3$. 2-Pentanone is a five-carbon ketone, with the double-bonded oxygen on the second carbon. This means it has a condensed structural formula of $CH_3CH_2CH_2COCH_3$.

(b) Translate the condensed structural formula into a complete structural drawing, giving you

```
       H   H   H   O   H
       |   |   |   ‖   |
   H — C — C — C — C — C — H
       |   |   |       |
       H   H   H       H
```

(c) Any five-carbon ketone or aldehyde would be an acceptable isomer, as long as it still has the formula $C_5H_{10}O$. 3-pentanone and pentanal are two possibilities, both shown below.

            Pentanae   3-pentanone

```
       H   H   H   H   O                   H   H   O   H   H
       |   |   |   |   ‖                   |   |   ‖   |   |
   H — C — C — C — C — C — H           H — C — C — C — C — C — H
       |   |   |   |                       |   |       |   |
       H   H   H   H                       H   H       H   H
```

Part VI
Getting Wet and Dirty: Labs

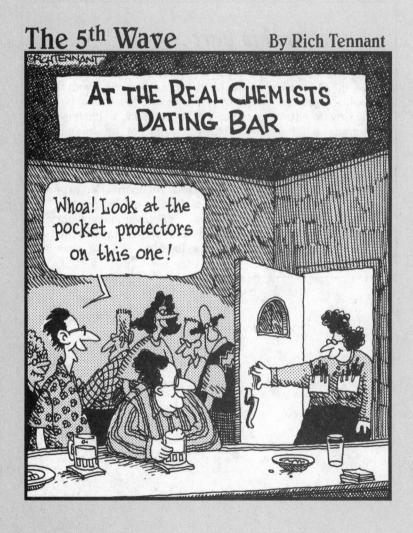

The 5th Wave   By Rich Tennant

AT THE REAL CHEMISTS DATING BAR

Whoa! Look at the pocket protectors on this one!

In this part . . .

Studying chemistry without spending time in a lab is a bit like reading up on the history of skydiving from the confines of an armchair in your basement. Something is missing from the experience. For this reason, AP chemistry courses include extensive labwork as part of the standard curriculum. Although the specific experiments may vary from one high school course to another, they all conform to the same set of guidelines and principles, and most of them are drawn from an official set of recommended experiments put forth by the AP chemistry Powers That Be. In this part, we give you a brief review of common laboratory equipment and techniques, and a complete survey of the Official List of AP chemistry experiments.

Chapter 27

Knowing Your Way around the Lab: Equipment and Procedures

In This Chapter
▶ Getting familiar with common laboratory equipment
▶ Reviewing common laboratory techniques

Chemistry is a laboratory science. Although there are such things as purely "theoretical" chemists who do all their reactions on computers, other chemists tend to make fun of these types behind their backs. Most chemistry still involves things like beakers, tubes, and interesting odors (beyond those possibly emanating from the chemists themselves). This state of affairs is not lost on the creators of the AP exam, who expect you to know your basic laboratory equipment and techniques. As you'll know if you're taking an AP chemistry course, lab is an important component. This chapter will help you to review the equipment and procedures that you've employed to create interesting odors of your own.

Gearing Up: An Overview of Common Lab Equipment

Figure 27-1 shows all of the lab equipment that an AP chemistry test-taker will need to be familiar with. You will not be directly tested on your knowledge of lab equipment, however the AP chemistry exam will often describe and/or diagram experimental setups so you will need to be able to recognize and understand the purpose of each of the pieces of equipment below.

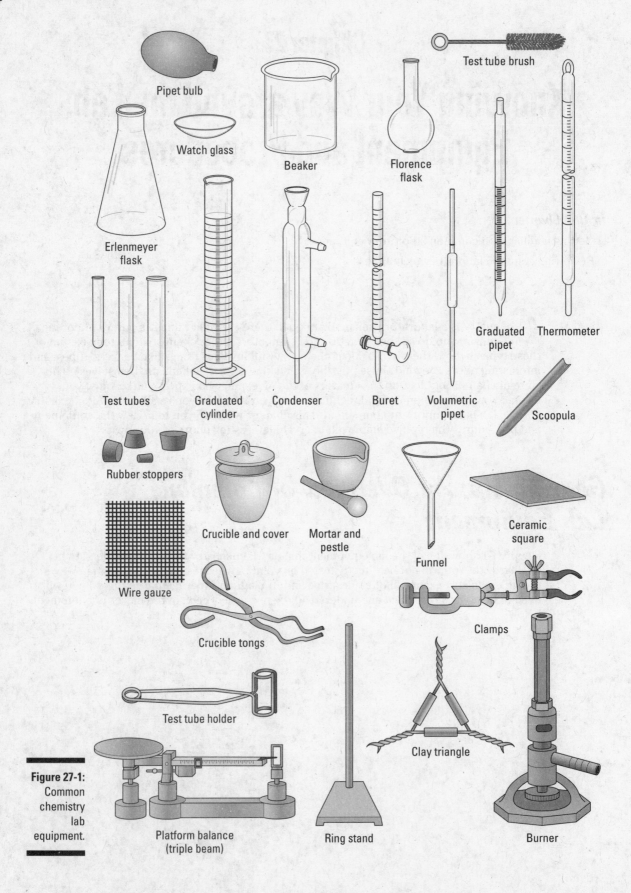

Pipet bulb

Test tube brush

Watch glass

Beaker

Florence flask

Erlenmeyer flask

Test tubes

Graduated cylinder

Condenser

Buret

Volumetric pipet

Graduated pipet

Thermometer

Scoopula

Rubber stoppers

Crucible and cover

Mortar and pestle

Funnel

Ceramic square

Wire gauze

Crucible tongs

Clamps

Test tube holder

Clay triangle

Figure 27-1:
Common chemistry lab equipment.

Platform balance (triple beam)

Ring stand

Burner

Figure 27-1 shows you what the equipment looks like, and the list below tells you how each piece of lab equipment functions:

- **Balance:** Used for obtaining the masses of solid and liquid samples

- **Beaker:** A flat-bottomed, cylindrical piece of glassware used for mixing and heating compounds

- **Bunsen burner:** Attached to a gas line and lit to provide heat for your experiments

- **Buret:** An extremely accurate device with a stopcock at the bottom used to measure volumes of reagents

- **Ceramic square:** Used to avoid burning the surface of your lab bench and incurring your chemistry teacher's wrath

- **Clamps:** Used to hold a variety of things in place, particularly test tubes

- **Clay triangle**: Used to hold a crucible while it is being heated

- **Condenser**: Used to collect vapors by condensing them into liquid as they contact the liquid-cooled inner surface of the condenser

- **Crucible:** A cup-shaped container capable of sustaining high temperatures. It is used to heat chemicals.

- **Crucible tongs:** Used to handle the hot crucible

- **Erlenmeyer flask:** Used to hold liquids. The small upper opening slows evaporation, so for some volatile liquids, a flask is a better choice than a beaker. The shape also makes it suitable for mixing and swirling liquids during a titration.

- **Florence flask:** A type of flask, generally round-bottomed, usually suspended and heated from below. Its shape makes it easy to swirl and mix liquids inside of it.

- **Funnel:** Used together with filter paper to filter precipitates out of solutions

- **Graduated cylinder:** Used to precisely measure volumes

- **Metal spatula:** Used to measure out solid substances

- **Mortar and pestle:** Used to grind sesame seeds for cooking and chemical compounds for chemistry experiments, though we recommend using a different set for each

- **Pipette bulb:** Used to transfer accurately measured amounts of liquid from one container to another

- **Rubber stoppers:** Used to close flasks or test tubes to prevent evaporation of liquids or escape of gases

- **Scoopula:** Another instrument used to transfer solids from one place to another

- **Test tube:** Cylindrical open-topped piece of glassware that comes in varying sizes

- **Thermometer:** Used to measure temperatures. Thermometers generally contain liquid mercury.

- **Watch glass:** A piece of glassware in the shape of a large contact lens used for evaporating liquids

- **Wire gauze:** Generally used as a surface for a beaker or flask to rest when being heated by a Bunsen burner

Following Protocol

What makes an effective laboratory "hunchback" is an intimate knowledge of two things. First, the hunchback knows the right tool for the job. Second, the hunchback knows how to use that tool. If you don't know the first thing, you may find yourself attempting to measure the pH of a solution by placing it dropwise onto the tip of your tongue (don't ever do this). If you know the first thing but not the second thing, you may find yourself staring in admiration at a stack of litmus paper that you have no idea how to use. To prevent both these kinds of embarrassing episodes, and to shore up your lab knowledge for the AP exam, make sure you thoroughly review this section.

Measuring mass, volume, and density

Mass and volume are by far the most frequently measured quanities in your basic chemistry lab. From mass and volume, you can calculate density. Alternately, you can use density to calculate mass or volume.

- **Mass:** You measure mass in the laboratory using a balance. Most laboratory balances are extremely accurate, capable of measuring masses with a precision of hundredths of grams.

- **Volume:** Techniques for measuring volume vary depending on what type of substance you are measuring, with liquids being the easiest to measure. Graduated cylinders are generally used to make approximate volume measurements for liquids. Pipettes and burets are used to measure precise volumes of liquids. Obtaining the volume of a solid can be somewhat tricky and, in some cases, impossible to do without altering the substance. If the substance does not react with water, you can submerge it in water and measure the volume of water displaced by it. Luckily, in most experiments it is mass of solid used that matters and not volume. You cannot measure the volume of a gas because gases have no definite volume. Recall that it is a fundamental property of gases that they expand to fill the entire volume available to them. However, the amount of gas can be obtained by measuring its pressure when held in a known volume.

- **Density:** Mass and volume measurements can also be used to determine the density of a substance using the formula density=mass÷volume. Density is a characteristic physical property for solids and liquids and can often be looked up in a table.

Determining melting point

Melting is the phase transition between the solid and liquid states and is equivalent to the freezing point. Because melting point is a characteristic physical property of a solid, you can use melting point to help to determine the identity of an unknown solid or to verify that an experiment has been successful and you have isolated the desired product.

You can use several different types of apparatus for measuring melting points, but they all have the same general setup. A sample of the solid in question is placed in a thin glass tube and heated slowly from the bottom. The experimenter watches the solid through a magnifying glass until it reaches the melting point and records it. Pure crystalline solids have very distinct melting points and the phase transition between solid and liquid occurs in a few seconds, giving a very accurate melting point determination. Impurities lower the melting point of a pure solid, by an amount depending on their concentration. So it is vital to have a pure sample if you are using melting point to confirm an identity. Amorphous solids such as glass and plastic do not have precise melting points and so chemists make special types of melting measurements for them.

Determining boiling point and doing distillation

The boiling point of a substance is slightly more difficult to measure than the melting point. This is because it is the phase transition between a liquid and a gas and gases are difficult to see so you need a way of verifying that your liquid is at its boiling point and not merely very, very hot. One convenient phenomenon to help you in measuring a boiling point is the tendency for a liquid to maintain a constant temperature during a phase transition. All of the energy being added to the solution goes into changing its phase rather than into heating it.

Distillation is a process that can aid you not only in determining a precise boiling point for a liquid but also in purifying that liquid. Say you have a mixture of two different liquids in a flask, one with a lower boiling point than the other. If you could somehow trap the gas being released as the first liquid boils you would have isolated the pure first substance. This is the basic purpose of a distillation apparatus such as the one shown in Figure 27-2.

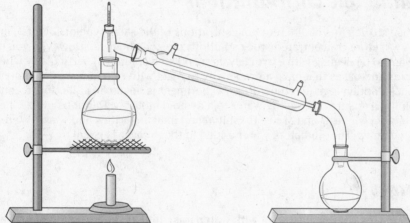

Figure 27-2:
A simple
distillation
apparatus.

The flask on the left is being heated, while the one on the right is cool. Connecting the two is a piece of lab equipment called a *condenser*. The inner tube in a condenser connects the first flask to the second, while cold water is run through the outer tube (this water is kept isolated and does not come into direct contact with the chemicals in the experiment). This causes the gas that is escaping from the boiling liquid in the first flask to condense in the inner tube of the condenser, which is angled downward so that it will collect in the second flask.

As the experimenter, you watch a thermometer connected to the first flask. When the first liquid begins to boil and condense in the condenser, the temperature of the liquid will remain constant until all of the lower boiling point liquid has escaped. Once it has all escaped, the condenser will collect no more liquid and the temperature in the first flask will again begin to rise. You then turn off the heat and disconnect your apparatus, having isolated a pure substance in your second flask. If your mixture contains more than two liquids or you wish to isolate a liquid with a higher boiling point, you can continue this process, switching out the second flask and isolating one substance at a time.

Dealing with solids in precipitation and filtration

In chemistry, you will often create solid precipitates in aqueous solutions. You separate the solution from the precipitate using a process called *filtration*, which involves pouring your solution/precipitate mixture through a funnel lined with filter paper. The filter paper allows the solution to pass through, but catches the precipitate. The precipitate should be washed to remove impurities.

Positive spin: Centrifugation

Centrifugation is another nifty way of separating compounds, this time according to their density. Centrifuges are quite expensive and most high school chemistry labs are not equipped with them, but you will often find reference to centrifugation in your chemistry text or on the AP exam. A centrifuge is essentially a big, extremely fast-spinning trough. A mixture in the centrifuge will separate according to density as the denser components migrate away from the center of the apparatus and the less dense migrate toward it.

Thinning out with dilutions

Rather than storing lots of different concentrations of the same type of solution, most chemistry labs will store concentrated stock solutions of each chemical. However, you don't always want to be dealing with extremely concentrated solutions, particularly when they are strong acids and bases that have a propensity to react with skin and are generally unpleasant. Often, a dilution is required before an experiment is begun. In a dilution, a concentrated stock solution is diluted by adding water to a desired molarity. Dilutions utilize the equation $M_i \times V_i = M_f \times V_f$, where M_i and M_f are the initial and final molarities and V_i and V_f are the initial and final volumes. The amount of water added in these cases is equal to $V_f - V_i$.

Drying off

Because so many compounds react with water, it is often desirable to eliminate water from certain substances. Since so many substances are soluble in water, it is often very difficult to get rid of. In many cases, you can simply heat the substance you wish to rid of water to higher than 100 degrees C and wait for the water to evaporate. However, high temperatures can also cause decomposition of the substance you wish to isolate so sometimes another method is needed.

This alternative method takes advantage of substances called *desiccants* that draw in water. If you leave your sample in a sealed container also containing a desiccant, over time the desiccant will absorb all of the water from your sample. Desiccants are slow acting, and your sample will generally have to be left in the container overnight to complete the drying process. Desiccants are capable of absorbing only a certain amount of water, so it is important that your sample have as much water as possible removed from it before being sealed with the desiccant. Silica gel-based desiccants are common in chemistry labs because they can be easily regenerated by heating. Most commercially available dessicants are prepared with additives so that they change color as they become more hydrated.

Making pH measurements

When doing acid/base reactions in the laboratory, you will often wish to know the pH of the solution you are dealing with. There are several ways to measure pH. Often, a material known as litmus paper is used. Litmus paper changes color as it reacts with solutions of different pH. Proper technique is to place a drop of solution onto the paper with a stir rod. When the color of the paper develops, you match its color to the pH scale that comes with the box of litmus paper. Many labs now use "pH paper" or "pH strips," that change to different colors over a range of pH values.

Many chemists prefer to use a pH indicator such as phenolphthalein instead of litmus paper, however, probably because it makes the solution itself turn pink when a pH above 9 is reached. Universal indicators can provide a range of color changes against a scale. pH meters have now become quite cheap and small, so a probe dipped in a solution can give you a direct digital readout of the pH in the solution.

Using spectroscopy/spectrometry to measure concentration

Spectroscopy is a technique for separating light into its component colors. *Spectrometry* means making measurements of the separated light. Sophisticated chemistry labs use spectrophotometers to create a graph showing the absorption and/or emission at each wavelength of the visible spectrum. However, as is the case with centrifuges, not all high school chemistry labs are equipped with spectrophotometers. If you are asked about spectroscopy on the AP chemistry exam, you will be required to know that it relies on the separation of light into its component colors. Recall from Chapter 3 that the colors absorbed or emitted by a pure element are unique to that element and are related to its atomic structure.

In addition, you will be expected to know that a spectrometer can be used to measure the concentration of a colored substance by using Beer's Law, and measuring the amount of light transmitted by the solution. The solution is poured into a clean cell and placed in the spectrometer. By comparing this to a cell containing a standard amount of the substance, the spectrometer can determine an unknown concentration by a simple ratio. It is important that no other light scattering (such as by a fingerprint on the cell) happens, as it will affect the result.

Chapter 28

Taking the Tour: Recommended AP Experiments

. . . .

In This Chapter

▶ Surveying the range of College Board-recommended AP chemistry experiments

▶ Highlighting key concepts and calculations from the labs

. . . .

Goggles are the geeky, timeless emblems of a chemistry lab. But before you laugh off this chapter, remember that the laboratory is where chemistry really takes place. That's why the College Board has put forth an extensive list of recommended AP chemistry labs, many of them interrelated. Any actual AP chemistry course will not include the full range of these labs — but you'll probably have done a significant fraction of them. To expose yourself to the pith of any labs your class skipped, and to shore up your understanding of those you did perform, review the following summary of the recommended labs, 22 things you might have done behind goggles. Examples offered for individual experiments may or may not exactly describe what you have done in your own lab classes but reflect the essential principles of the labs, and data like those you might encounter in lab are included in the examples. As you go through the steps of an experiment, keep alert to possible sources of error. Each measurement you make (massing a sample, measuring a temperature, etc.) carries some error with it. That error propagates through every subsequent calculation that you make using those measurements, and results in error (uncertainty) in any final experimental values.

Determination of the Formula of a Compound

By combusting a metal sample of known mass, you form a metal oxide with an unknown formula. Taking the mass of the oxide allows you to calculate the ratio of oxygen to metal within the oxide, and thereby determine the formula.

Example:

3.65g Mg is combusted within a 16.26g crucible. After combustion completes, the mass of the sample-containing crucible is 22.38g. What is the formula of the oxide?

Follow these steps to find the formula of this metal oxide:

1. Mass of crucible + combusted sample = 22.38g

2. Mass of combusted sample = 22.38g – 16.26g = 6.12g

3. Increase in mass of sample upon combustion = 6.12g – 3.65g = 2.47g

   Because combustion is essentially a combination reaction with oxygen, any added mass derives from oxygen. The final sample therefore contains the original amount of Mg:

4. Moles Mg = 3.65g Mg × (1mol Mg / 24.3g Mg) = 0.150mol Mg

5. Moles O added = 2.47g O × (1mol O / 16.0g) = 0.154mol O

6. Moles Mg : moles O ≈ 1:1

   The formula of the oxide is MgO.

Determination of the Percentage of Water in a Hydrate

By heating a hydrated ionic compound of known mass, you drive off hydrating water molecules. Often, this process can be followed visually by observing the colored hydrate compound lose its color. Taking the mass of the dehydrated compound allows you to calculate the percent of water in the original hydrated compound.

Example:

5.91g of hydrated copper (II) sulfate, $CuSO_4 \cdot nH_2O$, is heated within a 15.73g crucible until its original blue color changes entirely to white. The sample should be cooled and massed, and then heated and cooled again until the mass remains constant indicating all the water has been removed. The mass of the sample-containing crucible after dehydration is 19.48g. What is the apparent percentage of water in the hydrated compound?

Here are the steps to determine the apparent percent water in the original compound:

1. Mass of sample + crucible = 5.91g + 15.73g = 21.64g

2. Mass of dehydrated sample + crucible = 19.48g

3. Mass of dehydrated water = 21.64 – 19.48g = 2.16g

4. Moles of dehydrated water = 2.16g H_2O × (1mol H_2O / 18.02 g H_2O) = 0.120mol H_2O

   Because the mass that remains in the crucible after dehydration is $CuSO_4$ (not the hydrate), you can calculate the original moles of hydrated compound:

5. Moles $CuSO_4 \cdot nH_2O$ = moles $CuSO_4$

6. Mass remaining in crucible = 19.48g – 15.73g = 3.75g

7. Moles $CuSO_4$ = 3.75 g $CuSO_4$ × (1mol $CuSO_4$ / 159.6g $CuSO_4$) = 0.0235mol $CuSO_4$

8. Moles dehydrated water : mole $CuSO_4$ = 0.1120 : 0.0235 = 5.1: 1 ≈ 5:1

   So, given a plausible amount of experimental error, the original hydrate appeared to have the formula $CuSO_4 \cdot 5H_2O$.

   To calculate the mass percent of water in the original compound, you simply divide the mass of the hydrating waters (5.1 × 18.02g) by the mass of the whole hydrated compound and multiply by 100%:

   (91.9g / 249.7g) × 100% = 36.8%

Determination of the Molar Mass by Vapor Density

By evaporating a volatile liquid such that it fills a flask of known volume, and then measuring the mass of recondensed liquid, you determine the vapor density (mass of compound / volume of flask) of the compound. Then, you use the ideal gas law to determine the molar mass of the compound.

Example:

4.00g of an unknown volatile liquid is loaded into a flask of 312mL interior volume. The flask is capped with a single-hole stopper to allow controlled escape of evaporating material. The stoppered flask is immersed in a 373K water bath, causing the volatile liquid to begin to evaporate. Once the last visible trace of liquid evaporates, the flask is plunged into an ice bath, recondensing any volatile liquid that remains as vapor within the flask. By comparing the mass of the flask before and after the heating/cooling procedure, you calculate that 0.854g of recondensed liquid remains. What is the apparent molar mass of the volatile liquid?

Follow these steps to find the apparent molar mass:

1. The experiment takes place in atmospheric conditions, so the pressure ≈ 1.0atm.

2. At the moment the flask was immersed in an ice bath to recondense the volatile liquid, the moles of the sample filled the 312mL volume (0.312L). You know that the temperature of the original water bath was 373K and the pressure ≈ 1.0atm. Using the ideal gas law, you can calculate the moles of gas in the flask under those conditions:

   $PV = nRT$

   $n = (PV) / (RT) = (1.0\text{atm} \times 0.312\text{L}) / (0.0821 \text{ Latm mol}^{-1} \text{ K}^{-1} \times 373\text{K}) = 0.010\text{mol}$

3. Because the recondensed liquid had mass 0.854g, we calculate the apparent molar mass as follows:

   Molar mass ≈ 0.854g / 0.010mol = 85 g mol^{-1}

Determination of the Molar Mass by Freezing Point Depression

You dissolve a known mass of an unknown, nonelectrolyte solute into a known mass of solvent. Then you measure the freezing point of the resulting solution and compare it to the freezing point of the pure solvent. By calculating the change in the freezing point, you determine the molar mass of the unknown solute based on the colligative property of freezing point depression.

Constraining the unknown solute as an nonelectrolyte ensures that each mole of dissolved compound contributes one mole of dissolved particles, as opposed to the more complicated cases of acids, bases, and salts.

Example:

You dissolve 1.32g of an unknown, nonelectrolyte compound into 50.0g of benzene solvent. You measure the freezing point of the solution to be 2.6°C. What is the apparent molar mass of the solute?

Here are the steps to determine the apparent molar mass of this solute:

1. Freezing point depression depends on the molality (m) of the solution and on the solvent according to the freezing point depression equation, $\Delta T_f = K_f \times m$.

   The parameter K_f is the freezing point depression constant and varies from solvent to solvent. For benzene, $K_f = 5.12\ °C\ m^{-1}$.

   The parameter ΔT_f is the change in freezing point compared to that of the pure solvent, not the freezing point temperature itself. The normal freezing point of pure benzene is 5.5 °C.

   Substituting in known values, you get

   $$(5.5\ °C - 2.6\ °C) = (5.12\ °C\ m^{-1}) \times m$$

2. Solving for m gives you 0.57 molality, or 0.57 moles solute per kg solvent. Because you dissolved the unknown solute into 50g = 0.050kg solvent, you can calculate the apparent moles of solute:

   $$0.57\text{mol kg}^{-1} = n\ /\ 0.050\ \text{kg}$$

3. Solving gives you 0.029 moles. Those moles had a mass of 1.32g, so the apparent molar mass is 1.32g / 0.029mol = 46 gmol^{-1}.

Determination of the Molar Volume of a Gas

You react a known mass of a known metal with acid within a flask such that the reaction evolves hydrogen gas H_2. The volume of gas is measured by its ability to displace water within a connected eudiometer, a graduated column partially filled with water. By measuring the volume of gas evolved, you can use the ideal gas law to calculate the apparent molar volume of hydrogen gas. In so doing, you must account for the contribution of water vapor within the eudiometer to the total pressure.

Example:

You react 0.027g Mg with concentrated HCl within a stoppered flask that is connected by a tube to a eudiometer. As the reaction proceeds, hydrogen gas bubbles from the mixture into the eudiometer. By observing the water displacement, you estimate that the reaction evolved 26.15mL of H_2. The temperature of the lab is 22°C, a temperature at which the vapor pressure of water is 0.030atm. What is the apparent molar volume of the hydrogen gas? How do the moles of H_2 estimated by this method compare to the number of moles expected from the reaction?

To compare the experimentally estimated number of moles to the expected number, follow these steps:

1. Set yourself on solid ground by writing a balanced equation for the reaction:

   $$Mg(s) + 2HCl(aq) \rightarrow Mg^{2+}(aq) + 2Cl^-(aq) + H_2(g)$$

2. Given the stoichiometry of the balanced reaction equation, you can calculate a theoretical yield of H_2 gas from the starting mass of Mg:

   $$0.027\text{g Mg} \times (1\text{mol Mg} / 24.3\text{g Mg}) \times (1\text{mol } H_2 / 1\text{ mol Mg}) = 1.1 \times 10^{-3}\text{ mol } H_2$$

   Potentially, 1.1×10^{-3} mol H_2 gas could coexist with water vapor within the eudiometer tube.

3. You can estimate the moles of water in the displaced volume within the tube (26.15mL = 0.02615L) by using the ideal gas law and the given vapor pressure for water at 22 °C, 0.030atm:

    $PV = nRT$

    $n = (PV) / (RT) = (0.030\text{atm} \times 0.02615\text{L}) / (0.0821 \text{ Latm mol}^{-1} \text{K}^{-1} \times 295\text{K})$

    $n = 3.2 \times 10^{-5} \text{ mol } H_2O$

4. Raoult's law states that the partial pressure of a substance in the vapor pressure over a sample is proportional to the mole fraction of that substance within the sample. So, you can estimate the fractional pressure exerted by water vapor trapped in the eudiometer tube:

    Fractional pressure $\approx (3.2 \times 10^{-5} \text{ mol } H_2O) / (3.2 \times 10^{-5} \text{ mol } H_2O + 1.1 \times 10^{-3} \text{ mol } H_2)$

    Fractional pressure ≈ 0.028

5. So, the pressure that caused the displacement within the eudiometer derived not only from hydrogen gas but also from about 3 mole percent water vapor. Using this information, you can recalculate the moles of hydrogen within the eudiometer, including a correction for the partial pressure contributed by water vapor:

    $n = [(1.0\text{atm} - 0.028\text{atm}) \times (0.02615\text{L}) / (0.0821 \text{ Latm mol}^{-1} \text{K}^{-1} \times 295\text{K})]$

    $n = 1.0 \times 10^{-3} \text{ mol } H_2$

6. The volume occupied by this amount of hydrogen $\approx 0.02615\text{L} \times (1 - 0.028) \approx 0.025\text{L}$

7. So, the apparent molar volume of the hydrogen $\approx 0.025\text{L} / 1.0 \times 10^{-3} \text{ mol} = 25 \text{ L mol}^{-1}$

8. Finally, the percent error between the actual moles H_2 produced versus those estimated from the balanced equation calculates as:

    $[(1.1 \times 10^{-3}\text{mol} - 1.0 \times 10^{-3}\text{mol}) / 1.1 \times 10^{-3}\text{mol}] \times 100\% = 9\%$

Standardization of a Solution Using a Primary Standard

To reliably prepare working solutions from stock solutions, it helps to be able to measure the concentration of stock solutions by reacting them with "primary standards" of very certain concentration. This process is called *standardization*. To standardize solutions in this way, you must know the balanced reaction equation for the reaction undergone between the primary standard and sample solutions, and you must have some means to assess the extent of the reaction, such as a color change.

Analysis of the data from a standardization titration is essentially identical to the analysis used for acid-base titrations. Both experiments employ titration. Both require you to make use of a balanced reaction equation. Both often use a color change to indicate the endpoint of a titration. A common standard substance is potassium acid phthallate (or KHP for short), which can be prepared very pure, is solid at room temperature, and so easily weighed to obtain an exact mass for reaction. For an example of the types of calculations used in a standardization experiment, see "Determination of the Equilibrium Constant for a Chemical Reaction," later in this chapter.

Determination of Concentration by Acid-Base Titration, Including a Weak Acid or Base

To reliably determine the concentration of acid or base solutions, you can use the solution of unknown concentration in a neutralization reaction against an acid or base of highly certain concentration. To use this method, you must know the balanced reaction equation of the neutralization reaction and you must have a means to assess the extent of the reaction, such as the change in color of an indicator. If you know the number of acid or base equivalents contributed by your sample compound (HCl contributes one acid equivalent, H_2SO_4 contributes two acid equivalents, for example), you can determine the molar concentration of the compound. If you do not know the number of equivalents per mole of your compounds, you can still determine the normality, N, of your sample solution. (You may or may not have worked with "normality" in your AP class, but the AP exam does not include it.)

Example:

You wish to determine the concentration of a solution of ascorbic acid, $C_6H_8O_6$, a weak acid. You know that ascorbic acid is monoprotic, contributing one mole of acid equivalent per mole of compound. To determine the concentration of your solution, you titrate 50.0mL of ascorbic acid solution with a 0.10M NaOH standard solution. To follow the progress of the neutralization reaction, you use a colorimetric indicator, phenolphthalein. The phenolphthalein changes from colorless to pink when you have added 19.6mL of standard to the ascorbic acid sample solution. What is the apparent concentration of the sample solution?

To determine the apparent concentration of the original ascorbic acid solution, follow these steps:

1. To begin the analysis, you must know the stoichiometry of the neutralization reaction, as expressed in a balanced reaction equation:

   $$C_6H_8O_6(aq) + NaOH(aq) \rightarrow C_6H_7O_6Na(aq) + H_2O(l)$$

   Expressed as a net ionic equation,

   $$C_6H_8O_6(aq) + OH^-(aq) \rightarrow C_6H_7O_6^-(aq) + H_2O(l)$$

   In other words, the soichiometry of the interaction between NaOH and ascorbic acid is 1:1.

2. Phenolphthalein, the colorimetric indicator added to the titration solution, changed from colorless to pink at the equivalence point, indicating that the moles of acid in solution equal the moles of base. The color changed at 19.6mL (0.0196L) added 0.10M NaOH.

3. The moles of base added at the equivalence point is simply the concentration of the standard NaOH solution multiplied by the volume added:

   $$\text{Moles base} = 0.10 \text{ mol L}^{-1} \times 0.0196L = 2.0 \times 10^{-3} \text{ mol}$$

4. Because moles base = moles acid at the equivalence point, the original 50.0mL (0.0500L) ascorbic acid solution apparently contained 2.0×10^{-3} mol ascorbic acid. So, the molar concentration of the ascorbic acid solution is

   $$\textit{Molarity} = (2.0 \times 10^{-3} \text{ mol}) / (0.0500L) = 4.0 \times 10^{-2} \text{ } M$$

Determination of a Concentration by Oxidation-Reduction Titration

To determine the concentration of a solution of an oxidizing agent or reducing agent, you can use the sample solution in an oxidation-reduction reaction against a reducing agent or oxidizing agent of highly certain concentration. To use this method, you must know the balanced reaction equation of the redox reaction and you must have a means to assess the extent of the reaction, such as a color change.

For example, here is a balanced equation for a redox reaction between hydrogen peroxide and permanganate:

$$5H_2O_2(aq) + 6H^+(aq) + 2MnO_4^-(aq) \rightarrow 5O_2(g) + 2Mn^{2+}(aq) + 8H_2O(l)$$

In this reaction, the permanganate ion, MnO_4^-, itself acts as the indicator because it has an intense purple color. So, as the reaction proceeds and permanganate is reduced, the purple color disappears. When the reaction is complete, the solution entirely loses purple color, becoming colorless.

Analysis of the data from a oxidation-reduction titration is essentially identical to the analysis used for standardization and acid-base titrations. The key elements are the use of a color change to indicate a reaction endpoint and the use of a balanced reaction equation to make correct stoichiometric calculations. The fact that the reaction in question is an oxidation-reduction reaction doesn't really change the method of analysis. So, to see an example of the types of calculations used in a redox titration experiment, see determination of concentration by acid-base titration, earlier in this chapter.

Determination of Mass and Mole Relationships in a Chemical Reaction

This experiment is not typically performed on its own but rather as a necessary part of other experiments. The basic principle shows up in two main ways. First, you use a measured mass to calculate the apparent number of moles in a sample. Second, you use a measured number of moles to calculate the apparent mass of a sample. The measured quantities may be determined directly or indirectly, depending on the experiment. In either case, the apparent ratio of the mass of a substance to the corresponding number of moles represents the experimental molar mass ($g\ mol^{-1}$) of the substance.

Determination of the Equilibrium Constant for a Chemical Reaction

The equilibrium constant, K_{eq}, for a reaction is a ratio of the concentrations of products to reactants when the reaction has achieved equilibrium. So, you can determine the equilibrium constant by measuring the concentrations of reactants and products in an equilibrium solution. To use this method, you must know the balanced reaction equation for the reaction in question, and you must have a reliable means to measure the reactant and product concentrations.

Example:

You wish to determine the equilibrium constant for the ionization reaction of ascorbic acid, $C_6H_8O_6$, a weak acid. You know that ascorbic acid is monoprotic, contributing one mole of acid equivalent per mole of compound. Because one product of the ionization reaction is H^+, you choose to measure the concentration of reactants and products by measuring the pH of a standardized, 1.00×10^{-2} M, solution of ascorbic acid. You observe the pH to be 3.08. What is the apparent K_{eq} of ascorbic acid?

To find the apparent K_{eq} of ascorbic acid by this method, do the following:

1. To calculate the K_{eq} from a set of measured concentrations, you must have access to a balanced reaction equation. For the ionization reaction of ascorbic acid, the equation is

   $$C_6H_8O_6(aq) \rightarrow C_6H_7O_6^-(aq) + H^+(aq)$$

   Or to be more rigorously correct,

   $$C_6H_8O_6(aq) + H_2O(l) \rightarrow C_6H_7O_6^-(aq) + H_3O^+(aq)$$

2. For simplicity, we'll use the first equation to set up an expression for K_{eq}:

   $$K_{eq} = \text{products / reactants} = [C_6H_7O_6^-] \times [H^+] / [C_6H_8O_6]$$

   So, if you can measure the concentrations of $C_6H_8O_6$, $C_6H_7O_6^-$, and H^+ in a solution of ascorbic acid at equilibrium, you can use those values to solve for K_{eq}.

3. Because the reaction in question is an acid ionization reaction, measuring the pH of an equilibrium solution is a clever way to obtain the needed data. Prior to ionizing, the ascorbic acid in solution exists at 1.00×10^{-2} M concentration. Each mole of acid that ionizes depletes that initial concentration and adds one mole of ascorbate (the conjugate base) and one mole of H^+ to solution. In other words, the equilibrium mixture contains the following concentrations:

   $$[C_6H_8O_6] \quad = \quad (1.00 \times 10^{-2} - x) \, M$$
   $$[C_6H_7O_6^-] \quad = \quad x \, M$$
   $$[H^+] \quad = \quad x \, M$$

   All concentrations are expressed in terms of x, which means that if you can measure x, then you know all concentrations and can calculate the equilibrium constant.

4. The concentration of H^+ can be calculated from the measured pH:

   $$pH = -\log[H^+] \quad \text{therefore} \quad 10^{-pH} = [H^+]$$

   The pH of the solution is measured to be 3.08, so $[H^+] = 10^{-3.08} = 8.3 \times 10^{-4} M$.

5. So, $x = 8.3 \times 10^{-4} M$. Knowing this, you can substitute into the expression for K_{eq} and solve:

   $$K_{eq} = (8.3 \times 10^{-4} \times 8.3 \times 10^{-4}) / (1.00 \times 10^{-2} - 8.3 \times 10^{-4}) = 7.5 \times 10^{-5}$$

Determination of Appropriate Indicators for Various Acid-Base Titrations and Determining pH

This experiment is often performed as part of a larger acid-base titration lab. The core concept is that common colorimetric pH indicators work best within limited pH ranges. Different indicators work best in different pH ranges. The reason for this is that indicator dyes are simply molecules that can exist in protonated or deprotonated form, with each form exhibiting a different color. The affinity of a given dye molecule for a proton, H^+, is expressed by either of two related quantities: the K_a or the pK_a of the molecule.

The K_a is the acid dissociation constant, which is simply the equilibrium constant for the ionizing dissociation reaction of the molecule, $HA \leftrightarrow H^+ + A^-$. The pK_a is simply the negative logarithm of the K_a, $pK_a = -\log K_a$.

The pK_a is a convenient quantity to use for thinking about the effective pH range of an indicator dye. The pK_a represents the pH at which one half the indicator molecules exist in protonated form and one half exist in deprotonated form. Beyond about one pH unit above or below its characteristic pK_a, an indicator dye is outside of its useful range — no discernible changes in color can be detected this far from the pK_a. So, it is best to use indicator dyes with pK_a values as close as possible to the expected pH range of an experiment. If no reasonable pH range can be estimated prior to the experiment, it may be necessary to conduct the experiment multiple times, in the presence of indicators covering a range of pK_a values.

Here are the most common indicator dyes alongside their respective pKa values:

Indicator	*pKa*	*Low pH Color*	*High pH Color*
Methyl orange	3.7	Red	Orange
Bromophenol blue	4.0	Yellow	Purple
Methyl red	5.1	Red	Yellow
Bromothymol blue	7.0	Yellow	Blue
Phenolphthalein	7.9	Colorless	Pink
Thymol blue	9.3	Yellow	Blue

Determination of the Rate of a Reaction and Its Order

You determine the rate at which reactant is converted into product and, by varying the initial concentrations of reactants across a series of trials, you gather data that allow you to determine the reaction order. The overall reaction order is the sum of the orders of the individual reactants. Reaction order is a measure of the sensitivity of the reaction rate to changes in concentration.

Example:

You measure the initial rate of the following (unbalanced) reaction — the acid-catalyzed iodination (addition of iodine) of acetone — under four different sets of reactant concentrations.

$$C_3H_6O + I_2 \rightarrow C_3H_6IO + H^+ + I^-$$

You observe the following results.

$[C_3H_6O] / M$	$[I_2] / M$	$[H^+] / M$	Initial rate / $M\,s^{-1}$
0.050	0.050	0.050	1.2×10^{-7}
0.050	0.100	0.050	1.2×10^{-7}
0.050	0.200	0.100	2.4×10^{-7}
0.100	0.200	0.100	4.8×10^{-7}

What is the reaction order of each species (C_3H_6O, I_2 and H^+) and what is the overall reaction order? What are the rate law and the apparent rate constant? To determine the rate law for each reactant, observe the following:

1. To determine rates like those listed in the previous table, you need a means of detecting the progress of the reaction, which typically means having some way to measure the appearance of product or the disappearance of reactant. Often, the simplest and most precise way to measure reaction progress is spectrophotometry, as described in the determination of electrochemical series, later in this chapter.

   Because rates are often so dependent on reactant concentrations, it is usually advisable to measure "initial rates," which are fast, linear portions of the early reaction, before depletion of reactants appreciably slows the observed rate.

2. Once you have a set of initial rates at varying concentrations of reactants and/or products, you can seek patterns in the variation of rate with concentration. In the example data, it is clear that doubling the concentration of I_2 had no effect on rate when other concentrations remained constant, suggesting that the reaction is zero order in I_2.

3. Determining that I_2 concentrations do not impact rate allows you to focus exclusively on C_3H_6O and H^+ concentrations. Doubling the concentration of H^+ doubles the observed rate suggesting that the reaction is first order in H^+.

4. Maintaining the increased concentration of H^+ while doubling the concentration of C_3H_6O causes a further doubling of the rate, suggesting that the reaction is first order in C_3H_6O. You can convert the individual reaction orders to an overall reaction order and a rate law with a known rate constant, k, as follows:

   1. Because the reaction is first order in two reactants, the overall reaction order is 2.

   2. The reaction order is reflected in the rate law for the reaction, Rate = $k[C_3H_6O][H^+]$.

   3. To determine the value of the rate constant, k, simply substitute in any set of data from any of the individual rate experiments:

      Rate = $k[C_3H_6O][H^+]$

      k = Rate / ($[C_3H_6O][H^+]$) = $(4.8 \times 10^{-7}\ M\,s^{-1})$ / $(0.100M \times 0.100M)$ = $4.8 \times 10^{-5}\ M^{-1}\,s^{-1}$

Determination of the Enthalpy Change Associated with a Reaction

This lab involves *calorimetry*, the measurement of heat transfer associated with a process (such as a chemical reaction). One common reaction employed in this lab is the dissolution reaction of a solute into water, although many other types of reaction can be used. At constant pressure, the absorption or release of heat that accompanies a chemical reaction is equivalent to the reaction enthalpy, ΔH_{rxn}. Reactions that release heat are exothermic and have negative reaction enthalpy. Reactions that absorb heat are endothermic and have positive reaction enthalpy. By measuring the change in temperature of a reaction system of known mass and composition, you can calculate the molar reaction enthalpy, the amount of heat released or absorbed per mole of reaction.

Example:

You add 100.mL of water to a well-insulated (probably polystyrene foam) cup. You observe that the water temperature is 22.0°C. You then add 2.70g solid NaOH to the water, cap the cup with an insulated lid, and swirl gently to promote dissolution. While the NaOH dissolves, you monitor the temperature of the solution via a thermometer extending through a perforation in the lid. When the temperature levels off for approximately one half minute, indicating that the dissolution reaction has completed, you record the final temperature as 28.5°C. What is the apparent heat of solution, ΔH_{soln}, of sodium hydroxide?

Because the reaction performed in this lab is a dissolution reaction, the reaction enthalpy is also known as the "heat of solution." In other words, $\Delta H_{soln} = \Delta H_{rxn}$. The dissolution reaction is

$$NaOH(s) + H_2O(l) \rightarrow Na^+(aq) + OH^-(aq) \qquad\qquad \Delta H_{soln} = ?$$

The units of the heat of solution are joules per mole, J mol^{-1}. So, if you can measure the heat flow that accompanies the dissolution reaction, you can divide that heat by the moles of NaOH dissolved to find ΔH_{soln}.

The key experimental trick is realizing that as NaOH dissolves, heat flows to or from the water in which it is dissolving. We can neglect any heat flow to or through the polystyrene cup as it is such a good insulator. So, $q_{NaOH} = -q_{water}$. You can measure the change in the heat content of the water by measuring a change in the temperature of the solution (assuming that the mass of the solute is much smaller than the mass of the solvent — a fair assumption in dilute solutions).

To find the apparent heat of solution of sodium hydroxide from the data in this experiment, do the following:

1. To translate between temperature and heat flow, you need the constant-pressure calorimetry equation,

   $$q = m \times C_p \times \Delta T = m \times C_p \times (T_{final} - T_{initial})$$

   Here, m is the mass of the water, and C_p is the specific heat capacity of the water.

2. Because the density of water is 1.00 g mL^{-1}, you can calculate the mass of water in the experiment:

   Mass of water = 100mL $H_2O \times$ (1.00g / mL H_2O) = 100.g

3. The specific heat capacity of water is 4.186 J g^{-1} °C^{-1}. So, you can calculate the heat transfer during dissolution as

   $q = (100.g) \times (4.186$ J g^{-1} °C$^{-1}) \times (28.5$ °C $- 22.0$ °C$) = 2.72 \times 10^3$ J

4. So, as the NaOH dissolved, it released 2.72×10^3 J of heat energy into the water. In other words, $q_{NaOH} = -2.72 \times 10^3$ J. This amount of heat was released by 2.70g NaOH. You can convert to moles NaOH by using the molar mass:

   Moles NaOH = 2.70g NaOH $\times$ (1mol NaOH / 40.0g NaOH) = 0.0675mol NaOH

5. So, the apparent heat of solution of NaOH is:

   $\Delta H_{soln} = q_{NaOH}$ / mol NaOH $= -2.72 \times 10^3$ J / 0.0675mol $= -4.03 \times 10^4$ J mol^{-1}.

Separation and Qualitative Analysis of Anions and Cations

This lab is essentially about the descriptive chemistry of ionic compounds and their aqueous solutions. Qualitative analysis involves testing for the presence or absence of a compound, as opposed to measuring the precise concentration of that compound. Compounds can often be identified based on the formation of insoluble precipitates or colored compounds. Sometimes it can be difficult to test for the presence of one particular "target" ion within a solution of many ions, because the "nontarget" ions can interfere with intended reactions. For example, Cd^{2+} can be detected in aqueous solution by the formation of a yellow precipitate upon addition of S^{2-}. However, if Pb^{2+} or Cu^{2+} happen to be in the sample solution, a black precipitate forms upon addition of S^{2-}, which makes analysis for Cd^{2+} impossible. In cases like these, it is desirable to be able to selectively separate interfering ions from a solution while preserving other ions targeted for analysis. In other cases, the goal is simply to analyze a mixture of ions for the total ion composition of the solution. Separation schemes typically depend on the differential solubility of different ionic compounds.

Example:

You are given a solution that may contain Cl^- anion, I^- anion, both, or neither. To determine which of the four possibilities is true, you may use the following reagents: water, nitric acid (HNO_3), silver nitrate ($AgNO_3$), and ammonia (NH_3). In addition, you have a standard array of glassware and filtration devices.

To separate the anions in the sample solution with the materials provided, do the following:

1. First, acidify the unknown chloride/iodide solution by dropwise addition of HNO_3 until the solution is mildly acidic. Because Cl^- and I^- are weakly basic, addition of acid reduces their solubility and makes them easier to separate.

2. Next, slowly add $AgNO_3$ until any precipitation is complete. Although silver nitrate is soluble, other silver salts, like $AgCl$ and AgI, are insoluble. Filter and wash the precipitate with water. Examine the precipitate for color. $AgCl$ is white, and AgI is yellow.

3. Next, resuspend the precipitate in a small amount of water and add NH_3 and $AgNO_3$. Mix well. The added silver nitrate promotes precipitation of silver salts due to the common ion effect. At this point, any iodide anion present in solution should precipitate as yellow silver iodide, AgI. Separate the supernatant from any AgI precipitate.

4. The ammonia that remains in the supernatant forms a cationic complex with silver, $Ag(NH_3)_2^+$. After dropwise addition of HNO_3 to reduce anion solubility, any Cl^- in solution precipitates as white $AgCl$ salt.

Synthesis of a Coordination Compound and Its Chemical Analysis

"Coordination compounds" or "coordination complexes" are structures in which a central metal atom is surrounded by a host of nonmetallic groups called *ligands*. The ligands are bound to the metal by "coordinate covalent" bonds. These kinds of bonds are particularly common between transition metals and partners that possess lone pairs of electrons. The lone pair containing atom acts as an electron donor (a Lewis base), giving both electrons to a bond with the metal, which acts in turn as an electron acceptor (a Lewis acid). Coordination compounds are often intensely colored and can have properties that are quite different than those of the free metal.

This lab, in which you synthesize a coordination compound, is often combined with the two labs that follow it on this list: gravimetric determination and colorimetric or spectrophotometric determination. In keeping with that practice, we'll stick with one "example" system through these three labs.

Example:

You synthesize a copper-centered coordination compound by the addition of ammonia to a blue, aqueous solution of copper (II) sulfate. When ammonia is added to the copper (II) sulfate, the blue solution turns a deep purple color, indicating that you have successfully formed a coordination complex. To separate the complexes from the surrounding solution, you add ethanol to selectively precipitate the complex. This addition results in the formation of crystals of your coordination complex. Using a vacuum filtration apparatus, you repeatedly wash your crystals, rinsing them with an ammonia-ethanol mixture. You allow the crystals to dry overnight. Presumably, you have formed a complex that contains ammonia and/or sulfate ligands surrounding a copper cation center. In the following two labs, you'll analyze the compound gravimetrically and spectrophotometrically.

Analytical Gravimetric Determination

Gravimetric analysis entails measurements of mass to determine the composition of a compound. Here, we extend the copper coordination complex example from the synthesis of a coordination compound in the previous experiment.

Example:

Based on the chemicals used in the synthesis, the copper coordination complex likely includes ammonia and/or sulfate ligands, although the number of each type of ligand around the copper cation center is unknown. So, at this point, the formula of the complex is

$$Cu_a(NH_3)_b(SO_4)_c \cdot nH_2O$$

In this formula, a, b and c represent the stoichiometric amounts of copper, ammonia, and sulfate, respectively. The "$\cdot nH_2O$" indicates that some number n of water molecules may hydrate the complex.

To analyze the composition of your complex, do the following steps:

1. Dissolve 1.00g of your purified crystals in nitric acid, HNO_3, which causes the following reaction:

   $$Cu_a(NH_3)_b(SO_4)_c \cdot nH_2O + xH^+ \rightarrow aCu^{2+} + bNH_4^+ + cSO_4^{2-} + nH_2O$$

2. To this solution, add lead (II) acetate, $Pb(C_2H_3O_2)_2$, to selectively precipitate sulfate anion as lead sulfate:

   $$Pb^{2+}(aq) + SO_4^{2-}(aq) \rightarrow PbSO_4(s)$$

3. Carefully filter the white $PbSO_4$ precipitate and allow it to dry overnight before measuring its mass. From the mass of $PbSO_4$ you can determine the number of moles of sulfate recovered. Within error, this number should be equivalent to the number of moles of sulfate in the original 1.00g of crystals.

4. To determine the amount of ammonia in your compound, you perform an acid-base titration (covered in more detail in previously in the chapter). Briefly, you dissolve 1.00g of your crystals in 50.0mL water, add a pH indicator, and titrate with acid. By doing this, you can indirectly determine the amount of ammonia in your compound. By observing the endpoint of the titration, you can calculate the moles of ammonia that must have been present in the original sample, because the color change of the titration reflects the progress of this reaction:

   $$bNH_3(aq) + xH^+(aq) \rightarrow bNH_4^+(aq)$$

Within error, the number of moles of ammonia detected by your titration should be equivalent to the number of moles of ammonia in the original 1.00g of crystals.

The analysis for copper content of the coordination complex is done spectrophotometrically, as described in determination of electrochemical series, found below.

Colorimetric or Spectrophotometric Analysis

Spectrophotometric analysis measures the amount of a component within a mixture by exploiting the ability of the component to absorb electromagnetic radiation of a specific wavelength. Colorimetric analysis is simply spectrophotometric analysis using wavelength in the range of visible light. In practice, these analyses most often make use of Beer's law, a mathematical relationship between the concentration of a substance and the absorption of light of a certain wavelength by that substance:

$$A = abc$$

You may have encountered the equivalent equation $A = \varepsilon lc$. This equation is absolutely identical to the one given above, but simply uses the variable ε for a and the variable l for b.

To properly use Beer's Law, you must clearly understand the meaning of each variable.

The variable A is the absorbance of a specific wavelength of light. The wavelength used is typically one that is strongly absorbed by the compound of interest but not well absorbed by other components in the sample.

The variable a is the molar absorptivity, an experimentally determined constant that depends on the compound of interest and on the wavelength of light used for analysis. Typically, the units of a are M^{-1} cm^{-1}, but other units can be used. The key criterion is that the units used in a match those used in the variables b and c.

The variable b is the pathlength, the distance of the path of light as it travels through the sample. Most often, the pathlength is 1cm, but this value can change in the case of strongly or weakly absorbing samples. Whatever the pathlength, it must match the units used in the variable a.

The variable c is the concentration of the compound of interest. Typically, the units of c are molar (that is, M, mol L^{-1}), but that can change in special cases. Whatever the concentration units, they must math those used in the variable a.

> **Example:**
>
> To continue the analysis of the copper coordination complex synthesized and gravimetrically analyzed in the previous two experiments, you subject the compound to spectrophotometric analysis to determine its copper content.

To subject the coordination compound to spectrophotometric analysis, you must first estimate the molar absorptivity of your copper compound:

1. Calibrate a spectrometer at 645nm wavelength by using a series of Cu^{2+} standard solutions, diluted as necessary in $1M$ HNO_3.

2. Dissolve 0.50g of coordination compound crystals in 20.0mL $1M$ HNO_3 and transfer a sample of the solution to a 1cm-pathlength quartz cell.

   If the molar absorptivity of your compound, estimated by the Cu^{2+} calibration, is 5.2 M^{-1} cm^{-1}, and the absorbance reading of your dissolved sample is 0.47, what is the apparent copper concentration in the dissolved sample? What is the apparent molar mass of the compound in your crystals?

To determine the apparent copper concentration of your sample, do the following:

1. The apparent concentration of Cu^{2+} in the sample can be directly determined by using Beer's Law:

    $A = abc$ therefore $c = A / (ab)$

    $c = 0.47 / (5.2\ M^{-1}\ cm^{-1} \times 1\ cm) = 9.0 \times 10^{-2} M\ Cu^{2+}$

2. Assume for the moment that each coordination complex centers on a single Cu^{2+} ion (a reasonable working assumption), then the concentration of coordination complex in your sample is also $9.0 \times 10^{-2} M$.

You can proceed from these results to estimate the molar mass of your copper-centered coordination complex as follows:

1. You prepared the spectrophotometer sample by dissolving 0.50g compound into 20.0mL (0.0200L) $1M$ HNO_3. Using this information, you can estimate the moles of coordination compound in the sample:

    $9.0 \times 10^{-2} mol\ L^{-1} \times 0.0200L = 1.8 \times 10^{-3} mol$

2. So, 1.8×10^{-3} moles of compound appear to have been present in the 0.50g sample you dissolved. Therefore, the apparent molar mass of the compound is

    $0.50g / 1.8 \times 10^{-3} mol \approx 280\ g\ mol^{-1}$

Separation by Chromatography

Chromatography is a useful method for separating the components of a mixture. Although many variations of chromatography allow for quantitative analysis of the components of a mixture, the most frequent uses of chromatography in AP chemistry are qualitative, revealing how many components are in a mixture or determining whether a certain component is in the mixture or not.

In a chromatographic analysis, the sample to be analyzed is included in a "mobile phase," a fluid mixture that passes over a "stationary phase," a surface or porous medium that interacts with the mobile phase as it passes. Separation of components within the sample depends on those components interacting differently with the stationary phase. Some components of the sample may be more strongly attracted to the stationary phase than others, and these strongly attracted components take longer to be swept along by the flowing mobile phase. Separations can be based on charge, size, polarity, or other properties.

One common way to characterize the separated components of a mixture is the retention factor, R_f. To calculate an R_f, you must measure two quantities: the distance covered by the fastest-moving part of the mobile phase (often called the "solvent front") and the distance covered by the component in question. The retention factor is simply the ratio of the two quantities.

R_f = (distance of solvent front) / (distance of sample)

Preparation and Properties of Buffer Solutions

Buffer solutions are mixtures of conjugate acid-conjugate base pairs that resist changes in pH. Typically weak acids and their conjugate bases are used as buffers. Essentially, as extra acid or base is added to a buffer solution, the proportion of conjugate acid to conjugate base shifts in response, thereby minimizing the effect of the acid or base addition. The ability of a buffer solution to resist pH change depends on the concentration of the buffer compounds in solution and on the proportion of acid and base conjugate forms (that is, the ratio of HA to A^-). A buffer is at maximum buffering capacity when $[HA] = [A^-]$. Under this optimal-buffering condition, the pH of a buffer solution equals the pK_a of the weak conjugate acid. Buffers tend to be useful within about 1 pH unit of their characteristic pK_a.

The Henderson-Hasselbach equation describes the relationship between pH, pK_a and the relative concentrations of conjugates acid and base in a buffer solution:

$$pH = pK_a + \log([A^-]/[HA])$$

Example:

You are preparing an acetic acid/acetate buffer solution. At room temperature, acetic acid has $pK_a = 4.76$. You have 500.mL of $0.25M$ acetic acid. You want the final pH of your solution to be 5.00. What mass of sodium acetate salt must you add to achieve the desired pH?

To determine the mass of sodium acetate you must add to achieve a solution with pH 5.00, do the following:

1. Substitute into the Henderson-Hasselbach equation your known values for pH and pK_a, and then solve for the desired ratio of conjugate base to conjugate acid:

   $5.00 = 4.76 + \log([A^-]/[HA])$

   $0.24 = \log([A^-]/[HA])$

   $10^{0.24} = [A^-]/[HA] = 1.74$

2. So, for every mole of acetic acid (the conjugate acid, HA) in solution you want to add 1.74 moles acetate (the conjugate base, A^-).

   Moles acetic acid = $(0.25\text{mol L}^{-1}) \times (0.500\text{L}) = 0.13\text{mol}$

   Moles acetate = $1.74 \times 0.13\text{mol} = 0.23\text{mol}$

3. So, you want to add 0.23mol sodium acetate, $C_2H_3O_2Na$. The molar mass of anhydrous (that is, not hydrated) sodium acetate is 82.0 g mol^{-1}.

   $0.23 \text{ mol } C_2H_3O_2Na \times 82.0 \text{ g mol}^{-1} = 19 \text{ g } C_2H_3O_2Na$

Determination of Electrochemical Series

In this lab, you use your knowledge of electrochemical cells to explain and predict observed voltage potentials in cells of differing composition. Key tasks are to identify the anode and cathode of a cell and to use a table of standard reduction potentials to predict the voltage potential of a cell given its composition.

By measuring the voltage potentials of a series of half-cells, all against the same counterpart half-cell, you can construct an electrochemical series, a ranked list of redox reactions in order of reduction potential. Tables of standard reduction potentials represent just such a series, one in which the constant counterpart half cell is the standard hydrogen electrode:

$$2H^+(aq) + 2e^- \rightarrow H_2(g) \qquad E^\circ = 0 \text{ (by definition)}$$

Example:

A series of half-cell solutions is tested against a constant half-cell counterpart of $1.0M$ copper (II) nitrate, $Cu(NO_3)_2$, with a solid copper electrode. In all trials, an ammeter indicates that electrons flow toward the copper electrode from the opposing electrode. The following voltage potentials are observed:

$1.0M$ lead (II) nitrate, $Pb(NO_3)_2$ 0.47 V

$1.0M$ zinc (II) nitrate, $Zn(NO_3)_2$ 1.10 V

$1.0M$ iron (II) nitrate, $Fe(NO_3)_2$ 0.78 V

Given that E° = +0.34 V for the reduction of copper (II) cation, which divalent cation is the strongest reducing agent: lead (II), zinc (II) or iron (II)?

To determine the strongest reducing agent from the given data, observe the following:

1. The observed voltage potentials are standard cell potentials:

    $$E^\circ_{cell} = E^\circ_{red}(\text{cathode}) - E^\circ_{red}(\text{anode})$$

2. Because electron flow was toward the copper electrode, you know that the copper electrode was the cathode in each case.

3. Because each observed cell potential is positive and larger than +0.34 V, you know that each metal tested has a negative standard reduction potential.

    The most negative of the standard reduction potentials must belong to zinc (II) cation. Based on the data given, zinc (II) cation, Zn^{2+}, is the strongest reducing agent.

To calculate the standard reduction potential of the zinc (II) cation, do the following:

1. Substitute into the equation for standard cell potential:

    $$E^\circ_{cell} = E^\circ_{red}(Cu^{2+}) - E^\circ_{red}(Zn^{2+})$$

    $$1.10 \text{ V} = +0.34 \text{ V} - E^\circ_{red}(Zn^{2+})$$

    $$E^\circ_{red}(Zn^{2+}) = -0.76 \text{ V}$$

2. Because zinc (II) has the most negative standard reduction potential of the three metal ions tested, it is the most difficult to reduce and is therefore the strongest reducing agent.

Measurements Using Electrochemical Cells and Electroplating

This lab requires you to combine your knowledge of electrochemical cells with basic stoichiometry and gravimetric analysis. Electroplating occurs at a cathode when electrons reduce metal cations in solution, causing them to deposit onto the cathode. By measuring the mass of a cathode before and after an electrochemical reaction in which electroplating occurs, you can directly determine the mass of metal deposited onto the cathode during the reaction and compare that mass with one expected from calculations.

Example:

Identical carbon electrodes, each with a mass of 5.50g are immersed into a 1.0M solution of nickel (II) chloride, $NiCl_2$. The poles of the electrodes are connected to a battery, and current flows through the cell for 30. minutes at 1.5 amperes. What is the expected mass of the cathode after 30 minutes?

To determine the expected mass of the cathode, make the following observations and calculations:

1. Deposition of Ni(s) happens at the cathode, where the following reaction takes place:

   $Ni^{2+} (aq) + 2e^- \rightarrow Ni(s)$

   So, it takes 2 moles of electrons to electroplate 1 mole of nickel.

2. Calculate the total moles of electrons that flow through the cathode during the experiment. This calculation requires you to use the unit of the Faraday, F. One Faraday is charge associated with one mole of electrons. Each electron has 1.602×10^{-19} Coulombs of charge, so:

   $F = (6.022 \times 10^{23} mol^{-1})(1.602 \times 10^{-19}\, C) = 9.649 \times 10^4\, C\, mol^{-1}$

3. The experiment lasted for 30 minutes (1.8×10^3 seconds) at 1.5 amperes. One ampere is one Coulomb per second ($C\, s^{-1}$), so the total moles of electrons are

   $Moles\ e^- = (1.8 \times 10^3\, s) \times (1.5\, C\, s^{-1}) \times (1\, mol\ e^- / 9.649 \times 10^4\, C) = 2.8 \times 10^{-2}\, mol\ e^-$

4. Calculate the mass of nickel that you expect to electroplate as a result of transferring 2.8×10^{-2} moles of electrons into the $NiCl_2$ solution:

   $Mass\ Ni(s) = (2.8 \times 10^{-2}\, mol\ e^-) \times (1 mol\ Ni / 2 mol\ e^-) \times (58.7g / 1 mol\ Ni) = 0.82g$

5. Because the initial mass of the carbon electrode was 5.50g, the expected final mass (after electroplating) is 5.50g + 0.82g = 6.32g.

Synthesis, Purification, and Analysis of an Organic Compound

Typically, this lab involves a relatively straightforward synthesis — one with few steps. The synthesis is followed by a purification of the synthesized compound. In organic synthesis, you typically calculate a percent yield following each major step to track the progress of your reactions. Whatever the chemical details of the compound synthesized, the principles of "yield" are the same:

Percent yield = 100% × (actual yield) / (theoretical yield)

Actual yield is simply the amount of synthesized compound actually obtained after purification. Theoretical yield is the (never-actually-obtained) amount of compound that would result if each step of synthesis occurs with perfect efficiency and if there is no loss of compound between synthetic steps or during the purification.

Part VII
Measuring Yields:
Practice Tests

The 5th Wave   By Rich Tennant

Kenny didn't need the distraction, but it was just his luck that the Zorlocks' invasion of Earth would begin JUST as he opened his AP Chemistry booklet.

In this part . . .

By the time you make it to this part, chances are you've had your fill of Theory. It's time for Practice. However you feel about chemistry, you definitely want to turn the AP chemistry exam into a showcase for your skill and mastery. When you stride out of the exam hall, you want to be wearing a smile that says, "Nice try, guys, but I was ready for you." For that tasty moment to occur, you need practice. This part gives you that practice. Herein, you'll find two full-length practice tests. In case you don't know *every* answer, we've got answer keys with complete explanations. Earn your test-day smile here.

Chapter 29

Practice Test One

....

In This Chapter

▶ Trying your hand at 75 multiple-choice questions

▶ Taking on six free-response questions

....

*O*n the AP exam, you will be given a list of formulas and constants. You may use your Cheat Sheet (which mimics the AP formula list) and a periodic table for this exam, but do not use your calculator until the section specifically says that you may. Make sure to time yourself to get a sense of your pacing.

Answer Sheet

1. Ⓐ Ⓑ Ⓒ Ⓓ Ⓔ	26. Ⓐ Ⓑ Ⓒ Ⓓ Ⓔ	51. Ⓐ Ⓑ Ⓒ Ⓓ Ⓔ
2. Ⓐ Ⓑ Ⓒ Ⓓ Ⓔ	27. Ⓐ Ⓑ Ⓒ Ⓓ Ⓔ	52. Ⓐ Ⓑ Ⓒ Ⓓ Ⓔ
3. Ⓐ Ⓑ Ⓒ Ⓓ Ⓔ	28. Ⓐ Ⓑ Ⓒ Ⓓ Ⓔ	53. Ⓐ Ⓑ Ⓒ Ⓓ Ⓔ
4. Ⓐ Ⓑ Ⓒ Ⓓ Ⓔ	29. Ⓐ Ⓑ Ⓒ Ⓓ Ⓔ	54. Ⓐ Ⓑ Ⓒ Ⓓ Ⓔ
5. Ⓐ Ⓑ Ⓒ Ⓓ Ⓔ	30. Ⓐ Ⓑ Ⓒ Ⓓ Ⓔ	55. Ⓐ Ⓑ Ⓒ Ⓓ Ⓔ
6. Ⓐ Ⓑ Ⓒ Ⓓ Ⓔ	31. Ⓐ Ⓑ Ⓒ Ⓓ Ⓔ	56. Ⓐ Ⓑ Ⓒ Ⓓ Ⓔ
7. Ⓐ Ⓑ Ⓒ Ⓓ Ⓔ	32. Ⓐ Ⓑ Ⓒ Ⓓ Ⓔ	57. Ⓐ Ⓑ Ⓒ Ⓓ Ⓔ
8. Ⓐ Ⓑ Ⓒ Ⓓ Ⓔ	33. Ⓐ Ⓑ Ⓒ Ⓓ Ⓔ	58. Ⓐ Ⓑ Ⓒ Ⓓ Ⓔ
9. Ⓐ Ⓑ Ⓒ Ⓓ Ⓔ	34. Ⓐ Ⓑ Ⓒ Ⓓ Ⓔ	59. Ⓐ Ⓑ Ⓒ Ⓓ Ⓔ
10. Ⓐ Ⓑ Ⓒ Ⓓ Ⓔ	35. Ⓐ Ⓑ Ⓒ Ⓓ Ⓔ	60. Ⓐ Ⓑ Ⓒ Ⓓ Ⓔ
11. Ⓐ Ⓑ Ⓒ Ⓓ Ⓔ	36. Ⓐ Ⓑ Ⓒ Ⓓ Ⓔ	61. Ⓐ Ⓑ Ⓒ Ⓓ Ⓔ
12. Ⓐ Ⓑ Ⓒ Ⓓ Ⓔ	37. Ⓐ Ⓑ Ⓒ Ⓓ Ⓔ	62. Ⓐ Ⓑ Ⓒ Ⓓ Ⓔ
13. Ⓐ Ⓑ Ⓒ Ⓓ Ⓔ	38. Ⓐ Ⓑ Ⓒ Ⓓ Ⓔ	63. Ⓐ Ⓑ Ⓒ Ⓓ Ⓔ
14. Ⓐ Ⓑ Ⓒ Ⓓ Ⓔ	39. Ⓐ Ⓑ Ⓒ Ⓓ Ⓔ	64. Ⓐ Ⓑ Ⓒ Ⓓ Ⓔ
15. Ⓐ Ⓑ Ⓒ Ⓓ Ⓔ	40. Ⓐ Ⓑ Ⓒ Ⓓ Ⓔ	65. Ⓐ Ⓑ Ⓒ Ⓓ Ⓔ
16. Ⓐ Ⓑ Ⓒ Ⓓ Ⓔ	41. Ⓐ Ⓑ Ⓒ Ⓓ Ⓔ	66. Ⓐ Ⓑ Ⓒ Ⓓ Ⓔ
17. Ⓐ Ⓑ Ⓒ Ⓓ Ⓔ	42. Ⓐ Ⓑ Ⓒ Ⓓ Ⓔ	67. Ⓐ Ⓑ Ⓒ Ⓓ Ⓔ
18. Ⓐ Ⓑ Ⓒ Ⓓ Ⓔ	43. Ⓐ Ⓑ Ⓒ Ⓓ Ⓔ	68. Ⓐ Ⓑ Ⓒ Ⓓ Ⓔ
19. Ⓐ Ⓑ Ⓒ Ⓓ Ⓔ	44. Ⓐ Ⓑ Ⓒ Ⓓ Ⓔ	69. Ⓐ Ⓑ Ⓒ Ⓓ Ⓔ
20. Ⓐ Ⓑ Ⓒ Ⓓ Ⓔ	45. Ⓐ Ⓑ Ⓒ Ⓓ Ⓔ	70. Ⓐ Ⓑ Ⓒ Ⓓ Ⓔ
21. Ⓐ Ⓑ Ⓒ Ⓓ Ⓔ	46. Ⓐ Ⓑ Ⓒ Ⓓ Ⓔ	71. Ⓐ Ⓑ Ⓒ Ⓓ Ⓔ
22. Ⓐ Ⓑ Ⓒ Ⓓ Ⓔ	47. Ⓐ Ⓑ Ⓒ Ⓓ Ⓔ	72. Ⓐ Ⓑ Ⓒ Ⓓ Ⓔ
23. Ⓐ Ⓑ Ⓒ Ⓓ Ⓔ	48. Ⓐ Ⓑ Ⓒ Ⓓ Ⓔ	73. Ⓐ Ⓑ Ⓒ Ⓓ Ⓔ
24. Ⓐ Ⓑ Ⓒ Ⓓ Ⓔ	49. Ⓐ Ⓑ Ⓒ Ⓓ Ⓔ	74. Ⓐ Ⓑ Ⓒ Ⓓ Ⓔ
25. Ⓐ Ⓑ Ⓒ Ⓓ Ⓔ	50. Ⓐ Ⓑ Ⓒ Ⓓ Ⓔ	75. Ⓐ Ⓑ Ⓒ Ⓓ Ⓔ

Multiple-Choice Questions (90 Minutes)

CALCULATORS MAY NOT BE USED

Questions 1 through 3 refer to the following reaction types:

(A) Combustion

(B) Single Replacement

(C) Double Replacement

(D) Combination

(E) Acid/Base

1. Characteristic products of this reaction type are carbon dioxide and water.

2. The activity series of metals determines whether or not these reactions will take place.

3. Titrations take place by way of this type of reaction.

Questions 4 through 8 refer to the following elements:

(A) Na

(B) Mg

(C) Al

(D) S

(E) Cl

4. Is the heaviest metal

5. Is the most electronegative

6. Has the largest first ionization energy

7. Has the largest jump between second and third ionization energies

8. Has the largest atomic radius

Questions 9 through 11 refer to the following atomic models:

(A) Thompson Model

(B) Bohr Model

(C) Rutherford Model

(D) Quantum Mechanical Model

(E) Boyle Model

9. This method was developed via the famous gold foil experiment.

10. This model represents our best modern understanding of the actual structure of the atom.

11. Only one type of charge came in particle form in this atomic model.

Questions 12 through 14 refer to the following gas laws:

(A) Boyle's Law

(B) Charles' Law

(C) Graham's Law

(D) The Combined Gas Law

(E) The Ideal Gas Law

12. To use this law, you must assume that temperature is constant.

13. This law can be used to derive many of the other gas laws.

14. This law described the phenomenon of effusion.

15. Which of the following species is not planar?

(A) CO_2

(B) Cl_2

(C) BF_3

(D) CH_4

(E) NO_3^-

16. A compound has a formula X_2O. Which of the following species could be X?

(A) Be

(B) Mg

(C) Fe

(D) Li

(E) Sr

Go on to next page

17. Zn(s) + 2AgNO3(aq) Æ 2Ag(s) + Zn(NO3)2(aq)

    According to the reaction represented above, about how many grams of Zn must go into this reaction to produce 1.0mol of Ag?

    (A) 17g

    (B) 25g

    (C) 33g

    (D) 65g

    (E) 130g

18. A 0.25M solution has an [H$^+$] of 4.2×10^{-6}. What is its pH?

    (A) 5.00

    (B) 5.37

    (C) 6.00

    (D) 6.27

    (E) 7.00

19. Which of the following bonds is the most polar?

    (A) C-H

    (B) C-N

    (C) C-O

    (D) C-F

    (E) N-O

20. Which of the following is a correct representation of the electron configuration for molybdenum?

    (A) $1s^2 2s^2 2p^6 3s^2 3p^6 4s^2 3d^{10} 4p^6 5s^2 4d^4$

    (B) [Ar]$5s^2 4d^4$

    (C) [Ar] $5s^1 4d^5$

    (D) Kr] $5s^1 4d^5$

    (E) [Kr] $5s^2 5d^4$

21. Which of the following K_a values would be most suitable to buffer a solution with a pOH of 4.30?

    (A) 6.4×10^{-3}

    (B) 8.2×10^{-4}

    (C) 5.1×10^{-5}

    (D) 1.8×10^{-9}

    (E) 2.0×10^{-10}

22. Which of the following organic compounds contains 2 pi bonds?

    (A) $CH_3CHCHCHCH_3$

    (B) CH_3CH_2COOH

    (C) $CH_3CH_2CH_2CH_2OH$

    (D) $CH_3CHCHCH_2CH_3$

    (E) Benzene

23. Which of the following does NOT favor the formation of products in the reaction below?

    $C(s) + H_2O(g) \rightarrow CO(g) + H_2(g)$

    (A) Increasing the concentration of H_2O

    (B) Removing H_2 as it is formed

    (C) Increasing the pressure

    (D) Increasing the volume of the reaction container

    (E) Adding a catalyst

24. What are the possible values of the quantum number l for an atom with a principal quantum number of 3?

    (A) 0, 1, 2

    (B) –2, –1, 0, 1, 2

    (C) 0, 1, 2, 3

    (D) –3, –2, –1, 0, 1, 2, 3

    (E) –½ and ½

25. Which of the following is NOT a property of an ideal gas?

    (A) Its molecules occupy no volume.

    (B) Its particles do not interact with one another.

    (C) Its molecules are in constant random motion.

    (D) Collisions between molecules are completely inelastic.

    (E) There are no intermolecular forces acting between molecules.

Go on to next page

26. Compound A combines with chlorine as ACl_2. Which of the following is likely to be its electron configuration?

    (A) $1s^2 2s^2 2p^6 3s^1$

    (B) $1s^2 2s^2 2p^6 3s^2$

    (C) $1s^2 2s^1$

    (D) $1s^2 2s^2 2p^6$

    (E) $1s^2 2s^2 2p^6 3s^2 3p^1$

27. Which of the following substances contains both ionic and covalent bonds?

    (A) CH_3CH_2F

    (B) $NaCl$

    (C) BF_3

    (D) CH_3COOH

    (E) NH_4Cl

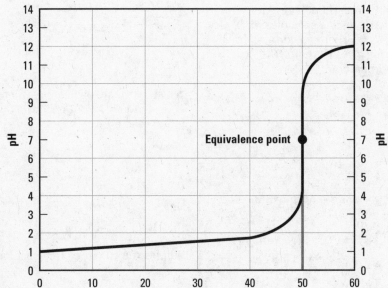

Figure 29-1:
Titration curve.

28. Which of the following matches the titration curve shown in Figure 29-1?

    (A) A strong acid is titrated into a weak base.

    (B) A strong acid is titrated into a strong base.

    (C) A strong base is titrated into a weak acid.

    (D) A strong base is titrated into a strong acid.

    (E) A weak base is titrated into a weak acid.

29. Miscible liquids are best separated from one another by

    (A) filtration.

    (B) condensation.

    (C) centrifugation.

    (D) distillation.

    (E) precipitation.

Go on to next page

30. HNO2(aq) + OH–(aq) Æ NO2–(aq) + H2O(l)

    In the reaction above, which species acts as the Lewis base?

    (A) HNO_2

    (B) OH^-

    (C) NO_2^-

    (D) H_2O

    (E) This is not an acid-base reaction.

31. One mole of a pure hydrocarbon undergoes complete combustion, creating six moles of water and five moles of carbon dioxide. Which of the following gives the empirical formula for that hydrocarbon?

    (A) C_2H_6

    (B) C_3H_8

    (C) C_4H_{10}

    (D) C_5H_{12}

    (E) C_6H_{14}

32. How many valence electrons does phosphorous have?

    (A) 3

    (B) 4

    (C) 5

    (D) 6

    (E) 7

33. The molecule PCl_5 has what type of geometry according to VSEPR theory?

    (A) Linear

    (B) Trigonal Planar

    (C) Trigonal bipyramidal

    (D) Tetrahedral

    (E) Octahedral

34. $4Ag(s) + O_2(g) + 4H^+(aq) \rightarrow 4Ag^+(aq) + 2H_2O(l)$

    $O_2(g) + 4H^+(aq) + 4e^- \rightarrow 2H_2O(l)$
    $E^\circ_{red} = 1.23V$

    $Ag^+(aq) + e^- \rightarrow Ag(s)$ $\qquad E^\circ_{red} = 0.80V$

    What is the EMF of the voltaic cell driven by the above reaction?

    (A) $-0.43V$

    (B) $0.43V$

    (C) $0.80V$

    (D) $1.23V$

    (E) $2.03V$

35. Which of the following ions will have the largest atomic radius according to periodic table trends?

    (A) N^{3-}

    (B) O^{2-}

    (C) F^-

    (D) Na^+

    (E) Mg^{2+}

36. What is the electron configuration of O^{2-}?

    (A) $1s^2 2s^2 2p^2$

    (B) $1s^2 2s^2 2p^4$

    (C) $1s^2 2s^2 2p^6$

    (D) $1s^2 2s^2 3p^2$

    (E) $1s^2 2s^2 3p^6$

37. The net ionic equation for the reaction when solutions of calcium chloride and sodium carbonate are mixed is

    (A) $CaCl_2(aq) + Na_2CO_3(aq) \rightarrow CaCO_3(s) + 2NaCl(aq)$.

    (B) $CaCl_2(aq) + Na_2CO_3(aq) \rightarrow CaCO_3(aq) + 2NaCl(aq)$.

    (C) $Ca^{2+}(aq) + 2Cl^-(aq) + Na^+(aq) + CO_3^{2-}(aq) \rightarrow CaCO_3(s) + 2Na^+(aq) + 2Cl^-(aq)$.

    (D) $Ca^{2+}(aq) + 2Cl^-(aq) + Na^+(aq) + CO_3^{2-}(aq) \rightarrow Ca^{2+}(aq) + CO_3^{2-}(aq) + 2Na^+(aq) + 2Cl^-(aq)$.

    (E) $Ca^{2+}(aq) + CO_3^{2-}(aq) \rightarrow CaCO_3(s)$.

Go on to next page

38. An oxide of sulfur is found to contain 40% sulfur. What is its identity?

    (A) SO

    (B) SO_2

    (C) SO_3

    (D) S_6O_2

    (E) S_7O_2

39. $Mg(s) + 2HCl(aq) \to MgCl2(aq) + H2(g)$

    According to the reaction above, how many grams of HCl are needed to produce 11.2L of hydrogen gas at STP?

    (A) 18.3g

    (B) 36.5g

    (C) 38.2g

    (D) 46.9g

    (E) 73.0g

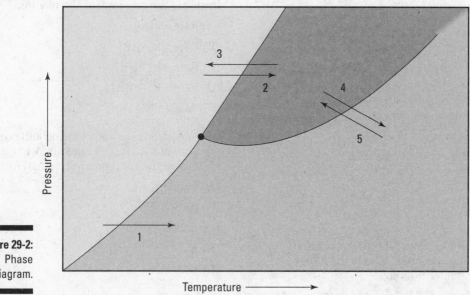

Figure 29-2: Phase diagram.

40. Which of the arrows on the phase diagram in Figure 29-2 represents the process of sublimation?

    (A) Arrow 1

    (B) Arrow 2

    (C) Arrow 3

    (D) Arrow 4

    (E) Arrow 5

Questions 41 through 42 refer to the following unbalanced half-reaction

$MnO_4^- + H^+ \to Mn^{2+}$

---

41. How many electrons will appear on the left-hand side in the final, balanced half-reaction?

    (A) 2

    (B) 3

    (C) 4

    (D) 5

    (E) 6

42. How many moles of water will need to appear on the righthand side of the equation to balance out the H and O on the left-hand side?

    (A) 2

    (B) 3

    (C) 4

    (D) 5

    (E) 6

Go on to next page

43. Compared to light with low energy, high energy light has

    (A) a shorter wavelength.

    (B) a smaller frequency.

    (C) a larger amplitude.

    (D) a faster speed.

    (E) X-rays.

44. Which of the following salts forms an acidic solution when dissolved in water?

    (A) NaCl

    (B) Na_2SO_4

    (C) KCl

    (D) KBr

    (E) NH_4I

45. Which of the following functional groups is characteristic of an amine?

    (A) –CHO

    (B) –COOH

    (C) –OH

    (D) $–NH_2$

    (E) $–NH_3$

46. If you dilute 20mL of a 1M NaOH stock solution with 30mL of water, what is the molarity of your diluted solution?

    (A) 0.3M

    (B) 0.4M

    (C) 0.5M

    (D) 0.6M

    (E) 0.7M

Experiment	Initial [A]	Initial [B]	Initial rate of formation of C
1	0.02	0.01	2.0×10^{-6}
2	0.04	0.01	2.0×10^{-6}
3	0.04	0.02	4.0×10^{-6}
4	0.02	0.04	1.6×10^{-5}

47. The data in the table above were obtained for the reaction $A + B \rightarrow C$. Which of the following is the rate law for the reaction?

    (A) Rate = k[A]

    (B) Rate = k[A][B]

    (C) Rate = k $[A][B]^2$

    (D) Rate = k $[A]^2[B]$

    (E) Rate = k[B]

48. When aqueous barium chloride and potassium sulfate are mixed together, what is the identity of the solid precipitate that forms?

    (A) $Ba(SO_4)_2$

    (B) Ba_2SO_4

    (C) $BaSO_4$

    (D) KCl

    (E) K_2Cl

49. Arrange these species in order of decreasing oxidation number:

    Fluoride, hydrogen bonded to a nonmetal, oxygen not in a peroxide, magnesium cation

    (A) Magnesium < oxygen< hydrogen < fluoride

    (B) Fluoride < magnesium < hydrogen < oxygen

    (C) Magnesium < hydrogen < oxygen < fluoride

    (D) Magnesium < hydrogen < fluoride < oxygen

    (E) Hydrogen < oxygen < fluoride < magnesium

Go on to next page

50. Which of the following polyatomic ions has the greatest amount of negative charge?

    (A) Nitrate

    (B) Sulfate

    (C) Phosphate

    (D) Permanganate

    (E) Ammonium

51. What formula could be expected for a binary compound of potassium and sulfur?

    (A) KS

    (B) K_2S

    (C) KS_2

    (D) K_3S

    (E) K_2S_3

52. What is the pK_a of a substance with a K_b of 1×10^{-6}?

    (A) 5

    (B) 6

    (C) 7

    (D) 8

    (E) 9

53. Li+(aq) + e− Æ Li(s)E° = −3.04

    $Zn^{2+}(aq) + 2e^- \rightarrow Zn(s)$ E° = −0.76

    $F_2(g) + 2e^- \rightarrow 2F^-(aq)$ E° = 2.87

    Based on the standard electrode potentials given above, which of the following is the strongest oxidizing agent?

    (A) Li^+

    (B) Li

    (C) Zn

    (D) F_2

    (E) F^-

54. Which of the following represents a reaction that can occur at the cathode of an electrolytic cell?

    (A) $Cu^{2+}(aq) + 2e^- \rightarrow Cu(s)$

    (B) $Sn^{2+}(aq) + 2e^- \rightarrow Sn(s)$

    (C) $Au^{3+}(aq) + 3e^- \rightarrow Au(s)$

    (D) $Fe^{3+}(aq) + e^- \rightarrow Fe^{2+}(aq)$

    (E) $Br_2(l) + 2e^- \rightarrow 2Br^-(aq)$

55. Which of the following organic compounds is the most soluble in water?

    (A) 2-fluoropentane

    (B) Benzene

    (C) Ethene

    (D) Propanol

    (E) Pentane

56. Which of the following pairs of compounds are not a conjugate acid/base pair?

    (A) H_3O^+ and H_2O

    (B) H_2O and OH^-

    (C) NH_3 and NH_2^-

    (D) CH_3COOH and CH_3COO^-

    (E) H_3PO_4 and PO_4^{3-}

57. The half-life of ^{125}I is about 60 days. A company in the United States has ordered that a sample of this radioisotope be shipped to their lab from a lab in China. The sample is sent by boat on a trip that takes 120 days. If the lab needs to receive 1.5g of the radioisotope, how much should be sent from the lab in China?

    (A) 1.0g

    (B) 1.5g

    (C) 1.75g

    (D) 3.0g

    (E) 6.0g

58. Which of the following compounds cannot form hydrogen bonds?

    (A) CH_3OH

    (B) HI

    (C) NH_3

    (D) HF

    (E) H_3O^+

59. What is the theoretical yield of water for a reaction between 6.00g of H_2 and 160.g O_2?

    (A) 53.4g

    (B) 98.0g

    (C) 108g

    (D) 120.g

    (E) 134g

Go on to next page

60. At constant temperature, if the volume of a gas is halved, its pressure will be

 (A) halved.

 (B) remain the same.

 (C) doubled.

 (D) tripled.

 (E) quadrupled.

61. The hybridization of the carbon atoms in ethene can be described as

 (A) sp.

 (B) sp^2.

 (C) sp^3.

 (D) sp^3d.

 (E) sp^3d^2.

62. Which of the following operations should have a final answer containing four significant figures?

 (A) 12.2×13.51

 (B) $(62.315)^2$

 (C) $0.023 + 1.311$

 (D) $1.010 - 11.623$

 (E) $64.5 \div 3.2$

63. Which of the following elements is not isoelectronic with the others?

 (A) S^{2-}

 (B) Cl^-

 (C) Ar

 (D) K^+

 (E) Mg^{2+}

64. If 149g of KCl is dissolved in 400.g of water, what is the new freezing point of the water $(K_f = 1.86°C/m)$

 (A) 255K

 (B) 282K

 (C) −0.930°C

 (D) 9.30°C

 (E) 109°C

65. 10.0mL of 1.00M NaOH will create a solution of pH = 7 when mixed with which of the following (assume the solution is 1.00M)?

 (A) 3.33mL HNO_3

 (B) 3.33mL CH_3COOH

 (C) 5.00mL HCl

 (D) 5.00mL H_2SO_4

 (E) 30.0mL H_3PO_4

66. How many lone electron pairs are in the molecule H–Cl?

 (A) one

 (B) two

 (C) three

 (D) four

 (E) six

67. Which of the following substances is insoluble in water?

 (A) K_2S

 (B) AgCl

 (C) $(NH_4)_3PO_4$

 (D) $Sr(OH)_2$

 (E) Li_2SO_4

68. How many moles of nitric acid are in 5.00L of nitric acid solution with a molarity of 0.250?

 (A) 0.250

 (B) 1.25

 (C) 2.50

 (D) 4.75

 (E) 5.25

69. An element undergoes alpha decay to form Lead-206. What is the identity of that element?

 (A) Mercury-206

 (B) Thalium-207

 (C) Lead-209

 (D) Bismuth-209

 (E) Polonium-210

Go on to next page

70. A 2.00L container holding oxygen at 1.50atm pressure is joined to a 3.00L container holding nitrogen at 0.500atm. What will the partial pressure of oxygen be when the gases mix and equilibrate in the new 5.00L volume?

    (A) 0.300atm

    (B) 0.450atm

    (C) 0.600atm

    (D) 1.25atm

    (E) 2.00atm

71. What is the molarity of a solution of NaOH if 3.00mL of the solution must be titrated with 5.00mL of 0.750M H_2SO_4 to reach the equivalence point?

    (A) 0.250M

    (B) 0.750M

    (C) 1.25M

    (D) 2.50M

    (E) 7.50M

72. What is the $[H^+]$ of a solution with a pOH of 4.50?

    (A) 3.1×10^{-4}

    (B) 5.6×10^{-5}

    (C) 4.2×10^{-9}

    (D) 3.2×10^{-10}

    (E) 7.1×10^{-11}

73. A reaction is exothermic. Which of the following must also be true?

    (I) The reaction is spontaneous.

    (II) ΔH is negative.

    (III) Reactants have lower energy than products.

    (A) I only

    (B) II only

    (C) I and II

    (D) I, II, and III

    (E) None of the above

74. Which of the following is not a likely decay product for the decay of iodine-131?

    (A) Antimony-127

    (B) Tellurium-131

    (C) Iodine-131

    (D) Xenon-131

    (E) Cesium-135

75. Polonium-218 undergoes an alpha decay followed by a beta minus decay. What is the end product of this reaction?

    (A) Thallium-214

    (B) Lead-214

    (C) Polonium-218

    (D) Bismuth-214

    (E) Astatine-218

STOP DO NOT TURN THE PAGE UNTIL TOLD TO DO SO.
DO NOT RETURN TO A PREVIOUS TEST.

Free-Response Questions

Part A (55 minutes)

YOU MAY USE YOUR CALCULATOR

CLEARLY SHOW ALL STEPS YOU TAKE TO ARRIVE AT YOUR ANSWER. It is to your advantage to do this, because you will receive partial credit for partially correct responses. Make sure to pay attention to significant figures.

Answer questions 1, 2, and 3. The score weighting for each question is 20 percent.

$C_6H_5COOH(aq) + H_2O(l) \rightarrow H_3O^+(aq) +$
$C_6H_5COO^-(aq)$ $K_a = 6.17 \times 10^{-5}$

---

1. Benzoic acid acid dissociates in water according to the reaction above.

   (a) Write the equilibrium constant expression for the reaction.

   (b) Calculate the molar concentration of $C_6H_5COO^-$ in a 0.0100M benzoic acid solution.

   (c) Write a balanced neutralization reaction equation for the reaction of benzoic acid and calcium hydroxide.

   (d) After adding 10.0mL of $5.00 \times 10^{-6} M$ $Ca(OH)_2$ to 90.0mL of an unknown concentration of benzoic acid, the pH of the solution is 5.26. Calculate each of the following:

      (i) The $[H^+]$ of the solution after the addition of $Ca(OH)_2$.

      (ii) The $[OH^-]$ of the solution after the addition of $Ca(OH)_2$.

      (iii) The initial molarity of the benzoic acid solution before the addition of $Ca(OH)_2$.

   (e) State whether the solution at the equivalence point of this titration is acidic, basic, or neutral. Explain your reasoning.

2. Reaction A: $N_2(g) + 3H_2(g) \rightarrow 2NH_3(g)$
   $\Delta H^° = -92kJ \; \Delta S^° = -198 \; J/K$

   Reaction B: $N_2F_4(g) \rightarrow 2NF_2(g)$
   $\Delta H^° = 85kJ \; \Delta S^° = 198 \; J/K$

   (a) Identify each reaction as exothermic or endothermic. Explain your reasoning.

   (b) In each reaction, which side is more ordered: the reactants or the products? Explain your reasoning.

   (c) Calculate the Gibbs free energy change for each reaction and identify it as spontaneous or nonspontaneous under standard conditions. Explain your reasoning.

   (d) How will the Gibbs free energy for each of these reactions change with increasing temperature?

   (e) What is the value of the equilibrium constant for reaction A at 298K?

3. A voltaic cell is constructed based on the redox reaction between cadmium and tin (both to their 2+ ions).

   (a) Write the half-reactions that occur and identify them as occurring at the anode or the cathode.

   (b) Calculate the value of the standard cell potential.

   (c) Sketch and label a diagram of the cell, indicating the anode, cathode, salt bridge, and voltmeter.

   (d) Draw an arrow on your diagram showing the direction that electrons flow through the wire.

   (e) Write a balanced net ionic equation for the spontaneous cell reaction.

   (f) Calculate the value of the standard free energy change for the reaction.

 STOP DO NOT TURN THE PAGE UNTIL TOLD TO DO SO.
DO NOT RETURN TO A PREVIOUS TEST.

Part B (40 minutes)

CALCULATORS MAY NOT BE USED

Answer question 4 below. The score weighting for this question is 10 percent.

---

4. For each of the following three reactions, in part (i) write a balanced equation for the reaction and in part (ii) answer the question about the reaction. In part (i), coefficients should be in terms of lowest whole numbers. Assume that solutions are aqueous unless otherwise indicated. Represent substances in solution as ions if the substances are extensively ionized. Omit formulas for any ions or molecules that are unchanged by the reaction.

(a) Solid lead carbonate is added to an aqueous solution of ammonium bromide.

   (i) Balanced equation:

   (ii) Which product(s) is soluble, if any? Which product(s) is insoluble, if any?

(b) Calcium metal is exposed to oxygen gas.

   (i) Balanced reaction:

   (ii) What is the oxidation number of oxygen before and after the reaction?

(c) Hexane undergoes complete combustion in the presence of oxygen gas.

   (i) Balanced reaction:

   (ii) How would the products be different if hexane were combusted in air instead?

Answer questions 5 and 6 below. The score weighting for each question is 15 percent.

---

5. Answer each of the following questions.

(a) Draw a complete Lewis electron-dot diagram for carbon dioxide and for PO_4^{3-}.

(b) On the basis of your Lewis diagram from part (a), identify the hybridization of the central carbon atom in CO_2 and the central phosphorous atom in PO_4^{3-}.

(c) When carbon dioxide dissolves in water, a small fraction (at equilibrium) of the carbon dioxide reacts with water to form carbonic acid. Write out a complete, balanced equation for this reaction and identify the Lewis acid and the Lewis base in the reaction.

(d) Is CO_2 polar? Explain.

(e) What is the O-P-O bond angle in PO_4^{3-}?

6. Use chemical and physical principles to account for each of the following.

(a) An opened soda can becomes "flat" (loses its carbonation) over time.

(b) Most medicines designed to relieve heartburn and calm the stomach contain hydroxide.

(c) A pure elemental gas emits distinct colors of light when viewed through a prism and no two are alike.

(d) The water molecule has a tetrahedral electron pair arrangement but bent geometry.

Answers to Practice Test 1

Multiple choice

1. (A). Carbon dioxide and water are the characteristic products of a combustion reaction.

2. (B). The activity series of metals determines whether or not single replacement reactions will occur.

3. (E). Titrations are used in acid/base reactions.

4. (C). Sodium, magnesium, and aluminum are all metals, but aluminum has the greatest mass.

5. (E). Chlorine is the farthest right on the periodic table and is therefore the most electronegative.

6. (E). Chlorine has the highest first ionization energy because it has the strongest pull on its electrons. (Chlorine wants to gain electrons, not lose them.)

7. (B). All of the alkaline earth metals have extremely large jumps between their second and third ionization energies because after the second ionization, they achieve the electron configuration of a noble gas and are reluctant to relinquish it.

8. (A). Atomic radius increases to the left across a period and down in a family. All of these elements are in the same row so only the first trend applies. Sodium has the largest radius because it is the farthest left.

9. (C). Rutherford discovered the nucleus through his famous gold foil experiment, which led to the development of the Rutherford model of the atom.

10. (D). Our modern understanding of the atom is based on the Quantum Mechanical model.

11. (A). The Thompson model of the atom, often referred to as the "plum pudding model," consisted of negatively charged particles in a diffuse soup of positive charge.

12. (A). Boyle's law assumes constant temperature.

13. (E). The ideal gas law can be used to derive many of the other gas laws.

14. (C). Graham's law describes the phenomenon of effusion.

15. (D). CH_4 has a tetrahedral geometry and is therefore not planar.

16. (D). Oxygen has a 2– charge as an ion, so it will form a salt in a 1:2 ratio with any alkali metal, of which lithium is the only one on the list.

17. (C). Convert 1mol of Ag to grams of Zn using the mole ratios in the balanced equation to get your answer.

$$\frac{1.00\,molAg}{1} \times \frac{1molZn}{2molAg} \times \frac{65.4g}{1molZn} = 32.7\,gZn$$

18. (B). pH does not depend on compound molarity, only on [H⁺]. You do not have a calculator to take its logarithm, but you don't need one because the answers don't require you to be nit-picky. You know that a solution with an [H⁺] concentration that has a coefficient greater than 1 will have a pH value between its coefficient (neglecting the negative sign) and its coefficient minus 1. In this case, 5.37 is the only number between 6 and 5.

19. (D). Fluorine is the most electronegative element on the periodic table and therefore pulls most heavily on the electrons of a bound carbon atom. Nitrogen and oxygen, lying so closely together on the periodic table, are quite close in electronegativity and therefore form a nonpolar bond.

20. (D). Molybdenum is one of the transition metals with an exceptional electron configuration. In order to achieve greater stability, it will steal an electron from its 5s orbital in order to obtain a half-full 4d orbital.

21. (E). The best buffers are those whose pKa values are close to the desired pH. You're given the pOH here, so you need to subtract it from 14 to get a desired pH of 9.7. Here again, you cannot take a logarithm without a calculator but you know that you have a pH with a value between 9 and 10, which means that your desired exponent will be –10. Choice E is the only one that fits the bill.

22. (A). If you draw out the Lewis structures (see Figure 29-3) of the compounds given, you will find that choice A is the only one with two double bonds.

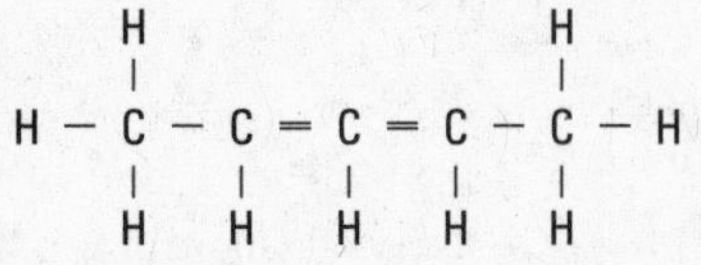

Figure 29-3:
A compound containing 2 pi bonds.

$$\begin{array}{ccccccccc} & & \overset{\displaystyle H}{|} & & & & & \overset{\displaystyle H}{|} & \\ H & - & C & - & C & = & C & = & C & - & C & - & H \\ & & | & & | & & | & & | & & | \\ & & H & & H & & H & & H & & H \end{array}$$

23. (C). Increasing the pressure in the reaction vessel will favor the side of the reaction with fewer moles of gas, which is the reactants side. Adding a catalyst promotes the formation of both products and reactants, so although the equilibrium concentration of product isn't altered by the catalyst, adding it nonetheless promotes product formation prior to attaining equilibrium.

24. (A). The quantum number l has values ranging from 0 to n-1. Since n is 3 in this case, you're left with three possible values: 0, 1, and 2.

25. (D). Collisions between the particles of an ideal gas are completely *elastic*, meaning that no energy is lost in the process of colliding.

26. (B). A compound that combines with chlorine in a 1:2 ratio is likely to be an alkaline earth metal, meaning its electron configuration should end with a full s orbital. B is the only choice that fits this criterion, as it is the electron configuration for magnesium.

27. (E). Ammonium is a polyatomic ion held together by covalent bonds, but its bond with chlorine is ionic.

28. (B). This is a steep curve that moves from low pH to high pH, with an equivalence point of 7. It must be a strong acid titrated with a strong base.

29. (D). Distillation is the best process for separating miscible liquids.

30. (B). A Lewis base donates a pair of electrons to a coordinate covalent bond. In this case, that pair of electrons comes from one of the lone pairs on the oxygen of OH⁻, so it is the Lewis base.

31. (D). The information provided should lead you to the formula $C_5H_{12} + 8O_2 \rightarrow 6H_2O + 5CO_2$ for the complete combustion of pentane.

32. (C). Phosphorous has 5 valence electrons.

33. (C). Phosphorous pentachloride has trigonal bipyramidal geometry.

34. (B). You calculate EMF, or E°_{cell}, by subtracting the reduction potential of the anode from that of the cathode. Oxidation happens at the anode, so the reaction $Ag^+(aq) + e^- \rightarrow Ag(s)$ must be the anode reaction. This leads you to an EMF of 1.23 – 0.80 = 0.43V.

35. (A). All of the ions listed have the same number of electrons, but nitrogen has the smallest number of protons, which means less positive charge to pull in the negatively charged electron cloud.

36. (C). O^{2-} has the regular electron configuration of oxygen plus an additional 2 electrons, giving it the electron configuration of neon.

37. (E). The complete reaction is answer choice A, however the net ionic reaction takes account of dissociation and eliminates spectators, which is answer choice E. $CaCO_3$ is also a carbonate and all carbonates are insoluble except for those containing ammonium or alkali metals so that product must be a solid, which choice E also shows.

38. (C). The easiest way to approach this problem is to use the answer choices and do rough calculations of the percent composition of sulfur in each one. Choice A gives you $32 \div 48$, which is too large. Choice B is also too large and choices D and E are too small. Choice C, which gives you a percent composition of $32 \div 80 = 0.4$, is the correct answer.

39. (B). Convert liters of H_2 to grams of HCl using the mole ratios in the balanced equation and the conversion 22.4L per mole of gas at STP.

$$\frac{11.2 L H_2}{1} \times \frac{1 mol H_2}{22.4 L} \times \frac{2 mol HCl}{1 mol H_2} \times \frac{36.5 g HCl}{1 mol HCl} = 36.5 g HCl$$

40. (A). Arrow 1 shows a transition from the solid to the gas regions on the phase diagram, which is the process of sublimation.

41. (D). Both questions 41 and 42 refer to the balanced half-reaction. Balance the excess oxygen on the lefthand side by adding 4mol H_2O to the right. Then balance the hydrogen on the left with the hydrogen atoms in the water on the right by adding the coefficient 8. Finally, balance charge by adding five electrons to the lefthand side, giving you a total charge of +2 on both sides, leaving you with. $MnO_4^- + 8H^+ + 5e^- \rightarrow Mn^{2+} + 4H_2O$.

42. (C). According to the equation you created to answer question 41, 4mol of water are needed.

43. (A). The higher the energy of light, the shorter its wavelength. Your spectroscopy equation $E = hc \div \lambda$ can be used to verify this inverse relationship.

44. (E). NH_4I is the only solution that will form a strong acid (HI) and a weak base NH_4OH when dissolved in water, leaving it with a low pH. All of the others will form both a strong acid and a strong base when they react with water.

45. (D). Primary amines have the functional group $-NH_2$.

46. (B). Use the formula $M_1 V_1 = M_2 V_2$ to solve for M_2. Be careful though! V_2 is the final diluted volume, which in this case is 50mL.

$$M_2 = \frac{1M \times 0.02L}{0.05L} = 0.4M$$

47. (E). When comparing experiments 1 and 2, the reaction rate does not change, indicating A is zero order. While comparing experiments 2 and 3, the rate doubles while B doubles indicating a first-order reaction.

48. (C). The balanced double replacement reaction between barium chloride and potassium sulfate is $BaCl_2(aq) + K_2SO_4(aq) \rightarrow BaSO_4(s) + 2KCl(aq)$. You know that the sulfate is the precipitate because barium is one of the elements whose presence makes a sulfate insoluble according to your solubility rules.

49. (D). The magnesium cation has an oxidation number of +2, hydrogen is +1 when bonded to a nonmetal, the fluoride ion is −1, and oxygen is always −2 except in peroxides.

50. (C). This question tests purely your ability to memorize the polyatomic ions and their charges. Phosphate has the greatest with a −3 charge.

51. (B). Potassium forms a +1 ion while sulfur forms a −2 ion, so they will combine in a 2:1 ratio.

52. (D). Use K_w and the given K_b to solve for K_a.

$$K_a = \frac{K_w}{K_b} = \frac{1 \times 10^{-14}}{1 \times 10^{-6}} = 1 \times 10^{-8}$$

Since the coefficient on K_a is 1, the pK_a is simply the exponent (neglecting the negative sign).

53. (D). Oxidizing agents have positive values of E°.

54. (B). This reaction requires the use of your chart of standard reduction potentials (which you will be given on the exam). Recall that reduction happens at the cathode of an electrolytic cell and choice B is the only answer with a negative reduction potential, so it is the only choice that is a reduction reaction.

55. (D). Remember the concept "like dissolves like." The intermolecular forces in propanol are the most similar to those of water.

56. (E). All of these pairs involve the loss of hydrogen ions, but choice E involves the loss of more than just a single hydrogen, so it is not a conjugate acid/base pair.

57. (E). 120 days amounts to two half-lives of the radioisotope, so they must send 6.0g total.

58. (B). O-H, N-H, and F-H bonds can all participate in hydrogen bonding, but I-H bonds cannot.

59. (A). The balanced reaction here is $2H_2 + O_2 \rightarrow 2H_2O$. Determine the limiting reagent by converting grams of hydrogen to grams of oxygen using the mole ratios from the balanced equation.

$$\frac{6.00 g H_2}{1} \times \frac{1 mol O_2}{2 mol H_2} \times \frac{16.0 g}{1 mol O_2} = 48.0 g O_2$$

This reveals that H_2 is the limiting reagent so now you need to convert grams of H_2 to grams of H_2O through a similar calculation.

$$\frac{6.00 g H_2}{1} \times \frac{1 mol H_2}{2.02 g H_2} \times \frac{2 mol H_2O}{2 mol H_2} \times \frac{18.0 g}{1 mol H_2O} = 108 g H_2O$$

60. (C). According to Boyle's law, pressure and volume are inversely proportional, so halving one should double the other.

61. (B). Both of the carbon atoms in ethene are sp^2 hybridized.

62. (C). Follow the rules for addition and subtraction; the answer contains the same number of decimal places as the least number of decimal places of the measurements used in the calculation. So, the calculation in answer C yields 1.334, with four significant figures.

63. (E). All of the elements listed have the same electron configuration as (are isoelectronic with) argon except for Mg^{2+}.

64. (A). Begin by calculating the molality of the solution by converting grams of KCl to moles and g of solvent to kg:

$$\frac{149 g KCl}{400. g} \times \frac{1 mol KCl}{74.5 g KCl} \times \frac{1000 g}{1 kg} = 5.00 m$$

Notice that KCl has a van't Hoff factor of 2; in other words, 1 mole of KCl dissolves into 2 moles of particles, K^+ and Cl^-. So, 5.00 molal KCl contributes a *solute particle molality* of 10.0. Multiply the particle molality by the K_f to get a freezing point depression of 18.6 degrees. This means that the new freezing point of the water is $0.00°C - 18.6°C = -18.6°C$ or 255K.

65. (D). 10.0mL of 1.00M NaOH will be neutralized by 10.0mL of a monoprotic acid, 0.500mL of a diprotic acid, or 3.33mL of a triprotic acid. D is the only answer choice that fits.

66. (C). There are three nonbonding pairs of electrons in the molecule HCl, which can be seen by drawing its Lewis dot structure (see Figure 29-4).

Figure 29-4:
Lewis dot
structure.

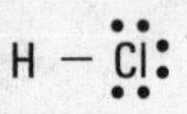

$$H - \overset{\displaystyle ..}{\underset{\displaystyle ..}{Cl}} :$$

67. (B). All chlorides, bromides, and iodides are soluble except those containing Ag, Pb, and Hg.

68. (B). Multiply the molarity by the volume to get the number of moles present. $5.00 \times 0.250 = 1.25$.

69. (E). When an element undergoes alpha decay, it loses 4 from its mass number and 2 from its atomic number, so the progenitor of Lead-206 must be polonium-210.

70. (B). The question requires you to use Boyle's Law, $P_1V_1 = P_2V_2$. You're asked to find a final pressure, so rearrange the gas law to the form $P_2 = (P_1V_1) / V_2$. Substituting in known values gives you $P_2 = (1.50\text{atm} \times 2.00\text{L}) / 5.00\text{L} = 0.600\text{atm}$.

71. (A). Convert the molarity of acid to the molarity of the base, taking account of equivalents and volumes.

$$\frac{0.750 mol H_2SO_4}{1L} \times \frac{5.00 \times 10^{-3}L}{1} \times \frac{2 mol NaOH}{1 mol H_2SO_4} \times \frac{1}{3.00 \times 10^{-3}L} = 2.50 m NaO1$$

72. (D). A solution with a pOH of 4.50 has a pH of 9.50, so the exponent on its H^+ concentration must be -10.

73. (B). Although most exothermic reactions are spontaneous, all are not. All exothermic reactions have a negative ΔH and the energy of their products is always lower than that of their reactants.

74. (E). Iodine-131 could undergo alpha decay to create antimony-127, beta plus decay to create tellurium-131, gamma decay to create iodine-131 (gamma decay does not change the identity of an atom), or beta minus decay to create xenon-131. Cesium-135 is the odd man out.

75. (D). An alpha decay of polonium-218 will change it into lead-214. Beta minus decay will not change the mass number, but it will increase the atomic number by one leaving bismuth-214.

Free response

1. (a) The equilibrium constant is constructed by taking the concentration of all of the products raised to the powers of their coefficients divided by the concentrations of all of the reactants (except for water) raised to the power of their coefficients. Because the reaction is an acid dissociation, the K_{eq} in this case is also the K_a.

$$K_{eq} = K_a = \frac{[C_6H_5COO^-][Br^-]}{[CH_3COOH]}$$

(b) $7.85 \times 10^{-4}M$. The given molarity is equivalent to $[C_6H_5COOH]$. When C_6H_5COOH dissociates in water, it creates 1 H_3O^+ molecule for every $C_6H_5COO^-$ molecule, so those two concentrations are the same, and we can say that they both equal x. The final concentration of C_6H_5COOH is $0.0100M - x$, which we can safely approximate to be $0.0100M$, leaving you with the equation

$$K_a = \frac{x^2}{0.0100} = 6.17 \times 10^{-5}$$

Solve the equation for x to get your answer.

(c) $2 C_6H_5COOH + Ca(OH)_2 \rightarrow 2H_2O + Ca(C_6H_5COO^-)_2$.

(d) This is clearly a titration problem.

   (i) $5.50 \times 10^{-6} M$. To calculate the $[H^+]$ you simply need to solve the equation $10^{-pH} = [H^+]$.

   (ii) $1.82 \times 10^{-9} M$. You know that $[H^+] = 5.50 \times 10^{-6} M$. Further, you know that in aqueous solution, $[H^+] \times [OH^-] = 10^{-14}$. So, you can substitute in your known value for hydrogen ion concentration and solve for the hydroxide ion concentration.

   (iii) $1.66 \times 10^{-6} M$ benzoic acid. The road to this answer is a bit long and winding, and has to do with understanding what happens when you titrate a weak acid with a strong base. Because the base is strong and the acid is weak, essentially all added base titrates away hydrogen ions from the weak acid — initally, at least. This 1-for-1 pattern holds through the "buffer region" of the weak acid, about 1 to 1.5 units above its pK_a. The pK_a of benzoic acid is $-\log(6.17 \times 10^{-5}) = 4.21$, so the final pH of 5.26 is within the buffer region. So, all added base acts to neutralize hydrogen ion from the weak acid. The total moles of base added in this case are:

$$(5.00 \times 10^{-6} M) \times (10.0 \times 10^{-3} L) \times (2\ OH^- / Ca(OH)_2) = 1.00 \times 10^{-7}\ \text{moles}\ OH^-$$

Because you are in the buffer region, you know that the original number of moles of weak acid equals the final number of moles *plus* the 1.00×10^{-7} moles that have been neutralized. That's fine, but what are the final moles of weak acid?

To find the final moles of weak acid, use your answers from part (b) and part (d-i). You know the $K_a = 6.17 \times 10^{-5}$. Further, you know the final concentrations of hydrogen ion and benzoate ion, $[H^+] = [C_6H_5COO^-] = 5.50 \times 10^{-6} M$. Plug these three known values into the equation for K_a, and you can solve for the final benzoic acid concentration, $[C_6H_5COOH] = 4.90 \times 10^{-7} M$. Multiply this value by the 0.100L final volume (10.0mL + 90.0mL) to obtain the final moles of benzoic acid, 4.90×10^{-8} moles. Now, add this final amount to the 1.00×10^{-7} moles that were neutralized, and you get the initial moles of benzoic acid, 1.49×10^{-7} moles. These moles were initially present in 90.0mL of solution. Calculate the initial molarity of benzoic acid by dividing 1.49×10^{-7} moles by 0.0900L to obtain $1.66 \times 10^{-6} M$. Phew.

(e) This is the titration of a weak acid with a strong base, so the solution will be basic at the equivalence point. The reason for this is that reaching the equivalence point involved the formation of a salt, calcium benzoate in this case, or $Ca(C_6H_5COO-)_2$. The benzoate anion of this salt is slightly basic, so the equivalence point solution will be slightly basic.

2. (a) Reaction A is exothermic because $\Delta H < 0$ and Reaction B is endothermic because $\Delta H > 0$.

(b) Reaction A has a negative entropy change (ΔS), which means that the products are more ordered than the reactants. This is consistent with the fact that Reaction A takes 4 moles of gas and converts it to 2 moles of gas. Reaction B has a positive entropy change, which means the reactants are more ordered than the products. This is consistent with the fact that Reaction B takes 1 mole of gas and converts it to 2 moles of gas. This also makes sense qualitatively because Reaction A is a combination reaction and Reaction B is a decomposition reaction.

(c) $\Delta G^\circ = -33 kJ$ for Reaction A and $\Delta G^\circ = 26 kJ$ for Reaction B. The Gibbs free energy change is governed by the equation $\Delta G^\circ = \Delta H - T\Delta S^\circ$. Here the temperature must be in K, and because they tell us that these reactions take place under standard conditions, we know that the temperature is 25°C or 298K. Plug the given values of ΔH and ΔS into the equation for each reaction to get your answers. Reaction A is spontaneous because $\Delta G^\circ < 0$ and Reaction B is nonspontaneous because $\Delta G^\circ > 0$.

(d) As temperature increases in the equation $\Delta G^\circ = \Delta H - T\Delta S^\circ$, the ΔG° for reaction A will become nonspontaneous, because the negative entropy change makes the term $-T\Delta S$ a positive quantity. Reaction B will become spontaneous.

(e) $K_{eq} = 1.6 \times 10^{-6}$. To answer this question, you must use the expression $\Delta G = \Delta G° + 2.303RT\log Q$. At equilibrium, $\Delta G=0$ and Q is the equilibrium constant, so the expression becomes $K_{eq} = 10^{(\Delta G/(2.303RT))}$. Plug in your known values, remembering to convert ΔG into joules (form kilojoules), and you calculate $K_{eq} = 1.6 \times 10^{-6}$.

3. (a) Your two half-reactions must be. $Cd^{2+} + 2e^- \rightarrow Cd$ and $Sn^{2+} + 2e^- \rightarrow Sn$. You must look up their standard reduction potentials to determine which occurs at the anode and which at the cathode. $Cd^{2+} + 2e^- \rightarrow Cd$ has an E° of –0.40 and $Sn^{2+} + 2e^- \rightarrow Sn$ has an E° of –0.14. The E° for Sn^{2+} is the higher of the two so it is reduced at the cathode. This means that Cd must be oxidized at the anode in the reaction $Cd \rightarrow Cd^{2+} + 2e^-$.

(b) The standard cell potential is calculated using the equation $E°_{cell} = E°_{red}(cathode) - E°_{red}$ (anode). Plug in the values you looked up in the table to get $E°_{cell} = -0.14V - (-0.40V) = 0.26V$.

(c) Be sure to label the anode, cathode, salt bridge, voltmeter, and the wire connecting the anode and cathode (see Figure 29-5).

Voltaic Cell

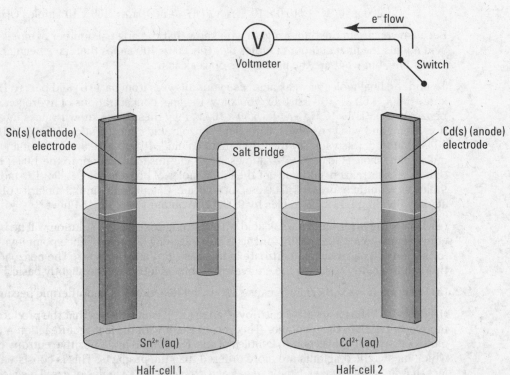

Figure 29-5:
Anode, cathode, salt bridge, voltmeter, and connecting wire.

(d) Electrons flow from anode to cathode.

(e) The net ionic equation for this cell is $Sn^{2+}(aq) + Cd(s) \rightarrow Cd^{2+}(aq) + Sn(s)$.

(f) Use the expression $\Delta G° = -nFE°$, where n is the number of moles of electrons exchanged in the net ionic equation (2 in this case) and F is Faraday's constant (96,500 C per mol e⁻). This gives you a $\Delta G°$ of $-2 \times 96,500 \times 0.26 = -50.2kJ$

4. (a) (i) Your balanced reaction should be $NH_4Br(aq) + PbCO_3(s) \rightarrow (NH_4)_2CO_3(aq) + PbBr_2(s)$. It is a relatively simple double replacement reaction as long as you have memorized your solubility rules and charges of the polyatomic ions. Eliminating spectators (NH_4^+), you should be left with $Br^-(aq) + PbCO_3(s) \rightarrow CO_3^{2-}(aq) + PbBr_2(s)$.

(ii) Carbonates are generally insoluble except for those containing ammonium and alkali metals so $(NH_4)_2CO_3$ is soluble. Bromides are generally soluble except for those containing Ag, Pb, or Hg, so $PbBr_2$ is insoluble.

(b) (i) This is a simple combination reaction of the form $2Ca(s) + O_2(g) \rightarrow 2CaO(s)$.

(ii) Pure elements have oxidation numbers of zero and oxygen has an oxidation number of –2 in the compound because it assumes its usual –2 charge.

(c) (i) The complete combustion of any hydrocarbon in the presence of oxygen alone will lead to the production of water and carbon dioxide. This reaction follows the equation $2C_6H_{14} + 19O_2 \rightarrow 14H_2O + 12CO_2$.

(ii) Nitrogen is also present in air and takes place in a combustion reaction, leading to the addition of N_2 among the reactants and NO_2 among the products.

5. See Figure 29-6.

(a) The carbon in CO_2 is sp hybridized while the phosphorous in PO_4^{3-} is sp^3 hybridized.

(b) The balanced reaction has the form $CO_2 + H_2O \rightarrow H_2CO_3$. In this reaction, the oxygen on the H_2O donates the lone pair to form a coordinate covalent bond with CO_2, so it is the Lewis base and CO_2 is the Lewis acid.

(c) No. The carbon-oxygen bonds of CO_2 are polar because O is more electronegative than C, and so the two oxygen atoms in CO_2 pull the electrons in their covalent bonds with carbon closer to them, leaving the oxygens with a slight negative charge and the carbon with a slight positive charge. However, the linear geometry of CO_2 opposes these polar bonds such that they cancel out and the molecule as a whole is nonpolar.

(d) PO_4^{3-} is tetrahedral, so the O-P-O bond angles are 109.5 degrees.

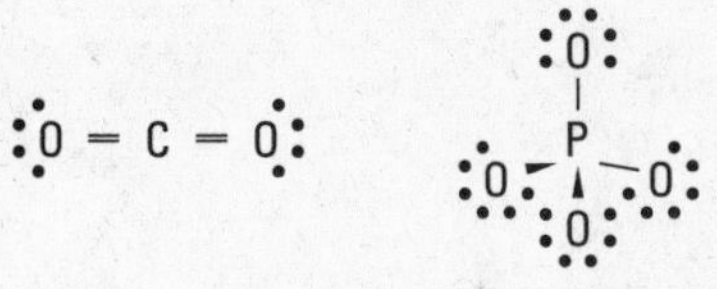

Figure 29-6:
Lewis
Structures.

6. (a) Carbon dioxide has a higher solubility in water at a higher pressure. Opening a can of soda lowers the pressure in the air in the headspace to 1atm and CO_2 begins to escape the solution.

(b) Heartburn and stomach upset are caused by an overproduction of stomach acid. Medications designed to relieve these symptoms contain hydroxide in order to fuel a neutralization of the stomach acid.

(c) No two atoms have the same available electron energy levels and since the colors emitted by an excited elemental gas depend on the energy difference between those levels, no two elements have the same emission spectrum.

(d) The two lone pairs on the oxygen atom of H_2O repel the hydrogen bonds, bending them downward slightly and changing the bond angles.

Chapter 30

Practice Test Two

In This Chapter
▶ Tackling 75 multiple-choice questions
▶ Responding to six free-response questions

*O*n the AP exam, you will be given a list of formulas and constants. You may use your Cheat Sheet (which mimics the AP formula list) and a periodic table for this exam, but do not use your calculator until the section specifically says that you may. Make sure to time yourself to get a sense of your pacing.

Answer Sheet

1. Ⓐ Ⓑ Ⓒ Ⓓ Ⓔ	26. Ⓐ Ⓑ Ⓒ Ⓓ Ⓔ	51. Ⓐ Ⓑ Ⓒ Ⓓ Ⓔ
2. Ⓐ Ⓑ Ⓒ Ⓓ Ⓔ	27. Ⓐ Ⓑ Ⓒ Ⓓ Ⓔ	52. Ⓐ Ⓑ Ⓒ Ⓓ Ⓔ
3. Ⓐ Ⓑ Ⓒ Ⓓ Ⓔ	28. Ⓐ Ⓑ Ⓒ Ⓓ Ⓔ	53. Ⓐ Ⓑ Ⓒ Ⓓ Ⓔ
4. Ⓐ Ⓑ Ⓒ Ⓓ Ⓔ	29. Ⓐ Ⓑ Ⓒ Ⓓ Ⓔ	54. Ⓐ Ⓑ Ⓒ Ⓓ Ⓔ
5. Ⓐ Ⓑ Ⓒ Ⓓ Ⓔ	30. Ⓐ Ⓑ Ⓒ Ⓓ Ⓔ	55. Ⓐ Ⓑ Ⓒ Ⓓ Ⓔ
6. Ⓐ Ⓑ Ⓒ Ⓓ Ⓔ	31. Ⓐ Ⓑ Ⓒ Ⓓ Ⓔ	56. Ⓐ Ⓑ Ⓒ Ⓓ Ⓔ
7. Ⓐ Ⓑ Ⓒ Ⓓ Ⓔ	32. Ⓐ Ⓑ Ⓒ Ⓓ Ⓔ	57. Ⓐ Ⓑ Ⓒ Ⓓ Ⓔ
8. Ⓐ Ⓑ Ⓒ Ⓓ Ⓔ	33. Ⓐ Ⓑ Ⓒ Ⓓ Ⓔ	58. Ⓐ Ⓑ Ⓒ Ⓓ Ⓔ
9. Ⓐ Ⓑ Ⓒ Ⓓ Ⓔ	34. Ⓐ Ⓑ Ⓒ Ⓓ Ⓔ	59. Ⓐ Ⓑ Ⓒ Ⓓ Ⓔ
10. Ⓐ Ⓑ Ⓒ Ⓓ Ⓔ	35. Ⓐ Ⓑ Ⓒ Ⓓ Ⓔ	60. Ⓐ Ⓑ Ⓒ Ⓓ Ⓔ
11. Ⓐ Ⓑ Ⓒ Ⓓ Ⓔ	36. Ⓐ Ⓑ Ⓒ Ⓓ Ⓔ	61. Ⓐ Ⓑ Ⓒ Ⓓ Ⓔ
12. Ⓐ Ⓑ Ⓒ Ⓓ Ⓔ	37. Ⓐ Ⓑ Ⓒ Ⓓ Ⓔ	62. Ⓐ Ⓑ Ⓒ Ⓓ Ⓔ
13. Ⓐ Ⓑ Ⓒ Ⓓ Ⓔ	38. Ⓐ Ⓑ Ⓒ Ⓓ Ⓔ	63. Ⓐ Ⓑ Ⓒ Ⓓ Ⓔ
14. Ⓐ Ⓑ Ⓒ Ⓓ Ⓔ	39. Ⓐ Ⓑ Ⓒ Ⓓ Ⓔ	64. Ⓐ Ⓑ Ⓒ Ⓓ Ⓔ
15. Ⓐ Ⓑ Ⓒ Ⓓ Ⓔ	40. Ⓐ Ⓑ Ⓒ Ⓓ Ⓔ	65. Ⓐ Ⓑ Ⓒ Ⓓ Ⓔ
16. Ⓐ Ⓑ Ⓒ Ⓓ Ⓔ	41. Ⓐ Ⓑ Ⓒ Ⓓ Ⓔ	66. Ⓐ Ⓑ Ⓒ Ⓓ Ⓔ
17. Ⓐ Ⓑ Ⓒ Ⓓ Ⓔ	42. Ⓐ Ⓑ Ⓒ Ⓓ Ⓔ	67. Ⓐ Ⓑ Ⓒ Ⓓ Ⓔ
18. Ⓐ Ⓑ Ⓒ Ⓓ Ⓔ	43. Ⓐ Ⓑ Ⓒ Ⓓ Ⓔ	68. Ⓐ Ⓑ Ⓒ Ⓓ Ⓔ
19. Ⓐ Ⓑ Ⓒ Ⓓ Ⓔ	44. Ⓐ Ⓑ Ⓒ Ⓓ Ⓔ	69. Ⓐ Ⓑ Ⓒ Ⓓ Ⓔ
20. Ⓐ Ⓑ Ⓒ Ⓓ Ⓔ	45. Ⓐ Ⓑ Ⓒ Ⓓ Ⓔ	70. Ⓐ Ⓑ Ⓒ Ⓓ Ⓔ
21. Ⓐ Ⓑ Ⓒ Ⓓ Ⓔ	46. Ⓐ Ⓑ Ⓒ Ⓓ Ⓔ	71. Ⓐ Ⓑ Ⓒ Ⓓ Ⓔ
22. Ⓐ Ⓑ Ⓒ Ⓓ Ⓔ	47. Ⓐ Ⓑ Ⓒ Ⓓ Ⓔ	72. Ⓐ Ⓑ Ⓒ Ⓓ Ⓔ
23. Ⓐ Ⓑ Ⓒ Ⓓ Ⓔ	48. Ⓐ Ⓑ Ⓒ Ⓓ Ⓔ	73. Ⓐ Ⓑ Ⓒ Ⓓ Ⓔ
24. Ⓐ Ⓑ Ⓒ Ⓓ Ⓔ	49. Ⓐ Ⓑ Ⓒ Ⓓ Ⓔ	74. Ⓐ Ⓑ Ⓒ Ⓓ Ⓔ
25. Ⓐ Ⓑ Ⓒ Ⓓ Ⓔ	50. Ⓐ Ⓑ Ⓒ Ⓓ Ⓔ	75. Ⓐ Ⓑ Ⓒ Ⓓ Ⓔ

Multiple-Choice Questions (90 Minutes)

CALCULATORS MAY NOT BE USED

Questions 1 through 3 refer to the following compounds:

(A) PI_3

(B) NO_2

(C) $Fe_2(SO_4)_3$

(D) MnO_2

(E) Li_2SO_4

---

1. Contains metal with oxidation number +1

2. Contains metal with oxidation number +3

3. Contains metal with oxidation number +4

Questions 4 through 8 refer to the following compounds, which may be used only once:

(A) C_2H_6

(B) C_2H_2

(C) $C_2H_5NH_2$

(D) C_2H_4

(E) C_2H_4O

---

4. Alkene

5. Aldehyde

6. Alkane

7. Amine

8. Alkyne

Questions 9 through 11 refer to the following types of bonding:

(A) Ionic bonding

(B) Covalent bonding

(C) Metallic bonding

(D) Sigma bonding

(E) Pi bonding

---

9. Characterized by orbital overlap that is symmetrical to a plane

10. Is the most polar form of bonding

11. Characterized by highly mobile electrons

Questions 12 through 14 refer to the following gases:

(A) HCl

(B) O_2

(C) NO

(D) NO_2

(E) CO

---

12. Diffuses at the slowest rate

13. Has the most weakly interacting molecules

14. At any given temperature and pressure, has particles with the highest average velocity

15. What volume is closest to that occupied by 56.0g of carbon monoxide gas at standard temperature and pressure?

(A) 2.00L

(B) 11.2L

(C) 34.0L

(D) 44.8L

(E) 68.0L

16. What is the conjugate base of ascorbic acid in the following reaction?

$C_5H_7O_4COOH + OH^- \leftrightarrow H_2O + C_5H_7O_4COO^-$

(A) $C_5H_7O_4COOH$

(B) OH^-

(C) H_2O

(D) $C_5H_7O_4COO^-$

(E) $C_5H_7O_4COOH^-$

Go on to next page

17. The following data list the results for three trials in which compounds W and X reacted to produce compound Y. Given these data, what is the overall reaction order?

Trial	[W] / M	[X] / M	Initial rate, $d[Y]/dt$ / M s-1
1	0.25	0.50	0.20
2	0.25	1.00	0.40
3	0.50	1.00	1.60

(A) 2

(B) 3

(C) 4

(D) 7

(E) 12

18. An aqueous solution of which of the following compounds is the best conductor of electricity?

(A) Glucose, $C_6H_{12}O_6$

(B) Ethanol, C_2H_6O

(C) Sodium hydroxide, NaOH

(D) Carbon monoxide, CO

(E) Carbon dioxide, CO_2

19. Which species does **NOT** have the following electron configuration?

$1s^2 2s^2 2p^6 3s^2 3p^6 3d^{10} 4s^2 4p^6$

(A) Rb^{+1}

(B) Sr^{+2}

(C) Kr

(D) Br^{-1}

(E) Se

20. A buffer solution of acetic acid and sodium acetate is prepared at pH 5.76. The pKa of acetic acid is 4.76. What is the ratio of acetate to acetic acid (acetate:acetic acid) in the buffer solution?

(A) 1:1

(B) 1:2

(C) 2:1

(D) 1:10

(E) 10:1

21. Consider the following balanced reaction equation:

$Cu + 2HCl \rightarrow CuCl_2 + H_2$

How does the oxidation number of copper change?

(A) 0 to +1

(B) 0 to +2

(C) +1 to –1

(D) +2 to –2

(E) –2 to 0

22. Which of the following is the formula for iron (III) sulfate?

(A) FeS

(B) Fe_2S_3

(C) $FeSO_4$

(D) $Fe_3(SO_4)_2$

(E) $Fe_2(SO_4)_3$

23. What is the approximate mass of nitrogen gas, N_2, that occupies a 45L container at 0.50atm and 273K?

(A) 7g

(B) 14g

(C) 22g

(D) 28g

(E) 56g

24. Which of the following is the shape of a molecule whose central atom is surrounded by two atoms and two lone pairs?

(A) Linear

(B) Bent

(C) Trigonal planar

(D) Trigonal bipyramidal

(E) Tetrahedral

Go on to next page

25. In the following figure of an electrochemical cell, what best represents events occurring at electrode #1 (in Figure 30-1)?

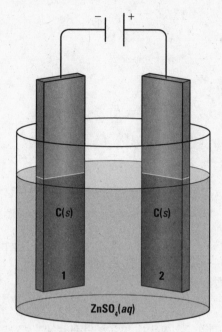

Figure 30-1:
Electro-
chemical
cell.

(A) Electrons travel out of electrode toward voltage source.

(B) Electrons travel out of electrode toward aqueous sulfate ions.

(C) Zinc metal electroplates onto electrode.

(D) Zinc metal donates electrons to electrode.

(E) Oxygen gas bubbles from electrode.

26. What volume of 0.25 H_3PO_4 neutralizes 500mL 1.5M KOH?

(A) 600mL

(B) 900mL

(C) 1L

(D) 3L

(E) 6L

27. Prepared at 0.5 molar concentration in water, which of the following solutions has the highest boiling point?

(A) Propanol

(B) Propanoic acid

(C) Iron (II) nitrate

(D) Iron (III) nitrate

(E) Nitric acid

Go on to next page

28. Which of the following statements about the energy diagram shown in Figure 30-2 is incorrect?

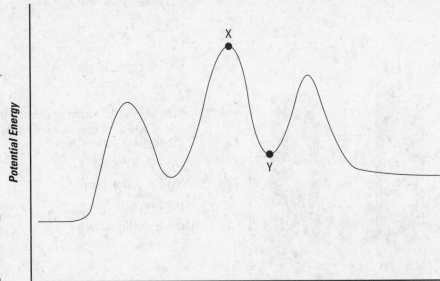

Figure 30-2:
Energy
diagram.

Reaction Progress

(A) The overall reaction is endothermic.

(B) The reaction has two intermediates.

(C) Point X represents a transition state.

(D) Raising the energy of point Y decreases the equilibrium concentration of product.

(E) Raising the energy of point X decreases the rate of product formation.

29. In the following reaction, which species are the Lewis acid and Lewis base, respectively?

$BF_3 + N(CH_3)_3 \rightarrow BF_3N(CH_3)_3$

(A) BF_3, $BF_3N(CH_3)_3$

(B) BF_3, $N(CH_3)_3$

(C) $BF_3N(CH_3)_3$, BF_3

(D) $BF_3N(CH_3)_3$, $N(CH_3)_3$

(E) $N(CH_3)_3$, BF_3

30. The free energy change for a reaction at 273K and 1atm is –35.7 kJ. If the equilibrium constant for the reverse of that reaction is e^x, then what is x?

(A) $-35700 / (8.314 \times 273)$

(B) $35700 / (8.314 \times 273)$

(C) $(8.314 \times 273) / -35700$

(D) $(8.314 \times 273) / 35700$

(E) $(\ln 35700) / (8.314 \times 273)$

31. If the ratio of the rate of effusion of an unknown gas to the rate of effusion for helium gas, He(g), is 0.35, what is the approximate molar mass of the unknown gas?

(A) 2.9g/mol

(B) 11g/mol

(C) 33g/mol

(D) 1.3×10^2g/mol

(E) 1.4×10^2g/mol

Go on to next page

32. Which of the following compounds is least soluble in water?

    (A) Sodium propanoate

    (B) Propanoic acid

    (C) Propanediol

    (D) Propanone

    (E) Propene

33. What is the EMF of the voltaic cell driven by the following reaction?

    $Zn(s) + Cu(NO_3)_2(aq) \rightarrow Zn(NO_3)_2(aq) + Cu(s)$

    The following data may help you arrive at your answer:

    $Cu^{2+}(aq) + 2e^- \rightarrow Cu(s)$ $\quad$ $E^0_{red} = 0.34V$

    $NO_3^-(aq) + 2H^+(aq) + e^- \rightarrow NO_2(g) + H_2O(l)$ $E^0_{red} = 0.78V$

    $Zn^{2+}(aq) + 2e^- \rightarrow Zn(s)$ $\quad$ $E^0_{red} = -0.76V$

    (A) −0.42V

    (B) 1.10V

    (C) 0.36V

    (D) −1.20V

    (E) 0.44V

34. Which of the following indicators would be the best choice to monitor a change that occurs at pH = 5.0?

    (A) Bromophenol blue, $pK_a = 4.0$

    (B) Phenolphthalein, $pK_a = 7.9$

    (C) Thymol blue, $pK_a = 9.3$

    (D) Methyl red, $pK_a = 5.1$

    (E) Methyl orange, $pK_a = 3.7$

35. Which element can have a mass number of 40 and 21 neutrons?

    (A) Calcium

    (B) Potassium

    (C) Argon

    (D) Neon

    (E) Promethium

36. The net ionic equation for the reaction of lead (II) nitrate with potassium iodide is

    (A) $KI(aq) + Pb(NO_3)_2(aq) \rightarrow KNO_3(aq) + PbI_2(aq)$.

    (B) $2KI(aq) + Pb(NO_3)_2(aq) \rightarrow 2KNO_3(aq) + PbI_2(aq)$.

    (C) $2KI(aq) + Pb(NO_3)_2(aq) \rightarrow 2KNO_3(aq) + PbI_2(s)$.

    (D) $2I^-(aq) + Pb^{2+}(aq) \rightarrow PbI_2(s)$.

    (E) $2K^+(aq) + 2I^-(aq) + Pb^{2+}(aq) + 2NO_3^-(aq) \rightarrow 2K^+(aq) + 2NO_3^-(aq) + PbI_2(s)$.

37. A sulfide of copper is found to contain 20% sulfur. What is the formula of the compound?

    (A) CuS

    (B) CuS_2

    (C) Cu_2S

    (D) Cu_2S_2

    (E) Cu_4S

38. At standard temperature and pressure, how many liters of H_2S are produced by the complete reaction of 45.4g of nickel (II) sulfide?

    $NiS(s) + 2HCl(aq) \rightarrow NiCl_2(aq) + H_2S(g)$

    (A) 0.500L

    (B) 11.2L

    (C) 41.0L

    (D) 49.0L

    (E) 91.0L

39. Which metal has the lowest ionization energy?

    (A) Lithium

    (B) Sodium

    (C) Calcium

    (D) Strontium

    (E) Cesium

Go on to next page

40. Which set of coefficients properly balances the equation for the combustion of ethane?

    __ethane(g) + __oxygen(g) → __carbon dioxide(g) + __water(g)

    (A) 1, 7, 2, 3

    (B) 1, 2, 1, 2

    (C) 2, 7, 4, 6

    (D) 1, 4, 1, 2

    (E) 1, 7, 4, 6

41. Which linear compound may exhibit cis-trans isomerism?

    (A) Butane

    (B) 1-butyne

    (C) 1-butene

    (D) 2-butene

    (E) 2-butyne

42. You dilute 100.mL of a stock NaCl solution by adding 400.mL water, creating a 0.200M NaCl working solution. What is the concentration of the stock solution?

    (A) 10.0M

    (B) 4.00M

    (C) 1.00M

    (D) 0.800M

    (E) 0.400M

43. At 298K, the K_{sp} for $CuCO_3$ is 2.5×10^{-10} and the K_{sp} for $BaCrO_4$ is 2.0×10^{-10}. Which of the following statements is true?

    (A) In saturated $CuCO_3$, the molar concentration of Cu^{2+} is 2.5×10^{-5}.

    (B) In saturated $BaCrO_4$, the molar concentration of Ba^{2+} is 2.0×10^{-10}.

    (C) In saturated $CuCO_3$, addition of Cu^{2+} increases $CuCO_3$ solubility.

    (D) $BaCrO_4$ is more soluble in water than $CuCO_3$.

    (E) In saturated $BaCrO_4$, the molar concentration of Ba^{2+} is 1.4×10^{-5}.

44. Which of the following compounds is possible?

    (I) $Fe_2(SO_4)_3$

    (II) $FeSO_4$

    (III) $Fe(OH)_3$

    (A) I only

    (B) II only

    (C) III only

    (D) I and III only

    (E) I, II, and III

45. 48g methane reacts with fluorine gas to produce 48g 1-fluoroethane and hydrogen gas. What is the percent yield? Percent yield = 100% × (actual yield / theoretical yield)

    (A) 33%

    (B) 66%

    (C) 50%

    (D) 100%

    (E) 300%

46. In the compound 2,3-dichloropropene-1-ol what are the hybridizations of carbons 1, 2, and 3, respectively?

    (A) sp, sp^2, sp^3

    (B) sp^3, sp^2, sp

    (C) sp^2, sp^2, sp^3

    (D) sp^2, sp^3, sp^3

    (E) sp, sp, sp^2

47. Based on the standard reduction potentials listed below, which is the strongest oxidizing agent?

    $Ni^{2+}(aq) + 2e^- \rightarrow Ni(s)$ $E^{\circ}_{red} = -0.23$

    $Fe^{2+}(aq) + 2e^- \rightarrow Fe(s)$ $E^{\circ}_{red} = -0.44$

    $Zn^{2+}(aq) + 2e^- \rightarrow Zn(s)$ $E^{\circ}_{red} = -0.76$

    $Sn^{2+}(aq) + 2e^- \rightarrow Sn(s)$ $E^{\circ}_{red} = +0.15$

    $2H^+(aq) + 2e^- \rightarrow H_2(g)$ $E^{\circ}_{red} = 0$

    (A) Ni^{2+}

    (B) Fe^{2+}

    (C) Zn^{2+}

    (D) Sn^{2+}

    (E) H^+

Go on to next page

48. The following reaction achieves equilibrium at 400K:

    $4H_2(g) + 2NO(g) \leftrightarrow 2H_2O(g) + N_2(g)$

    At constant temperature, increasing the volume of the reaction vessel has what effect?

    (A) Reaction rates increase.

    (B) Water condenses into liquid.

    (C) More mass shifts into the form of hydrogen gas.

    (D) Less mass shifts into the form of nitrogen monoxide.

    (E) There is no effect.

49. 15g of an unreactive crystalline compound are massed and set aside for later use. When the same sample is massed once more before dissolving into a solution, their new mass is 15.9g. What is the most likely explanation?

    (A) The compound degrades.

    (B) The compound reacts with itself.

    (C) The compound is amphoteric.

    (D) The compound is hygroscopic.

    (E) The compound is hydrophobic.

50. At 200.K, a certain reaction is spontaneous, having a free energy change of $-500.$ J mol^{-1} and an enthalpy change of -1.50×10^3 J mol^{-1}. At what temperature does the reaction become nonspontaneous?

    (A) 197K

    (B) 203K

    (C) 300K

    (D) 500K

    (E) The reaction is spontaneous at all temperatures.

51. The pK_a of a certain acid is 4.00. What is the K_b of that substance?

    (A) 1×10^{-10}

    (B) 1×10^{-4}

    (C) 1×10^{-3}

    (D) 1×10^3

    (E) 1×10^{10}

52. An electromagnetic wave travels through air with a wavelength of 300. nm. What is the frequency of the wave?

    (A) 6.00 s^{-1}

    (B) 9.00 s^{-1}

    (C) 9.00×10^{10} s^{-1}

    (D) 1.00×10^{-15} s^{-1}

    (E) 1.00×10^{15} s^{-1}

53. What amount of water is in a 2.5 molal NaCl solution made with 292.3g NaCl?

    (A) 8.6kg

    (B) 8.6kg

    (C) 5.0L

    (D) 2.0kg

    (E) 2.0L

Go on to next page

54. Moving from left to right, which of the following describes events in segments 2 and 4, respectively, of the heating curve shown in Figure 30-3?

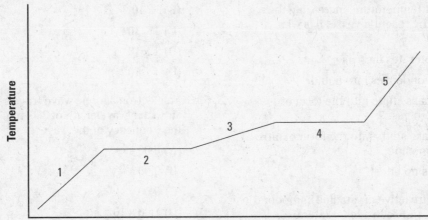

Figure 30-3:
Heating
curve.

(A) Kinetic energy remains constant, gas expands.

(B) Kinetic energy increases, gas expands.

(C) Kinetic energy increases, kinetic energy remains constant.

(D) Solid melts, gas expands.

(E) Kinetic energy remains constant in both segments.

55. Which of the following statements is false?

(A) In a closed system, energy can flow from system to surroundings.

(B) In a closed system, the amount of internal energy can change.

(C) The total difference in energy between system and surroundings is constant.

(D) In an adiabatic system, energy does not flow between system and surroundings.

(E) The energy of the universe minus the energy of the system can change.

56. Which of the following processes cannot cause solid water to become liquid?

(A) Adding energy

(B) Decreasing molecular motion

(C) Increasing pressure

(D) Decreasing pressure

(E) Decreasing potential energy

57. Which of the following is most likely to cause precipitation in a solution of $Ca(OH)_2$? The molar solubility of $MgCl_2$ is $5.7 mol L^{-1}$. The molar solubility of $Ca(OH)_2$ is $0.025 mol L^{-1}$.

(A) Add heat

(B) Add $MgCl_2$

(C) Add 2% $CH_3OH(aq)$

(D) Add base

(E) Add acid

58. The conjugate base of a weak acid is which of the following?

(A) A weaker base

(B) A stronger base

(C) A stronger acid

(D) Neutral

(E) A salt

59. At 300.K, a 100.L sample of He(g) at 2.00atm pressure consists of hown many moles?

(A) $6.00 / R$

(B) $1.50 \times R$

(C) $0.667 \times R$

(D) $0.667 / R$

(E) $1.50 / R$

Go on to next page

60. Which of the following properties of water will be affected by dissolved NaCl?

    (I)   Boiling point

    (II) Vapor pressure

    (III) Osmotic potential

    (A) I only

    (B) II only

    (C) I and II only

    (D) II and III only

    (E) I, II and III

61. For how long must a current of 2.00 amperes flow in order to transport 2.00 moles of charge?

    (A) 9.65 s

    (B) 9.65×10^4 s

    (C) 38.6 s

    (D) 38.6×10^4 s

    (E) Not enough information is given.

62. When mixed with 100.mL of 0.100M H_2SO_4, which of the following produces a solution with pH $\approx$ 7?

    (A) 50.0mL 0.2M NaOH

    (B) 100.mL 0.2M $Mg(OH)_2$

    (C) 50.0mL 0.4M CH_3OH

    (D) 50.0mL 0.4M KOH

    (E) 50.0mL 0.4M Na_2SO_4

63. The K_{eq} for the following reaction is 0.01:

    A + B $\leftrightarrow$ 2C

    If the concentrations of A and C are 5.0M and 1.0M, respectively, what must be the approximate concentration of B in an equilibrium mixture?

    (A) 5.0M

    (B) 4.0M

    (C) 1.0M

    (D) 20.M

    (E) 0.050M

64. Which of the following manipulations represents good laboratory practice?

    (I)   Adding 10mL 2M HCl to 1L water

    (II) Adding 1L water to 10mL 2M HCl

    (III) Adding 10mL 10M HCl to 10mL 10M NaOH

    (A) I only

    (B) II only

    (C) III only

    (D) I and II only

    (E) 1, II, and III

65. The thermodynamic critical point of a CO_2 is 304K and 73atm. At 310K and 80atm, what is the phase of CO_2?

    (A) Solid

    (B) Liquid

    (C) Gas

    (D) Mixture of solid, liquid, and gas

    (E) Supercritical fluid

66. A 1.0L container holding He(g) at 2.0atm pressure is joined to a 2.0L container holding CO_2(g) at 3.0atm pressure. Assuming constant temperature, what will the mole fraction of helium be when the gases equilibrate in the new volume?

    (A) 0.10

    (B) 0.25

    (C) 0.33

    (D) 0.67

    (E) 0.75

67. Solid zinc and silver nitrate undergo a single displacement reaction. If 4.00 moles of solid silver is formed, what mass of solid zinc was consumed?

    (A) 32.7g

    (B) 65.4g

    (C) 108g

    (D) 131g

    (E) 216g

Go on to next page

68. After resolving the components of a sample by using thin layer chromatography (TLC), a compound is visible on the TLC plate at retention factor (R_f) value 0.36. The solvent front had mobility of 8.4cm. What was the mobility of the compound?

    (A) 4.3×10^{-2} cm

    (B) 4.3 cm

    (C) 3.0 cm

    (D) 5.4 cm

    (E) 8.0 cm

69. Which molecule is most polar?

    (A) Carbon dioxide

    (B) Carbon tetrachloride

    (C) Ammonia

    (D) Ammonium cation

    (E) Sulfur hexafluoride

70. 400.mL of 0.025M NaOH are added to 100.mL of a strong monoprotic acid solution. The pH of the final solution is 7.0. What was the pOH of the 100.mL solution?

    (A) –1.00

    (B) 10.0^{-1}

    (C) 1.00

    (D) 10.0^{-13}

    (E) 13.00

71. What are the oxidation numbers of chromium in chromate and dichromate anions, respectively?

    (A) +8, +14

    (B) +8, +7

    (C) +7, +7

    (D) +6, +6

    (E) +4, +7

72. The pressure of a gaseous recation container is increased. The container holds hydrogen gas, nitrogen gas, and ammonia gas, which all participate in the reaction shown below. What are the possible effects?

    $$3H_2(g) + N_2(g) \leftrightarrow 2NH_3(g)$$

    (I) Reaction rates increase.

    (II) Reaction shifts toward products.

    (III) Reaction shifts toward reactants.

    (A) I only

    (B) II only

    (C) III only

    (D) I and II only

    (E) I, II, and III

73. Under equivalent conditions of pressure, temperature, and volume, what is the most likely ranking of the number of moles present in samples of methane(g), helium(g), and water(g)?

    (A) Helium > methane > water

    (B) Water > methane > helium

    (C) Helium > water > methane

    (D) Methane > water > helium

    (E) Water > helium > methane

74. Which of the following is an alkaline earth metal?

    (A) Mass number 88, atomic number 38

    (B) Mass number 86, atomic number 37

    (C) Mass number 88, atomic number 39

    (D) Mass number 84, atomic number 36

    (E) Mass number 137, atomic number 57

75. If uranium-238 emits an alpha particle, what will be the product?

    (A) Neptunium-234

    (B) Neptunium-238

    (C) Uranium-236

    (D) Thorium-234

    (E) Radium-236

STOP DO NOT TURN THE PAGE UNTIL TOLD TO DO SO.
DO NOT RETURN TO A PREVIOUS TEST.

Free-Response Questions

Part A (55 minutes)

YOU MAY USE YOUR CALCULATOR

CLEARLY SHOW ALL STEPS YOU TAKE TO ARRIVE AT YOUR ANSWER. It is to your advantage to do this, since you will receive partial credit for partially correct responses. Make sure to pay attention to significant figures.

> **Answer questions 1, 2, and 3. The score weighting for each question is 20 percent.**

1. Formic acid is a significant component of bee venom. Also known as methanoic acid, formic acid has an acid dissociation constant, K_a, of 1.80×10^{-4}.

    (a) Calculate the pOH of a $0.25M$ solution of formic acid.

    (b) Calculate the percent dissociation of the solution in part (a).

    (c) Calculate the pH of a solution prepared by mixing equal 1.00L volumes of $0.25M$ formic acid and $0.20M$ sodium methanoate.

    (d) Using only compounds already mentioned, what should be added to the solution in part (c) to produce a solution with maximum capacity to resist change in pH? Mention

       (i) The compound to be added.

       (ii) The mass of the compound to be added

2. A chemist dissolves 51.2g of an unknown, nonelectrolyte compound into 750.g of water. The freezing point depression constant (K_f) for water is 1.86.

    (a) Attempting to characterize the unknown, the chemist observes a 1.41 °C difference between the freezing point of the solution and that of pure water. What is the melting point of the solution?

    (b) Calculate the molal concentration of the unknown compound in the solution.

    (c) Calculate the molar mass of the unknown compound.

    (d) A sample of the solid compound is subjected to analysis for percent composition, yielding the following results: 40.0% carbon, 6.70% hydrogen, and 53.3% oxygen. What is the molecular formula of the unknown compound?

    (e) What is one possible Lewis structure for the unknown compound? Write the structure in the space below.

Go on to next page

3. Electrical current is passed through a $1.0M$ solution of HCl(aq) by means of two nonreactive electrodes immersed into the solution, with the electrodes connected to opposing terminals of a voltage source.

   (a) Sketch and label the diagram of the electrolytic cell, including the direction of electron flow, the half-reaction occurring at each electrode, labels for anode and cathode, and the voltage source.

   (b) At 4.5 amperes, how many Coulombs pass through the cell in 30. minutes?

   (c) How many moles of electrons pass through the cell in the period described in part (b)?

   (d) What number of moles of gas bubble pass from each electrode during the time described in part (b)?

   (e) Calculate the volume of gas (at standard temperature and pressure) that would bubble from the cell under a current of 3.0 amperes for 60 minutes.

   (f) What happens to the pH of the solution as current passes through it within the cell?

Part B (40 minutes)

CALCULATORS MAY NOT BE USED

> Answer question 4 below. The score weighting for this question is 10 percent.

> Answer questions 5 and 6 below. The score weighting for each question is 15 percent.

4. For each of the following three reactions, in part (i) write a balanced equation for the reaction and in part (ii) answer the question about the reaction. In part (i), coefficients should be in terms of lowest whole numbers. Assume that solutions are aqueous unless otherwise indicated. Represent substances in solution as ions if the substances are extensively ionized. Omit formulas for any ions or molecules that are unchanged by the reaction.

   (a) Potassium hydroxide is added to a solution of iron (III) sulfate.

   　(i) Write the balanced equation.

   　(ii) What would you observe about the reactants and/or products during this reaction?

   (b) Current is passed through molten nickel chloride.

   　(i) Write the balanced equation.

   　(ii) What occurs at the anode and cathode, respectively?

   (c) Magnesium oxide is added to water.

   　(i) Write the balanced equation.

   　(ii) What happens to the pH of the resulting solution as the reaction proceeds?

5. Give plausible scientific explanations for each of the following observations.

   (a) Oxygen concentrations in deep waters are sometimes higher than those in shallower waters.

   (b) Dissolving potassium hydroxide in water heats the solution while dissolving ammonium nitrate in water cools the solution.

   (c) $Cu(s)$ conducts electricity well but $CuCl_2(s)$ conducts electricity poorly.

6. A set of three vials contains three different organic compounds. Each compound contains only one kind of functional group, and each functional group is different from the others. None of the compounds has an ester or amide linkage, and none is an alkene or alkyne.

   (a) All of the compounds possess a carbonyl group. What kinds of compounds are these three?

   (b) Assuming that each of the three compounds contains four carbon atoms, is and is linear (not branched), draw Lewis structures for the three compounds.

   (c) Ethanol is added to each of the three vials. With which of the three compounds is ethanol most likely to react to produce an ester?

   (d) Draw the Lewis structure of the ester that would be produced in the reaction described in part (c).

Answers to Practice Test Two

Multiple choice

1. (E). As a group IA metal (that is, an alkali metal), lithium exclusively forms a +1 cation, the better to achieve a full valence shell.

2. (C). Although phosphorous has an oxidation number of +3 in PI_3, phosphorous isn't a metal. In iron (III) sulfate, the iron metal ions each have +3 charge, offsetting the $(3)(-2) = -6$ charge brought by the sulfate anions.

3. (D). Although nitrogen has a +4 oxidation number in NO_2, nitrogen isn't a metal. In manganese (IV) oxide, the manganese metal ion has +4 charge, offsetting the $(2)(-2) = -4$ charge brought by the oxygens.

4. (D). As 2-carbon hydrocarbon (too short for branching) with the general formula C_nH_{2n}, C_2H_4 can only be an alkene, $H_2C = CH_2$.

5. (E). As the only listed hydrocarbon with an oxygen-containing functional group, C_2H_4O must be an aldehyde. In this case, C_2H_4O has the structure H_3C-CHO.

6. (A). As 2-carbon hydrocarbon with the general formula C_nH_{2n+2}, C_2H_6 can only be an alkane, H_3C–CH_3.

7. (C). As the only listed hydrocarbon with an amine-containing functional group, $C_2H_5NH_2$ must be the amine. In this case, the compound has the structure H_3C-$CHNH_2$.

8. (B). As a 2-carbon hydrocarbon with the general formula C_nH_{2n-2}, C_2H_2 can only be an alkyne, HC-CH.

9. (E). Although sigma bonding is completely symmetrical around the axis of a sovalent bond, pi bonding is symmetrical about only a single plane through that axis, with overlapping p orbitals above and below the plane.

10. (A). In pure ionic bonding, electrons are entirely transferred from one atom or group to another. Because the separation of charge is so complete, the bond is extremely polar.

11. (C). Metallic bonding is characterized by a lattice of positively charged metal atom nuclei, around and through which migrates a "sea" of mobile electrons.

12. (D). Because the molar mass of NO_2 is the greatest among the compounds listed, its rate of diffusion is the smallest.

13. (B). Because the two atoms of molecular oxygen are identical, the bonds that connect them are entirely nonpolar. This apolarity results in very weak interactions between O_2 molecules.

14. (E). As the compound with the lowest molar mass among those listed, the average kinetic energy associated with a given temperature and pressure derives more from velocity in the case of CO than is the case with the other gases. Because kinetic energy — equal between gases at the same temperature — equals $(1/2)(mass)(velocity)^2$, the kinetic energy of the other gases derives more from mass.

15. (D). Because the molar mass of carbon monoxide is 28.0g/mol, the 56.0g described represents 2 moles CO. At STP, 1 mole of (ideal) gas occupies 22.4L, so 2 moles occupy 44.8L.

16. (D). Typically, conjugate acid/conjugate base pairs include a common molecular "chunk," differing only in the presence or absence of an H^+. In the given reaction, for example, H_2O is the conjugate acid of OH^-, a conjugate base. However, the conjugate base of ascorbic acid, $C_5H_7O_4COOH$, is ascorbate, $C_5H_7O_4COO^-$.

17. (B). Doubling the concentration of X doubles the reaction rate, suggesting the reaction is first-order in X. Doubling the concentration of W further quadruples the reaction rate, suggesting the reaction is second-order in W. These observations suggest the rate law, Rate = $k[W]^2[X]$. The overall reaction order is the sum of the individual exponents, $2 + 1 = 3$.

18. (C). Dissociated ions are necessary to carry current through a solution. Although many of the listed compounds are soluble in water, and a few dissociate into ions, only NaOH strongly dissociates into ions when dissolved in water.

19. (E). The given electron configuration is most obviously that of the noble gas, krypton. All the listed ions either shed or gain electrons to achieve the same stable, filled valence shell as krypton. Selenium, though it might gain two electrons to achieve the given configuration, is shown in its neutral, elemental form.

20. (E). Use the Henderson Hasselbach equation, pH = pK_a + log([A^-]/[HA]). The pK_a of 4.76 is 1 pH unit lower than the desired pH of 5.76. So, the term log([A^-]/[HA]) needs to equal +1. This, in turn, means that the ratio of A^- to HA must equal 10:1.

21. (B). On the reactant side, copper is shown in elemental form, so its oxidation number is 0. On the product side, copper participates in the compound $CuCl_2$. Because the oxidation number of chlorine within compounds is –1, and because the compound has two chloride anions, you know that the oxidation number of copper within the compound must be +2 to offset the negative charge.

22. (E). The Roman numeral within parentheses alerts you to the fact that iron in this compound has +3 charge. Next, recognize the sulfate anion, SO_4^{2-}. Finally, find the right numerical combination of iron cations and sulfate anions to result in a neutral ionic compound: $Fe_2(SO_4)_3$.

23. (D). Before finding the mass of nitrogen gas, you must find the number of moles. To do so, first recognize that at 273K, one mole of an ideal gas (like N_2) occupies 22.4L at 1atm. But the given pressure is half that — so one mole of gas occupies *twice* that volume, which is $2 \times 22.4L = 44.8L \approx 45L$. So, you've got about one mole of N_2 on your hands. The molar mass of N_2 is 28.0g.

24. (B). The four electron pairs represented by the two bound atoms and two lone pairs suggest a tetrahedral geometry. Because only two of the four orbital axes include actual bonds to actual atoms, the overall molecular shape resulting from the tetrahedral orbital geometry is "bent," just like water.

25. (C). The signs on the terminals of the voltage source clue you into the direction of electron flow within this electrolytic cell. Electrons flow away from the negative terminal, down into electrode 1, through the electrolyte solution and up through electrode 2 toward the positive terminal. So, at electrode 1, Zn^{2+} cations are reduced by the flow of electrons, electroplating solid zinc onto the electrode surface.

26. (C). The key idea in a neutralization reaction is that the total moles of acid must equal the total moles of base. So, you must provide enough acid to counteract (0.500L)(1.5M KOH) = 0.75mol KOH, where each KOH contributes a single OH^-. Because each mole of H_3PO_4 contributes three equivalents of acid, 0.25M H_3PO_4 effectively acts as 0.75M acid. Adding 1000mL (that is, 1.00L) of H_3PO_4 therefore neutralizes the base.

27. (D). This is a colligative properties question, requiring you to know that the more moles of solute you dissolve into a given mass of water, the more you'll elevate the boiling point. Because iron (III) nitrate, $Fe(NO_3)_3$, dissociates into four ionic particles, the given concentration of that solute contributes the most particles and leads to the highest boiling point.

28. (D). All the listed statements are correct except for the statement about point Y, which alters the energy level of an intermediate state. The equlibirium concentration of product depends on the energy difference between initial (reactant) and final (product) states.

29. (B). Lewis acids are electron acceptors, and Lewis bases are electron donors. The boron of BF_3 has two spots still empty in its valence shell. The nitrogen of $N(CH_3)_3$ has two available electrons in a lone pair. So, the nitrogen donates its lone pair to the boron, creating a coordinate covalent bond in a Lewis acid/base reaction.

30. (A). First, recognize that the free energy change for a reverse reaction has the same magnitude and opposite sign as the free energy change for the forward reaction. So, the free energy change for the reverse reaction is +35.7kJ = +35700 J. Second, recall the following relationship between the equilibrium constant and the free energy change:

$$K_{eq} = e^{-\Delta G/RT}$$

This means that $x = -(-\Delta G/RT) = -35700 / (8.314 \times 273)$.

31. (C). Recall Graham's law of diffusion and effusion: The ratio of the square roots of the molar masses equals the inverse of the rates of effusion or diffusion. Because the ration of the rates of effusion is 0.35, the ratio of the molar masses is $(1 / 0.35)^2 = 8.3$. Helium has a molar mass of 4.0g/mol. Solve for the molar mass of the unknown by multiplying 8.3×4.0g/mol = 33 g/mol.

32. (E). Choices A and B are both strongly polar because of the carboxylate/carboxylic acid group. Choices C and D are both moderately polar because of the alcohol and ketone groups, respectively. Choice E, an alkene, is simply nonpolar, and is therefore the least soluble in water.

33. (B). To understand the redox reaction involved here, view the reaction as a net ionic equation:

$$Zn(s) + Cu^{2+}(aq) \rightarrow Zn^{2+}(aq) + Cu(s)$$

Now you can see that the first and third listed standard reduction potentials are the relevant ones. Because copper is more easily reduced than zinc (as revealed by its more positive standard reduction potential), you know that copper reduction occurs at the cathode and zinc oxidation occurs at the anode. Finally, apply the equation for calculating a standard cell potential:

$$E^0_{cell} = E^0_{red}(cathode) - E^0_{red}(anode) = 0.34V - (-0.76V) = +1.10V$$

34. (D). Methyl red has a pK_a of 5.1, the closest to the target pH, and is therefore the most suitable.

35. (B). With a mass number of 40 and 21 neutrons, the element in question must have 19 protons, and therefore an atomic number of 19. Unambiguously, then, the element is potassium.

36. (D). Choice A is the balanced reaction equation. Choice E is the total ionic equation. But only choice D is the net ionic equation.

37. (C). Percent composition is all about mass. The only formula listed in which the mass of sulfur is 20% of the total mass of the compound is Cu_2S. Choice E is a red herring, because in that compound sulfur is 20% of the compound by atom count, not by mass.

38. (B). First, concentrate on the moles of H_2S produced by the given reaction. The molar mass of nickel (II) sulfide is 90.8g/mol, so 45.4g represents 0.500 mole of the compound. The stoichiometry of the reaction tells you that you'll therefore produce 0.500 mole of hydrogen sulfide. At standard temperature and pressure, 1 mole gas occupies 22.4L volume. One half of that quantity is 11.2L.

39. (E). Ionization energy increases overall upward and to the right within the periodic table. So, the element with the lowest ionization energy should hold the lowest, most leftward position. Of those elements listed, cesium best fits the bill.

40. (C). To answer this question, you must know the formulas for the common compounds listed, and you must be able to balance a reaction equation:

$$2C_2H_6 + 7O_2 \rightarrow 4CO_2 + 6H_2O$$

41. (D). Each compound listed shares a 4-carbon skeleton. In linear (not cyclic) compounds, geometric (cis-trans) isomerism doesn't occur about single bonds, so choice A is out. Both butyne compounds create *sp*-hybridized carbons with purely linear geometry. Double bonds are classic sites for cis-trans isomerism, but 1-butene is disqualified because one of the carbons participating in the double bond is a terminal carbon attached to two identical hydrogen atoms. So, only 2-butene, with its central double bond, offers the possibility of cis-trans isomerism.

42. (C). This problem requires you to use the dilution equation:

$$C_1 \times V_1 = C_2 \times V_2$$

Because you add 400mL to 100mL, notice that your final volume is 500mL, not 400mL.

43. (E). To answer this question, you must remember the definition of the solubility product constant. For the dissolution reaction $XY(s) \rightarrow X(aq) + Y(aq)$, $K_{sp} = [X] \times [Y]$. Each of the listed compounds undergoes simple dissociation to 2 counterions (Cu^{2+} and CO_3^{2-} on the one hand, Ba^{2+} and CrO_4^{2-} on the other hand), $[X] = [Y]$ in each case. So, the concentration of either ion in a saturated solution is simply the square root of the K_{sp}. This fact rules out choices A and B. The common ion effect rules out choice C. The given K_{sp} values reveal that choice D is wrong. Thus, by process of elimination and because the math works out properly, choice E is correct.

44. (E). Ionic compounds have neutral charge. Iron can take on variable charge, as Fe (II) cation or Fe (III) cation. Sulfate anion has –2 charge. Hydroxide anion has –1 charge. So, all three compounds are possible.

45. (B). To answer this question you must first generate a balanced reaction equation:

$$4CH_4 + F_2 \rightarrow 2C_2H_5Fl + 3H_2$$

If 48g of CH_4 react, then 48g / 16g / mol^{-1} = 3 moles methane react. Because only 2 moles of fluoroethane product are made for every 4 moles of ethane reactant, the theoretical yield of fluoroethane is 1.5 moles. The molar mass of fluoroethane is 48g/mol. So, the theoretical yield of fluoroethane is 1.5 mol × 48g/mol = 72g. The actual yield is 48g. Percent yield = 100% × (actual/theoretical) = 66%.

46. (C). Remember to take bonds to hydrogen into account. Carbons 1 and 2 each have three bond axes, and so are sp^2 hybridized. Carbon 3 has four bond axes, and so is sp^3 hybridized.

47. (D). Standard reduction potentials reflect ease of reduction. Tin (II) cation is most easily reduced, and therefore is mostly likely to oxidize other atoms. So, tin (II) cation is the strongest oxidizing agent listed.

48. (C). Equilibrium shifts to oppose any perturbation. At constant temperature, increasing volume decreases temperature. So, the equilibrium shifts mass in the way that increases pressure — in other words, it shifts to increase the total moles of gas. Because the reactant side has 6 moles gas to the product side's 3 moles of gas, equilibrium shifts to the left. Choice A is wrong because lower pressures decrease gas-phase reaction rates. Choice B is wrong because lower pressure favors the water vapor, not liquid water. Choice D is wrong because the equilibrium shifts to the left for the reasons already described. Choice E is just plain wrong.

49. (D). Because the compound is decribed as unreactive, choices A, B, and E are unlikely. Choice C may or may not be true, but has little relevance to the observed increase in mass. Hygroscopic compounds are those which interact so strongly with water that they absorb water vapor from the air, thereby increasing the overall mass of the sample.

50. (C). Begin this problem with the Gibbs equation: $\Delta G = \Delta H - T\Delta S$. Plugging in the given values for ΔG, ΔH and T, you can solve for ΔS, getting –5 J mol^{-1} K^{-1}. Next, recognize that the temperature at which the reaction moves from spontaneous to nonspontaneous is the one at which the free energy change is zero, so plug in "0" for ΔG, plug in the values for ΔH and ΔS, and solve for T. Doing so gives you 300K.

51. (A). Because $pK_a = -\log K_a$, it follows that $K_a = 10^{-pK_a}$. Using this equation, solve for the acid dissociation constant, $K_a = 1 \times 10^{-4}$. Next, remember that $K_a \times K_b = K_w = 1 \times 10^{-14}$. Using that fact, solve for the base dissociation constant, $K_b = 1 \times 10^{-10}$.

52. (E). For any wave, velocity = frequency × wavelength. For electromagnetic waves in air, velocity = 3.00×10^8 m/s. So, to solve for the frequency, simply divide 3.00×10^8 m/s by the wavelength, $300. \times 10^{-9}$ m. Doing so gives you 1.00×10^{15}.

53. (D). Molality = (moles solute particles) / (kilograms solvent). Start by determining the moles of NaCl present in the 292.3g sample. Using the molar mass of NaCl (58.45g/mol), this figure comes to 5.00 moles. To achieve 2.5 molality, you require 2kg water.

54. (E). Segments 2 and 4 are flat with respect to temperature — though heat is added, that heat goes into accomplishing the phase change of melting or vaporizing, not into increasing the temperature of the sample. So, within neither of these segments does kinetic energy (the underlying cause of temperature) change. Choice D is tempting because solid does melt in segment 2, but segment 4 does not include gas expansion, but rather the conversion of liquid at the boiling point into gas.

55. (C). All the statements are true except choice C. What remains constant is the *sum* of energy between system and surroundings.

56. (B). Water has some pretty anomalous properties, including regions of its phase diagram in which either increasing pressure or decreasing pressure could result in a change from solid to liquid state. However, nowhere in the phase diagram does decreasing temperature (that is, reducing the kinetic energy of molecular motions) cause a shift from solid to liquid.

57. (D). For most solid solutes, adding heat increases solubility. Adding acid increases calcium hydroxide solubility by converting dissociated hydroxide ions into water. Choices B and C could conceivably reduce solubility, but by far the best choice is D, because adding base (that is, effectively adding OH^-) reduces calcium hydroxide solubility by the common ion effect.

58. (B). Descriptive chemistry at its finest. The conjugate base of a weak acid is a stronger base. That is why the weak acid is weak.

59. (D). Use the ideal gas law for this one. Rearranged to solve for moles, n, the equation becomes $n = (PV) / (RT)$. Substituting in known values (and leaving R unsubstituted) gives you $(2.00\text{atm} \times 100.\text{L}) / (300.\text{K} \times R) = 0.667 / R$.

60. (E). Boiling point, vapor pressure, and osmotic potential are all colligative properties, properties that change as solute is added to solution..

61. (B). Current equals charge divided by time ($I = Q / t$), so to solve for time, you must divide the total charge by the current. There are 2.00 moles of charge, which means there are $2 \times F$ coulombs, or $2(9.65 \times 10^4$ coulombs). Divide this quantity by 2.00 amperes (2.00 coulombs per second) to get 9.65×10^4 seconds.

62. (D). A neutral solution (pH = 7) emerges when total moles of base and total moles of acid are equal. Each mole of sulfuric acid contributes two moles acid, so the original solution contains $(0.100\text{L})(0.100\text{mol L}^{-1})(2) = 0.0200$ moles acid. Using similar calculations on the choices, you'll find that the 50.0mL of 0.400M KOH offered by choice D does the trick.

63. (D). First, you must be able to construct the expression for the equilibrium constant of the reaction: $K_{eq} = [C]^2 / ([A] \times [B])$. Next plug in the known values for K_{eq}, [A] and [C], and solve for [B]. Doing so gives you 20.M.

64. (A). When diluting concentrated acid or concentrated base, always add the concentrated compound to a large volume of water, and not the other way around. The dilution may generate a lot of heat, and by adding to the large volume of water, that heat can be adequately dispersed.

65. (E). Temperatures and pressures above the critical point produce a supercritical fluid.

66. (B). The question is asking about a mole fraction — that of helium within the final mixture, to be precise. From the ideal gas law, you can calculate for the separate helium container that the moles of helium, nHe = (2.0atm)(1.0L) / (RT). Second, you can calculate the moles of carbon dioxide, nCO$_2$ = (3.0atm)(2.0L) / (RT). Because each question shares RT, you can set up the proportion (3.0 / nCO$_2$) = (1.0 / nHe). In other words, there are three times the moles of CO$_2$ as there are moles He. So, the moles fraction of He = 1.0 / (1.0 + 3.0) = 0.25.

67. (D). Start by writing a balanced reaction equation

$$Zn(s) + 2AgNO_3(aq) \rightarrow 2Ag(s) + Zn(NO_3)_2(aq)$$

From the equation you can see that two moles of solid silver form for every mole of solid zinc consumed. So, to produce four moles of solid silver required two moles of solid zinc. Because each mole of zince has 65.4g mass, the two mole total has mass 131g.

68. (C). To attack this problem, you need to recall that

$$R_f = (sample\ mobility) / (solvent\ mobility)$$

Substitute in the known values of R_f = 0.36 and solvent mobility = 8.4cm, and you can solve for the sample mobility, which is 3.0cm.

69. (C). Although many of the other compounds have polar covalent bonds, only ammonia has a geometric distribution of bonds and a lone pair whose polarities don't entirely cancel out.

70. (E). If the pH of the final solution is 7.0 (that is, neutral), then the moles of base added equal the moles of acid originally present in the strong acid solution. The moles of base added were (0.400L)(0.025mol/L) = 0.010mol OH$^-$. Because acid must have been equal to base for the final solution to be neutral, there must have been 0.010mol H$^+$ in the original 0.100L acid solution. In other words, [H$^+$] = 0.010mol / 0.100L = 0.10M. This means that the pH of that acid solution was pH = $-$log[0.10] = 1.00. Because pH + pOH = 14.00, you know that the pOH of the acid solution was 13.00.

71. (D). You must know the formulas of these polyatomic anions (CrO$_4^{2-}$ and Cr$_2$O$_7^{2-}$) to figure out the oxidation numbers of the metals. In each case, start with the oxygen atoms, each of which carries an oxidation number of -2. In order to offset much of this negative charge while still leaving a -2 formal charge, each chromium atom must carry on oxidation number of $+6$.

72. (D). Increased pressure can increase reaction rates as particles collide more frequently. Increasing pressure perturbs the equilibrium, which shifts to create lower pressure. In the case of the reaction shown, shifting toward products results in lower pressure because the products contain fewer moles of gas than the reactants.

73. (B). Water molecules have the greatest degree of intermolecular attraction, particularly from hydrogen bonding. Methane molecules lag far behind water in this respect, but can still experience a small intermolecular attraction due to London forces. Of the three substances listed, the atomic noble gas helium has the fewest intermolecular attractions. All other things being equal, the gas with the greatest attractive forces will have the greatest tendency to contract, so achieving any given pressure (at a given temperature) will require more molecules for that gas than for the others.

74. (A). Focus on the atomic number here — the mass numbers can only mislead you because they may represent different isotopes of different elements. Alkaline earth metals are in group IIA, and only atomic number 38, strontium, fits that bill.

75. (D). Alpha particles are helium nuclei, containing two protons and two neutrons. So, a nuclide that decays by alpha emission loses four mass number units and two atomic number units. By losing two atomic number units, the element is transformed into a different element entirely. The new mass number is 238 $-$ 4 = 234. The new atomic number is 92 $-$ 2 = 90. The new element is thorium, Th.

Free response

1. (a) pOH = 11.83. Make sure you understand the meaning of the K_a for this system by writing down the acid dissociation reaction. If the common name, formic acid, doesn't help you with the structure, the systematic name, methanoic acid, should do so.

$$CHOOH \leftrightarrow CHOO^- + H^+$$

$$K_a = ([CHOO^-] \times [H^+]) / [CHOOH] = 1.80 \times 10^{-4}$$

Making a 0.25mol/L formic acid solution begins the process of acid dissociation, in which some concentration, x, shifts from CHOOH form to equivalent concentrations, x, of each of the dissociated forms, CHOO$^-$ and H$^+$:

Initial concentration, mol/L	Change, mol/L	Final concentration, mol/L
CHCOOH, 0.25	–x	0.25–x
CHCOO$^-$, 0	+x	x
H$^+$, 0	+x	x

Substituting into the expression for K_a

$$K_a = (x^2) / (0.25 - x) = 1.80 \times 10^{-4}$$

This equation is a somewhat messy quadratic to solve under time pressure, and it is reasonable to assume that, because formic acid is a weak acid, 0.25–x ≈ 0.025. So, the equation simplifies to

$$1.80 \times 10^{-4} = (x^2) / (0.25)$$

Solving for x gives you 6.7×10^{-3} = x = [H$^+$].

$$pH = -\log[H^+] = 2.17$$

$$pH + pOH = 14.00, \text{ so } pOH = 11.83$$

(b) Percent dissociation = 2.6%. If the total concentration of formic acid and formate together is 0.25mol/L, and the concentration of formate alone is 6.7×10^{-3} mol/L, then the percent dissociation is

$$100\% \times (6.7 \times 10^{-3}) / (6.7 \times 10^{-3} + 0.25) = 2.6\%$$

(c) pH = 3.64. This is a buffer problem, for which you should use the Henderson-Hasselbach equation:

$$pH = pK_a + \log([formate]/[formic acid])$$

To calculate the pK_a: pK_a = –logK_a = 3.74

$$pH = 3.74 + \log(0.20 / 0.25) = 3.64$$

(d) (i) Sodium formate. Maximum buffering capacity occurs when the conjugate base and conjugate acid are present in equal concentrations. The initial concentrations are 0.20mol formate / 2.00L = 0.10mol/L formate; 0.25mol formic acid / 2.00L = 0.13mol/L formic acid. You must add enough solid sodium formate, CHCOONa, to result in 0.13mol/L formate, CHCOO$^-$.

(ii) 11g sodium formate

(0.10mol + x mol) / 2.00L = 0.13mol/L formate

Solving for x gives you 0.16mol formate (added as sodium formate, 68.0g/mol)

0.16mol × 68.0g/mol = 11g sodium formate

2. (a) −1.41 °C. Freezing point and melting point are the same temperature. The normal melting point of water is 0 °C, and adding solute lowers the freezing point.

(b) 0.758mol/kg

$$\Delta T_f = \textit{molality} \times K_f$$

Substituting in the known values gives you 1.41 °C = *molality* × 1.86. Solving for molality gives you 0.758mol/kg.

(c) 90.1g/mol

Molality = (moles solute particles) / (kilograms solvent)

Because the compound is a nonelectrolyte, you needn't worry that a mole of compound dissolves into multiple moles of ionic particles. So, you can substitute in the known values:

0.758mol/kg = (moles solute) / (0.750kg water)

Moles solute = 0.569 moles

Because the chemist added 51.2g of the compound, you can calculate the apparent molar mass as (mass solute added) / (moles solute) = 51.2g / 0.569 moles = 90.0g/mol.

(d) $C_3H_6O_3$. Begin by determining the empirical formula. Assume a 100.g sample of the compound. At the given percent composition, the sample will contain 40.0g carbon, 6.70g hydrogen, and 53.3g oxygen. Based on the gram atomic mass of each element, these values correspond to 3.33mol carbon, 6.67mol hydrogen, and 3.33mol oxygen. Dividing all mole values through by the smallest among them (3.33), you come to 1 mole carbon, 2 moles hydrogen, and 1 mole oxygen. These values correspond to the empirical formula CH_2O. Next, divide the apparent molar mass of the compound calculated in part (c) by the molar mass of the empirical formula: (90.0g/mol) / (30.0g/mol) = 3. The molecular formula of the compound is a multiple of 3 times the empirical formula:

$3 \times CH_2O = C_3H_6O_3$

(e) Figure 30-4 is a possible Lewis structure for the compound.

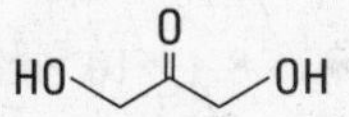

Figure 30-4: Lewis dot structure

Dihydroxyacetone, $C_3H_6O_3$

3. (a) Figure 30-5 is a labeled diagram of the electrolytic cell.

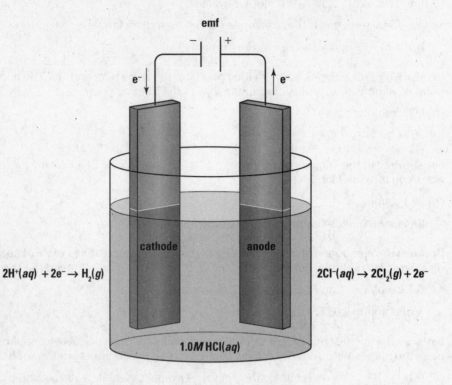

$2H^+(aq) + 2e^- \rightarrow H_2(g)$

$2Cl^-(aq) \rightarrow 2Cl_2(g) + 2e^-$

Figure 30-5:
Electrolytic
cell.

1.0M HCl(aq)

(b) 8.1×10^3 coulombs. Use the equation $Q = It$ (charge = current × time).

   Charge = (4.5 coulombs/s)(1.8×10^3 s) = 8.1×10^3 coulombs

(c) 8.4×10^{-2} mol electrons. To calculate the moles of charge, you must use Faraday's constant: 96485 C/mol e^-:

   Moles of electrons = (8.1×10^3 coulombs)(1mol / 96485 coulombs) = 8.4×10^{-2} mol

(d) 4.2×10^{-2} mol gas per electrode. Based on the half-reactions shown for the electrodes in part (a), 2 moles of electrons are required to liberate 1 mole of $H_2(g)$ and 1 mole of $Cl_2(g)$, by reduction and exidation, respectively.

   8.4×10^{-2} mol electrons × (1mol gas / 2mol electrons) = 4.2×10^{-2} mol gas

So, 4.2×10^{-2} mol gas are liberated at each electrode, for a total of 8.4×10^{-2} mol gas.

(e) 2.5L total gas. Again, start with $Q = It$, then convert to moles charge by using Faraday's constant, and then to moles gas by using the half-reaction equations (noting that 2 moles electrons frees 2 moles gas, if you count both electrodes), and finally to volume of gas by using 22.4L/mol.

   (3.0C/s)(3.6×10^3 s)(1mol e^-/96485C)(2mol gas/2mol e^-)(22.4L/1mol gas) = 2.5L gas (total)

(f) As the electrolysis continues, dissolved hydrochloric acid leaves the solution as H^+ is converted into H_2 and as Cl^- is converted into Cl_2. Because the concentration of acid decreases over time, the pH rises.

4. (a) (i) $3OH^-(aq) + Fe^{3+}(aq) \rightarrow Fe(OH)_3(s)$. Because the potassium cation and sulfate anions are ionized and unchanged on both sides of the reaction equation, they are omitted from the net ionic equation.

(ii) A solid precipitate forms as the reaction proceeds. The precise colors of the iron (III) sulfate solution and iron (III) hydroxide precipitate depend on their concentrations. The iron (III) sulfate solution typically has a greenish color, whereas the iron (III) hydroxide product would appear as a red-brown precipitate.

(b) (i) $Ni^{2+}(l) + 2Cl^-(l) \rightarrow Ni(s) + Cl_2(g)$. Molten salts are liquid, not aqueous. All reactants and products remain because each changes during the reaction.

(ii) During the course of this salt hydrolysis, solid nickel would deposit at the cathode and chlorine gas would bubble from the anode.

Cathode: $Ni^{2+}(l) + 2e^- \rightarrow Ni(s)$

Anode: $2Cl^-(l) \rightarrow Cl_2(g) + 2e^-$

(c) (i) $MgO(s) + H_2O(l) \rightarrow Mg(OH)_2(s)$. All reactants and products are shown because all change during the reaction.

(ii) During the course of the reaction, the metal oxide, initially a white powder, would dissolve into a clear aqueous solution and undergo the reaction shown in part (c). Although magnesium hydroxide is only sparingly soluble, some hydroxide ion would dissociate into solution. As the reaction progressed, the pH of the solution would progressively increase due to the appearance of hydroxide ion product.

5. (a) Although atmospheric oxygen occurs at the surface of a deep body of water, two important parameters change as you move to greater depths. First, the temperature decreases. Second, the pressure increases. Decreasing temperatures increase the solubility of gaseous solutes (like oxygen); at lower temperatures, the solutes have decreased kinetic energy and are less able to disrupt the solute–solvent interactions that hold them in solution. At increased pressures, gaseous solutes are more soluble in accordance with Henry's law.

(b) The tendency of a solute to heat or cool a solution while dissolving is quantified by the parameter, ΔH_{soln}, the molar heat of solution. If this parameter is negative, the solute releases heat into solution as it dissolves. If the parameter is positive, the solute absorbs heat from solution as it dissolves. The physical basis for differences in heats of solution lies in the collection of solute-solute, solute-solvent, and solvent-solvent interactions that govern the dissolution process. If, in dissolving, an excess of enthalpically favorable interactions must be disrupted, dissolution is accompanied by an absorption of heat — in other words, by cooling. If, in dissolving, an excess of enthalpically favorable interactions becomes available, dissolution is accompanied by a release of heat — in other words, by heating. Even when dissolution is enthalpically unfavorable, it may nevertheless occur if it is driven by entropic effects.

(c) Solid copper, $Cu(s)$, participates in metallic bonding. Solid copper (II) chloride, $CuCl_2(s)$ participates in ionic bonding. In metallic bonding, metal atoms pack into a crystalline lattice in which the electropositive nuclei sit amidst a sea of mobile electrons. The electrons move freely throughout the lattice because the energetic penalty for hopping from one nucleus to the next is negligible. So, when motivated to flow by an applied voltage, the mobile electrons do so. In ionic bonding, the crystalline lattice is composed of alternating ectropositive and electronegative bonding partners. One partner releases electrons and the other partner binds them. In order for electrons to flow under an applied voltage, the favorable electron-electronegative partner interactions would have to be disrupted, which is unfavorable.

6. (a) The three compounds are an aldehyde, a ketone, and a carboxylic acid.

(b) Here are the 4-carbon compounds:

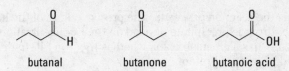

butanal   butanone   butanoic acid

(c) Ethanol is most likely to react with the carboxylic acid compound to form an ester. Esters are the typical products of dehydration reactions between carboxylic acids and alcohols.

(d) Here is a figure of the ester that would form:

ethyl butanoate

Part VIII
The Part of Tens

The 5th Wave · By Rich Tennant

In this part . . .

This part is a lean, hard-edged, white-knuckled distillation of the book. Big ideas, little details, key equations, and nasty traps to avoid. They're here, ready for your quick perusal. These lists are ideal for last-minute review, or for the occasional going-over to remind yourself of important things you've forgotten. Let your eyes linger on these pages when actual study just isn't going to happen. Paste these pages to the ceiling of your bedroom (we won't tell anyone). Make small copies of these pages, laminate them, and keep them in your wallet. Some of these suggestions may be a *bit* extreme, we concede, but do give the Part of Tens a look — you'll be rewarded for your time.

Chapter 31

Ten Math Skills to Bring with You to the Exam

*I*f you think that you're going to get away with doing chemistry and chemistry alone on the AP exam, you're sorely mistaken. Chemistry is a very mathematically oriented science, and you will be asked to do a great number of calculations as well as applying chemistry concepts. Listed in this chapter are the ten most essential AP chemistry math skills and some tips and tricks on avoiding errors. In particular, this chapter points out the skills necessary to build a natural intuition for the correctness of your answers.

Dimensional Analysis/Conversion Factors

You're guaranteed to use conversion factors, perhaps the most essential of all math skills for the purposes of AP chemistry, several times on every AP chemistry exam. The purpose of a conversion factor is just as its name implies — to change (convert) one unit to another. All conversion factors have the same essential property — their numerator and denominator are two different representations of the exact same quantity. For example, at STP 22.4L and 1mol of a gas are the exact same thing, so both the conversion factor 22.4L/1 mol and its companion 1mol/22.4L are equivalent to each other and both equal to 1 because their numerators and denominators are exactly equal quantities. So what's the purpose of multiplying something by 1? It allows you to manipulate units, getting rid of unwanted ones and bring in the ones that the problem asks for. In general, you will place the unit you wish to get rid of in the denominator and the one that you wish to change it to in the numerator of the conversion factor.

pH and Logs

pH calculations are notoriously difficult even for a seasoned Chemistry student such as yourself because they apply the difficult algebraic concept of the logarithm. It is essential that you realize that pH utilizes a base ten logarithm, which is the reverse operation of raising ten to a number. In other words, you can "undo" a logarithm by raising ten to both sides of a pH equation. This step leads you to the very useful formula $10^{-pH} = [H^+]$, which allows you to backsolve from pH to $[H^+]$. The other essential thing to realize about pH is that if the coefficient on the H^+

concentration is 1, then the negative power of ten to which it is raised is equivalent to the pH and no actual calculation is necessary. Also, any coefficient greater than 1 will have a pH between its power of ten and one less than its power of ten. For example, a solution with $[H^+]$ = 4.55×10^{-12} has a pH between 11 and 12. This knowledge can save you lots of time on multiple-choice questions or on free-response questions with which you cannot use a calculator.

Isolating a Variable

Another essential algebra skill to bring with you to the exam is the ability to isolate a variable in an equation. The AP exam provides you with a plethora of equations. However, it will be up to you to isolate each variable that you need to solve for (i.e. get it "by itself" on one side of the equation). We are not attempting to undermine your intelligence by pointing out this seemingly trivial concept. There is a subtlety to this skill that many AP chemistry exam-takers neglect — you are much more likely to get the correct answer if you isolate the variable that you need to solve for *before* you plug any numbers in. This minimizes algebraic and rounding errors, which can lose you points.

Significant Figures

After you get an answer on the AP exam, make sure that you provide the correct number of significant figures in your answer. Here is a summary of the rules for assigning significant figures:

- ✓ **Any nonzero digit is significant.** Thus, 6.42 seconds (s) contains three significant figures.

- ✓ **Zeros sandwiched between nonzero digits are significant.** Thus, 3.07 s contains three significant figures.

- ✓ **Zeros on the left side of the first nonzero digit are *not* significant.** Thus, 0.0642 s and 0.00307 s each contain three significant figures.

- ✓ **When a number is greater than 1, all digits to the right of the decimal point are understood to be significant.** Thus, 1.76 s has three significant figures, while 1.760 s has four significant figures. We understand that the *6* is uncertain in the first measurement, but is certain in the second measurement.

- ✓ **When a number has no decimal point, any zeros after the last nonzero digit *may or may not* be significant.** Thus, in a measurement reported as 1370 s, we cannot be certain if the "0" is a certain value, or if it is merely a placeholder. Be a good chemist. Report your measurements in scientific notation to avoid such annoying ambiguities.

- ✓ **Numbers resulting from *counting* (e.g., one kangaroo, two kangaroos, three kangaroos . . .) or from *defined quantities* (e.g., 60 seconds per 1 minute) are considered completely certain.** These values are understood to have an unlimited number of significant figures, consistent with their complete certainty.

- ✓ **When taking the log of a number, as when calculating pH or pOH from $[H^+]$ or $[OH^-]$, only the decimal portion of the answer applies toward the significant figure count (not the preceding integer).** For example, if $[H^+]$ = $0.0100M$ and pH = $-\log[H^+]$, then pH = 2.000. Why? 0.0100 contains three significant figures. Therefore, the decimal portion of the log answer (the mantissa) contains three significant figures. The preceding integer (the characteristic "2" in this case) does not count toward the significant figure total.

The final thing to burn into your brain about significant figures is that your final answer should always be rounded to the same number of significant figures as the least-precise number you were given in the problem. However, do not round any of the numbers you are given until the very end after you have plugged them into your equations in their full, precise glory.

Scientific Notation

You must be comfortable with scientific notation before taking the AP chemistry exam. Remember that the coefficient for a number in scientific notation must be greater than one and less than ten, that positive exponents mean very large numbers, and that negative exponents mean very small numbers. Also, you must be able to properly add, subtract, multiply, and divide numbers in scientific notation. Adding and subtracting is easily accomplished by adding or subtracting the coefficients *as long as the exponents of the two are the same*. So, you must adjust the values to have the same exponents *before* adding or subtracting. To multiply two numbers in scientific notation together, you must multiply their coefficients but *add* their exponents. To divide, you must divide their coefficients but *subtract* the exponent of the denominator from that of the numerator.

Percent Error, Percent Yield, and Percent Composition

The key percentage measurements in chemistry all have the exact same underlying concept: Remember, "percentage" is just asking how many would you have if you had a hundred of the total. Keep that in mind and you won't get lost. Divide the portion under consideration by the whole and you have the percentage you're looking for. For percent errors, subtract your result from the accepted value of the quantity (this value is the error in your measurement) and then divide that by the accepted value. Percent yield is calculated in an equivalent way: by subtracting the actual yield from the theoretical yield and then dividing by the theoretical yield. Unsurprisingly, percent composition is also quite similar, though even simpler. Divide the number of grams of each particular atom in one mole of a substance by its molar mass and you have its percent composition. For all three of these quantities, remember that your answer will be in decimal form so you will need to multiply by 100 to change it into percentage form.

The SI System of Units

Make sure that you're familiar with the SI system of units and, more importantly, have a sense of the relative size of each one. This familiarity can go a long way toward honing your intuition as to whether an answer seems to be about the right size based on what you know about the problem. The five SI base units that you will need to do chemistry problems, as well as their non-SI equivalents, are given in Table 31-1.

Table 31-1	SI Units		
Measurement	SI Unit	Symbol	Non-SI Unit
Amount of a substance	Mole	mol	
Length	Meter	m	feet, inch, yard, mile
Mass	Kilogram	kg	pound
Temperature	Kelvin	K	degree Celsius or Fahrenheit
Time	second	s	minute, hour

Metric System Prefixes

Nobody, not even a chemist, likes dealing with scientific notation if she doesn't have to. Metric system prefixes often appear in problems in lieu of scientific notation. For example, the size of the nucleus of an atom is roughly 10 femtometers across, which is a nicer way of saying 10×10^{-15} meters. The most useful of these prefixes are given in Table 31-2 below for you to memorize.

Table 31-2	The Metric System Prefixes		
Prefix	Symbol	Meaning	Example
Centi	c	10^{-2}	$1cm = 10^{-2}m$
Deci	d	10^{-1}	$1dm = 10^{-1}m$
Deco	D	10^{1}	$1Dm = 10^{1}m$
Femto	f	10^{-15}	$1fm = 10^{-15}m$
Kilo	k	10^{3}	$1km = 10^{3}m$
Main unit	1	1m	
Micro	μ	10^{-6}	$1\mu m = 10^{-6}m$
Milli	m	10^{-3}	$1mm = 10^{-3}m$
Nano	n	10^{-9}	$1nm = 10^{-9}m$
Pico	p	10^{-12}	$1 pm = 10^{-12}m$

Deducing Qualitative Trends from an Equation

It is important not only that you be adept at manipulating chemical equations, but that you also understand their underlying meaning if you wish to do well on the AP chemistry exam. In other words, when you look at an equation, you should always be thinking first and foremost about the relationship between variables. Are they inversely or directly proportional (in other words, does one increase as the other decreases, or do they increase or decrease together)? Recognizing trends like this can help you eliminate wrong answers in the multiple-choice section more quickly and can help to hone your intuition for correct and incorrect responses. A typical example is in solving for changes in ideal gas

If $P_1 V_1 = nRT_1$, then changing to new conditions gives you $P_2 V_2 = nRT_2$.

The Greek Letter Delta

The delta symbol (Δ) appears quite often in chemistry and always signifies that a quantity has changed from an initial to a final state. It can be interpreted to mean "change in" the variable that follows it. For example ΔT is the change in temperature from the initial to the final state and can be written as $T_f - T_i$.

Chapter 32

Ten Critical Types of Chemistry Problems

• •

In This Chapter

▶ Concentrating on the most important concepts

▶ Listing key kinds of calculations

▶ Breaking down exam questions into their component parts

• •

During the course of the AP exam, you'll be asked to answer a variety of types of chemistry questions. Some of these questions will be explicit and straightforward; in other words, you'll know for certain just what you're being asked. Other questions will be implicit and indirect; in other words, you'll have to unpack the question to determine just what you're being asked. Whether explicit or implicit, no matter what clothes they wear, the questions will fall into one or another of a limited number of categories. Here's an overview of those categories.

Periodic Table

Knowledge of the patterns of the periodic table is essential to do well on the AP chemistry exam; both explicitly and implicitly, questions will require you to refer to this basic chemical knowledge. Here are some of the contexts in which you might be called to consult the table:

✔ What is the subatomic composition of a particular atom?

✔ What is the ground state or excited electronic structure of an atom?

✔ What kind of compound, with what formula, will be formed if two elements react?

✔ Is a particular chemical species likely or unlikely?

✔ Will two compounds react or not?

✔ What kind of bond forms between the atoms of two elements?

✔ Is a particular atom likely to form an ion and if so, which ion?

✔ How many covalent bonds will a particular atom form?

Composition and Structure

Knowing the rules by which atoms assemble into compounds is key. Know how to do the following:

- ✔ Determine a percent composition for a compound, given the compound's formula
- ✔ Determine a compound's empirical formula, given its percent composition (and a molecular formula, given the compound's molar mass)
- ✔ Move easily between compound names and formulas
- ✔ Given a compound's formula, be able to

  - Determine whether the compound is ionic or molecular
  - Propose a likely covalent structure for a molecular compound, including possible double or triple bonds, and draw its Lewis dot structure
  - Propose a likely molecular shape, especially by using VSEPR principles
  - Explain the polarity of a molecule with respect to molecular shape and bond polarity

Solubility and Colligative Properties

Solubility is important because it can control the phases of reactants or products, and even determine whether a given reaction is possible. Almost all soutions on the AP exam have water as the solvent. Whether or not a substance dissolves in a given solvent is intimately connected to the structure and properties of both the substance and the solvent. In other words, be familiar with the following:

- ✔ How molecular structure, especially polarity, controls solubility and miscibility in different solvents
- ✔ How temperature affects the solubility of solid and gaseous solutes
- ✔ How pressure affects the solubility of gaseous solutes

Different chemistry problems require you to use different units of concentration. Some problems require you to convert between different units. So, you must be familiar with these different units, when it is appropriate to use each unit, and how to calculate dilutions:

- ✔ **Molarity:** Moles solute per liter solution — by far the most used
- ✔ **Molality:** Moles solute particles per kilogram solvent

  - To be used in colligative properties calculations

- ✔ **Partial pressure of an ideal gas:** Equals the mole fraction of the gas multiplied by the total pressure
- ✔ **Percent solution:** Mass percent, ppm, ppb, and volume percent — relatively less common

Colligative properties, the properties of a solution that depend on the number of solute particles, vary with molality. Molality can be used to calculate three key properties with which you should be familiar:

✔ Boiling point elevation

✔ Freezing point depression

✔ Molar mass of a solute

In addition, you should be able to describe the effects of added nonvolatile solute on the vapor pressure of solvent over a solution.

Gas Behavior and Phase Behavior

Don't enter the exam without knowing ideal gas behavior thoroughly and effortlessly. Here are the key relationships that may crop up in questions:

✔ **Charateristics of an ideal gas:** For instance, molecules have no volume, molecules are in random motion, molecules neither attract nor repel each other, molecules collide elastically

✔ **Meaning of gas temperature:** As the temperature of a gas increases, the distribution of kinetic energies of the particles gets broader, and the average kinetic energy increases.

✔ **Ideal Gas Law:** $PV = nRT$, the most complete statement of ideal gas behavior — forget this one at your peril.

✔ **Combined Gas Law:** $P_1V_1/T_1 = P_2V_2/T_2$, a useful combination of other laws that allows you to calculate responses to changes in pressure, volume, and temperature

- **Boyle's Law:** $PV = k$, which leads to $P_1V_1 = P_2V_2$

- **Charles's Law:** $V/T = k$, which leads to $V_1/T_1 = V_2/T_2$

- **Gay-Lussac's Law:** $P/T = k$, which leads to $P_1/T_1 = P_2/T_2$

✔ **Dalton's Law:** $P_{total} = P_1 + P_2 \ldots + P_n$

✔ **Graham's Law:** The ratio of the square roots of the molar masses equals the inverse of the rates of effusion or diffusion

The same concepts of kinetic energy and particle motion (in other words, kinetic molecular theory) that inform our model of gas behavior can also be used to help explain the movement of substances between gas, liquid, and solid phases. Be comfortable with the following:

✔ The microscopic description of solid, liquid, and gas phases, especially as it relates to particle motions and kinetic energy

✔ How the microscopic details of each phase correspond to macroscopic properties such as density, viscosity, and heat capacity

✔ How molecular structure and interparticle forces (like dipole-dipole and dispersion forces) factor into phase behavior, especially in condensed phases

Stoichiometry and Titration

Knowing the numerical relationships between reactants and products is the starting point for countless questions. So, be sure you can do the following:

- ✔ Balance reaction equations

- ✔ Use balanced reaction equations to construct conversion factors

- ✔ Move easily between units of moles and other units (mass, volume, and particles)

- ✔ Apply balanced reaction equations to determine the identity of a limiting reagent and calculate the theoretical, actual, and percent yield of a reaction

- ✔ Apply knowledge of stoichiometry to acid-base reactions, especially in accounting for acid/base equivalents within titration reactions

- ✔ Make predictions about whether a given titration reaction is at its equivalence point (equal amonts of acid and base reaction according to $H^+ + OH^- \leftrightarrow H_2O$)

- ✔ Move easily between molar concentrations of acid or base and related quantities of pH, pOH, K_a, K_b, and K_w

- ✔ Understand the concept of a buffered solution, when buffered solutions are effective, and be able to use the Henderson-Hasselbach equation

Equilibrium

The concept of equilibrium is core to understanding how chemical systems respond to change, and it figures prominently in a large percentage of questions. Here are some of the ways in which equilibrium might rear its head:

- ✔ Determine the direction in which a given reaction will shift, in response to the following

    - • increasing or decreasing the concentration of a reactant

    - • increasing or decreasing the concentration of a product

    - • increasing or decreasing temperature

    - • increasing or decreasing pressure or volume

- ✔ Given concentrations of reactants or products, calculate an equilibrium constant and vice versa

- ✔ Using an equilibrium constant or a reaction quotient and the concentrations of reactants or products to

    - • Determine whether a system is at equilibrium

    - • Determine the direction in which a nonequilibrium system will proceed

- ✔ Manipulate equilibrium constant expressions and values for coupled equilibria and for forward versus reverse reactions

- ✔ Apply equilibrium concepts to questions of solubility and acid-base behavior, especially with respect to K_{sp}, K_a, K_b and K_w

Thermodynamics and Thermochemistry

Thermodynamics deals quantitatively with differences in energy between reactants and products and states of substances, and thermochemistry deals specifically with thermal energy transfer as heat during chemical reactions. Reaction equilibrium can be predicted from knowledge of thermodynamic parameters enthalpy and entropy. Both of these topics are fundamental to chemical change. Have a solid grasp on the concepts and calculations associated with them.

✔ Move easily between equilibrium concentrations and the free energy change for a reaction by using $\Delta G° = -RT \ln K_{eq}$ and for nonequilibrium conditions using $\Delta G = \Delta G° - RT \ln Q$.

✔ Understand the interrelationships of free energy, enthalpy, entropy, and temperature in determining the spontaneity of a chemical reaction

- Know and be able to use the Gibbs equation, $\Delta G = \Delta H - T\Delta S$

- Be able to connect the concepts of enthalpy and entropy with behavior at the atomic and molecular level such as bonding and motion

✔ Understand and be able to differentiate the concepts of heat capacity, molar heat capacity, and specific heat capacity

✔ Be able to apply the concept of heat capacity and calorimetry to reactions in which heat is released or absorbed, especially by using $q = mC\Delta T$

✔ Apply Hess's Law to coupled systems of chemical reactions and determine a missing quantity from a Hess's Law cycle

Kinetics

Just as a grasp of thermodynamics is key to understanding spontaneity, having a grasp of kinetics is key to understanding the rates of reactions. Knowing one concept without the other is like walking on one stilt — you go nowhere and you fall down a lot.

✔ Understand the concept of reaction rate and how questions of rate differ from questions of equilibrium

✔ Given a set of experimental data, be able to formulate and use rate laws and understand the distinct roles of

- Rate constants

- Reactant concentrations

- Reaction orders

✔ Be able to apply the concepts of molecularity, elementary steps, and rate-determining steps to the mechanism of a chemical reaction

✔ Interpret reaction energy diagrams with regard to the kinetics and mechanism of a reaction

✔ Understand the concept of activation energy, E_a, and be able to apply it to questions of rate, especially by using the Arrhenius equation

Redox and Electrochemistry

Charges can change during chemical reactions, and when they do, they become critical players, altering the stoichiometry and energetics of the reaction. Oxidation-reduction reactions and electrochemistry are must-know categories for the exam, especially with respect to the following skills:

- Determining the oxidation number of all atoms in a balanced reacton equation and using those numbers to determine

    - Whether or not a given reaction involves redox

    - The identity of both oxidizing and reducing agents

- Decomposing redox reactions into oxidation and reduction half-reactions

- Using half-reactions to balance overall redox reaction equations (under acidic and basic conditions)

- Applying redox chemistry to electrochemical and electrolytic cells in order to

    - Determine the identity of anodes, cathodes

    - Determine the direction in which electrons and ions flow

    - Determine if and and at which electrode electroplating occurs

    - Calculate standard cell potentials, especially with regard to standard reduction potentials of half-reactions, electrical current and the Faraday, and free energy changes under standard conditions

    - Calculate cell potential under nonstandard conditions from concentrations and the standard cell potential by using the Nernst equation

Descriptive and Organic Chemistry

These topics are sometimes given short shrift in general Chemistry classes, but that doesn't mean they'll be absent from the AP exam — up to one-sixth of the material will test this information directly or indirectly. Be sure to focus on these items:

- Given the reactants, make reasonable predictions about the products of the most common kinds of reactions, to include

    - Single and double replacement/displacement

    - Synthesis/combination and decomposition

    - Combustion

- Be familiar with scenarios most likely to result in

    - Precipitation

    - Gas release

    - Redox

    - Formation of a colored compound

✔ Be able to move easily between the names for organic compounds and their structures, to include

- Alkanes, alkenes, and alkynes

- Cyclic and aromatic compounds, including existence of isomers and resonance structures

- Geometric isomers

- Compounds with functional groups, including halides, alcohols, ethers, and carboxylic acids

✔ Be able to make reasonable predictions about the properties of simple organic compounds (such as solubility, volatility, and reactivity)

✔ Be familiar with common organic reaction types, to include substitution, condensation, hydrolysis, addition, and elimination

Chapter 33

Ten Annoying Exceptions to Chemistry Rules

*E*xceptions seem like nature's way of hedging its bets. As such, they're annoying — why can't nature just go ahead and *commit?* But nature knows nothing of our rules, so it certainly isn't going out of its way to annoy you. Nor is it going out of its way to make things easier for you. Either way, seeing as you have to deal with exceptions, we thought we'd corral many of them into this chapter so you can confront them more conveniently.

However, one big caveat! This chapter is at the end for a reason. Don't get so hung up on the exceptions that you don't master the basics properly. Many AP students lose points on the exam because they omit describing the basic principles that underlie the exceptions. It is not enough to only understand the exceptions without quoting the "normal" rules or trends.

Hydrogen Isn't an Alkali Metal

In the field of psychology, Maslow's hierarchy of needs declares that people need a sense of belonging. It's a good thing hydrogen isn't a person because it belongs nowhere on the periodic table (which we describe in Chapter 3). Although hydrogen is usually listed atop Group IA along with the alkali metals, it doesn't really fit. Sure, hydrogen can lose an electron to form a +1 cation, just like the alkali metals, but hydrogen can also gain an electron to form hydride, H^-, especially when bonding to metals. Furthermore, hydrogen doesn't have metallic properties under normal conditions, but typically exists as the diatomic gas, H_2.

These differences arise largely from the fact that hydrogen has only a single $1s$ orbital and lacks other, more interior orbitals that could shield the valence electrons from the positive charge of the nucleus. (See Chapter 7 for an introduction to s and other types of orbitals.) Another factor is that H^+ is a bare proton with extremely compact charge density.

The Octet Rule Isn't Always an Option

An *octet,* as we explain in Chapter 5, is a full shell of eight valence electrons. The octet rule states that atoms bond with one another so as to acquire completely filled valence shells that contain eight electrons. It's a pretty good rule. Like most pretty good rules, it has exceptions:

- Atoms containing only 1*s* electrons simply don't have eight slots to fill. So, hydrogen and helium obey the "duet" rule.

- Certain molecules (like NO, for example) contain an odd number of valence electrons. In these cases, full octets aren't an option. Like people born with an odd number of toes, these molecules may not be entirely happy with the situation, but they deal with it.

- Atoms often attempt to fill their valence shells by covalent bonding (see Chapter 5). Each covalent bond adds a shared electron to the shell. But covalent bonding usually requires an atom to donate an electron of its own for sharing within the bond. Some atoms run out of electrons to donate, and therefore can't engage in enough covalent bonds to fill their shell octets. Boron trifluoride, BF_3, is a typical example. The central boron atom of this molecule can only engage in three B–F bonds, and ends up with only six valence electrons. This boron is said to be electron deficient. You might speculate that the fluorine atoms could pitch in a bit and donate some more electrons to boron. But fluorine is highly electronegative and greedily holds fast to its own octets. C'est la vie.

- Some atoms take on more than a full octet's worth of electrons. This is known as an *expanded octet.* These atoms are said to be *hypervalent* or *hypercoordinated.* The phosphorous of phosphorous pentachloride, PCl_5, is an example. These kinds of situations require an atom from period (row) 3 or higher within the periodic table. The exact reasons for this restriction are still debated. Certainly, the larger atomic size of period 3 and higher atoms allows more room to accommodate the bulk of all the binding partners that distribute around the central atom's valence shell.

Some Electron Configurations Ignore the Orbital Rules

Electrons fill orbitals from lowest energy to highest energy. This fact is true.

The progression of orbitals from lowest to highest energy is predicted by an Aufbau diagram. This isn't always true. Some atoms possess electron configurations that deviate from the standard rules for filling orbitals from the ground up. For Aufbau's sake, why?

Two conditions typically lead to exceptional electron configurations:

- First, successive orbital energies must lie close together, as is the case with 3*d* and 4*s* orbitals, for example.

- Second, shifting electrons between these energetically similar orbitals must result in a half-filled or fully filled set of identical orbitals, an energetically happy state of affairs.

Want a couple of examples? Strictly by the Aufbau, but not by the "energy" rules, chromium should have the following electron configuration:

$[Ar]3d^44s^2$

Because shifting a single electron from $4s$ to the energetically similar $3d$ level half-fills the $3d$ set and lowers the energy, the actual configuration of chromium is

$$[Ar]3d^54s^1$$

For similar reasons, the configuration of copper is not the expected $[Ar]3d^94s^2$, but instead is $[Ar]3d^{10}4s^1$. Shifting a single electron from $4s$ to $3d$ fills the $3d$ set of orbitals.

Strictly speaking, the above arguments only apply to isolated atoms. Once the atom is involved in bond formation in a compound, the whole energy level scheme has to be re-examined, and apparent filling order may change again.

Flip to Chapter 5 for full details on electron configurations and Aufbau diagrams.

One Partner in Coordinate Covalent Bonds Giveth Electrons; the Other Taketh

To form a covalent bond (as we explain in Chapter 5), each bonding partner contributes one electron to a two-electron bond, right? Not always. *Coordinate covalent bonds* are particularly common between transition metals and partners that possess lone pairs of electrons.

Here's the basic idea: Transition metals have empty valence orbitals. *Lone pairs* are pairs of nonbonding electrons within a single orbital. So, transition metals and lone-pair bearing molecules can engage in Lewis acid-base interactions (see Chapter 17). The lone-pair containing molecule acts as an electron donor (a Lewis base), giving both electrons to a bond with the metal, which acts as an electron acceptor (a Lewis acid). When this occurs, the resulting molecule is called a *coordination complex*.

The partners that bind to the metal are called *ligands*. Coordination complexes are often intensely colored and can have properties that are quite different than those of the free metal.

All Hybridized Electron Orbitals Are Created Equal

Different electron orbital types have grossly different shapes. Spherical s orbitals look nothing like lobed p orbitals, for example. So, if the valence shell of an atom contains both s- and p-orbital electrons, you might expect those electrons to behave differently when it comes to things like bonding, right? Wrong. If you attempt to assume such a thing, valence bond theory politely taps you an the shoulder to remind you that valence shell electrons occupy hybridized orbitals. These hybridized orbitals (as in sp^3, sp^2, and sp orbitals) reflect a mixture of the properties of the orbitals that make them up, and each of the orbitals is equivalent to the others in the valence shell.

Although this phenomenon represents an exception to the rules, it's somewhat less annoying than other exceptions because hybridization allows for the nicely symmetrical orbital geometries of actual atoms within actual molecules. VSEPR theory presently clears its throat to point out that the negative charge of the electrons within the hybridized orbitals causes those equivalent orbitals to spread as far apart as possible from one another. As a result, the geometry of sp^3-hybridized methane (CH_4), for example, is beautifully tetrahedral. However, this is messed up easily once total symmetry is upset. CH_3I is no longer a perfect tetrahedron. Sigh.

Check out Chapter 7 for the details on VSEPR theory and hybridization.

Use Caution When Naming Compounds with Transition Metals

The thing about transition metals is that the same transition metal can form cations with different charges. Differently charged metal cations need different names, so chemists don't get any more confused than they already are. These days, you indicate these differences by using Roman numerals within parentheses to denote the positive charge of the metal ion. An older method adds the suffixes –*ous* or –*ic* to indicate the cation with the smaller or larger charge, respectively. For example

Cu^+ = copper (I) ion or cuprous ion

Cu^{2+} = copper (II) ion or cupric ion

Metal cations team up with nonmetal anions to form ionic compounds. What's more, the ratio of cations to anions within each formula unit depends on the charge assumed by the fickle transition metal. The formula unit as a whole must be electrically neutral. The rules you follow to name an ionic compound must accommodate the whims of transition metals. The system of Roman numerals or suffixes applies in such situations:

$CuCl$ = copper (I) chloride or cuprous chloride

$CuCl_2$ = copper (II) chloride or cupric chloride

Chapter 7 has the full scoop on naming ionic and other types of compounds.

You Must Memorize Polyatomic Ions

Sorry, it's true. Not only are polyatomic ions annoying because they must be memorized, but they pop up everywhere. If you don't memorize the polyatomic ions, you'll waste time trying to figure out weird (and incorrect) covalent bonding arrangements when what you're really dealing with is a straightforward ionic compound. Here they are in Table 33-1.

Table 33-1	Common Polyatomic Ions
–1 Charge	*–2 Charge*
Dihydrogen phosphate ($H_2PO_4^-$)	Hydrogen Phosphate (HPO_4^{2-})
Acetate ($C_2H_3O_2^-$)	Oxalate ($C_2O_4^{2-}$)
Hydrogen Sulfite (HSO_3^-)	Sulfite (SO_3^{2-})
Hydrogen Sulfate (HSO_4^-)	Sulfate (SO_4^{2-})
Hydrogen Carbonate (HCO_3^-)	Carbonate (CO_3^{2-})
Nitrite (NO_2^-)	Chromate (CrO_4^{2-})
Nitrate (NO_3^-)	Dichromate ($Cr_2O_7^{2-}$)
Cyanide (CN^-)	Silicate (SiO_3^{2-})

−1 Charge	−3 Charge
Hydroxide (OH⁻)	Phosphite (PO_3^{3-})
Permanganate (MnO_4^-)	Phosphate (PO_4^{3-})
Hypochlorite (ClO⁻)	
Chlorite (ClO_2^-)	**+1 Charge**
Chlorate (ClO_3^-)	Ammonium (NH_4^+)
Perchlorate (ClO_4^-)	

Liquid Water Is Denser than Ice

Kinetic molecular theory, which we discuss in Chapters 9 and 11, predicts that adding heat to a collection of particles increases the volume occupied by those particles. Heat-induced changes in volume are particularly evident at phase changes, so liquids tend to be less dense than their solid counterparts. Weird water throws a wet monkey wrench into the works. Because of H_2O's ideal hydrogen-bonding geometry, the lattice geometry of solid water (ice) is very "open" with large empty spaces at the center of a hexagonal ring of water molecules. These empty spaces lead to a lower density of solid water relative to liquid water. So, ice floats in water. Although annoying, this watery exception is quite important for biology.

No Gas Is Truly Ideal

No matter what your misty-eyed grandparents tell you, there were never halcyon Days of Old when all the gases were ideal. To be perfectly frank, not a single gas is really, truly ideal. Some gases just approach the ideal more closely than others. At very high pressures, even gases that normally behave close to the ideal cease to follow the Ideal Gas Laws that we discuss in Chapter 9.

When gases deviate from the ideal, we call them *real gases*. Real gases have properties that are significantly shaped by the volumes of the gas particles and/or by interparticle forces.

Common Names for Organic Compounds Hearken Back to the Old Days

Serious study of chemistry predates modern systematic methods for naming compounds. As a result, chemists persistently address a large number of common compounds, especially organic compounds, by older, "trivial" names. This practice won't change anytime soon. A cynical take on the situation is to observe that progress occurs one funeral at a time. A less cynical approach involves serenely accepting that which you cannot change and getting familiar with these old-fashioned names. Table 33-2 lists some important ones; head to Chapter 25 for details on organic compounds.

Table 33-2	Common Names for Organic Compounds	
Formula	*Systematic Name*	*Common Name*
$CHCl_3$	Trichloromethane	Chloroform
H_2CO	Methanal	Formaldehyde
CH_2O_2	Methanoic acid	Formic acid
CH_3COCH_3	Propanone	Acetone
CH_3CO_2H	Ethanoic acid	Acetic acid
C_2H_4	Ethene	Ethylene
C_2H_2	Ethyne	Acetylene
C_3H_8O	Propan-2-ol	Isopropanol
$CH_3CH(CH_3)_2$	2-methylpropane	Isobutane

Index

• Q •

• R •

BUSINESS, CAREERS & PERSONAL FINANCE

Accounting For Dummies, 4th Edition*
978-0-470-24600-9

Bookkeeping Workbook For Dummies†
978-0-470-16983-4

Commodities For Dummies
978-0-470-04928-0

Doing Business in China For Dummies
978-0-470-04929-7

E-Mail Marketing For Dummies
978-0-470-19087-6

Job Interviews For Dummies, 3rd Edition*†
978-0-470-17748-8

Personal Finance Workbook For Dummies*†
978-0-470-09933-9

Real Estate License Exams For Dummies
978-0-7645-7623-2

Six Sigma For Dummies
978-0-7645-6798-8

Small Business Kit For Dummies, 2nd Edition*†
978-0-7645-5984-6

Telephone Sales For Dummies
978-0-470-16836-3

BUSINESS PRODUCTIVITY & MICROSOFT OFFICE

Access 2007 For Dummies
978-0-470-03649-5

Excel 2007 For Dummies
978-0-470-03737-9

Office 2007 For Dummies
978-0-470-00923-9

Outlook 2007 For Dummies
978-0-470-03830-7

PowerPoint 2007 For Dummies
978-0-470-04059-1

Project 2007 For Dummies
978-0-470-03651-8

QuickBooks 2008 For Dummies
978-0-470-18470-7

Quicken 2008 For Dummies
978-0-470-17473-9

Salesforce.com For Dummies, 2nd Edition
978-0-470-04893-1

Word 2007 For Dummies
978-0-470-03658-7

EDUCATION, HISTORY, REFERENCE & TEST PREPARATION

African American History For Dummies
978-0-7645-5469-8

Algebra For Dummies
978-0-7645-5325-7

Algebra Workbook For Dummies
978-0-7645-8467-1

Art History For Dummies
978-0-470-09910-0

ASVAB For Dummies, 2nd Edition
978-0-470-10671-6

British Military History For Dummies
978-0-470-03213-8

Calculus For Dummies
978-0-7645-2498-1

Canadian History For Dummies, 2nd Edition
978-0-470-83656-9

Geometry Workbook For Dummies
978-0-471-79940-5

The SAT I For Dummies, 6th Edition
978-0-7645-7193-0

Series 7 Exam For Dummies
978-0-470-09932-2

World History For Dummies
978-0-7645-5242-7

FOOD, GARDEN, HOBBIES & HOME

Bridge For Dummies, 2nd Edition
978-0-471-92426-5

Coin Collecting For Dummies, 2nd Edition
978-0-470-22275-1

Cooking Basics For Dummies, 3rd Edition
978-0-7645-7206-7

Drawing For Dummies
978-0-7645-5476-6

Etiquette For Dummies, 2nd Edition
978-0-470-10672-3

Gardening Basics For Dummies*†
978-0-470-03749-2

Knitting Patterns For Dummies
978-0-470-04556-5

Living Gluten-Free For Dummies†
978-0-471-77383-2

Painting Do-It-Yourself For Dummies
978-0-470-17533-0

HEALTH, SELF HELP, PARENTING & PETS

Anger Management For Dummies
978-0-470-03715-7

Anxiety & Depression Workbook For Dummies
978-0-7645-9793-0

Dieting For Dummies, 2nd Edition
978-0-7645-4149-0

Dog Training For Dummies, 2nd Edition
978-0-7645-8418-3

Horseback Riding For Dummies
978-0-470-09719-9

Infertility For Dummies†
978-0-470-11518-3

Meditation For Dummies with CD-ROM, 2nd Edition
978-0-471-77774-8

Post-Traumatic Stress Disorder For Dummies
978-0-470-04922-8

Puppies For Dummies, 2nd Edition
978-0-470-03717-1

Thyroid For Dummies, 2nd Edition†
978-0-471-78755-6

Type 1 Diabetes For Dummies*†
978-0-470-17811-9

* Separate Canadian edition also available
† Separate U.K. edition also available

Available wherever books are sold. For more information or to order direct: U.S. customers visit www.dummies.com or call 1-877-762-2974.
U.K. customers visit www.wileyeurope.com or call (0) 1243 843291. Canadian customers visit www.wiley.ca or call 1-800-567-4797.

 WILEY

INTERNET & DIGITAL MEDIA

AdWords For Dummies
978-0-470-15252-2

Blogging For Dummies, 2nd Edition
978-0-470-23017-6

**Digital Photography All-in-One
Desk Reference For Dummies, 3rd Edition**
978-0-470-03743-0

Digital Photography For Dummies, 5th Edition
978-0-7645-9802-9

**Digital SLR Cameras & Photography
For Dummies, 2nd Edition**
978-0-470-14927-0

**eBay Business All-in-One Desk Reference
For Dummies**
978-0-7645-8438-1

eBay For Dummies, 5th Edition*
978-0-470-04529-9

eBay Listings That Sell For Dummies
978-0-471-78912-3

Facebook For Dummies
978-0-470-26273-3

The Internet For Dummies, 11th Edition
978-0-470-12174-0

Investing Online For Dummies, 5th Edition
978-0-7645-8456-5

iPod & iTunes For Dummies, 5th Edition
978-0-470-17474-6

MySpace For Dummies
978-0-470-09529-4

Podcasting For Dummies
978-0-471-74898-4

**Search Engine Optimization
For Dummies, 2nd Edition**
978-0-471-97998-2

Second Life For Dummies
978-0-470-18025-9

**Starting an eBay Business For Dummies,
3rd Edition†**
978-0-470-14924-9

GRAPHICS, DESIGN & WEB DEVELOPMENT

**Adobe Creative Suite 3 Design Premium
All-in-One Desk Reference For Dummies**
978-0-470-11724-8

**Adobe Web Suite CS3 All-in-One Desk
Reference For Dummies**
978-0-470-12099-6

AutoCAD 2008 For Dummies
978-0-470-11650-0

**Building a Web Site For Dummies,
3rd Edition**
978-0-470-14928-7

**Creating Web Pages All-in-One Desk
Reference For Dummies, 3rd Edition**
978-0-470-09629-1

**Creating Web Pages For Dummies,
8th Edition**
978-0-470-08030-6

Dreamweaver CS3 For Dummies
978-0-470-11490-2

Flash CS3 For Dummies
978-0-470-12100-9

Google SketchUp For Dummies
978-0-470-13744-4

InDesign CS3 For Dummies
978-0-470-11865-8

**Photoshop CS3 All-in-One
Desk Reference For Dummies**
978-0-470-11195-6

Photoshop CS3 For Dummies
978-0-470-11193-2

Photoshop Elements 5 For Dummies
978-0-470-09810-3

SolidWorks For Dummies
978-0-7645-9555-4

Visio 2007 For Dummies
978-0-470-08983-5

Web Design For Dummies, 2nd Edition
978-0-471-78117-2

Web Sites Do-It-Yourself For Dummies
978-0-470-16903-2

Web Stores Do-It-Yourself For Dummies
978-0-470-17443-2

LANGUAGES, RELIGION & SPIRITUALITY

Arabic For Dummies
978-0-471-77270-5

Chinese For Dummies, Audio Set
978-0-470-12766-7

French For Dummies
978-0-7645-5193-2

German For Dummies
978-0-7645-5195-6

Hebrew For Dummies
978-0-7645-5489-6

Ingles Para Dummies
978-0-7645-5427-8

Italian For Dummies, Audio Set
978-0-470-09586-7

Italian Verbs For Dummies
978-0-471-77389-4

Japanese For Dummies
978-0-7645-5429-2

Latin For Dummies
978-0-7645-5431-5

Portuguese For Dummies
978-0-471-78738-9

Russian For Dummies
978-0-471-78001-4

Spanish Phrases For Dummies
978-0-7645-7204-3

Spanish For Dummies
978-0-7645-5194-9

Spanish For Dummies, Audio Set
978-0-470-09585-0

The Bible For Dummies
978-0-7645-5296-0

Catholicism For Dummies
978-0-7645-5391-2

The Historical Jesus For Dummies
978-0-470-16785-4

Islam For Dummies
978-0-7645-5503-9

**Spirituality For Dummies,
2nd Edition**
978-0-470-19142-2

NETWORKING AND PROGRAMMING

ASP.NET 3.5 For Dummies
978-0-470-19592-5

C# 2008 For Dummies
978-0-470-19109-5

Hacking For Dummies, 2nd Edition
978-0-470-05235-8

Home Networking For Dummies, 4th Edition
978-0-470-11806-1

Java For Dummies, 4th Edition
978-0-470-08716-9

**Microsoft® SQL Server™ 2008 All-in-One
Desk Reference For Dummies**
978-0-470-17954-3

**Networking All-in-One Desk Reference
For Dummies, 2nd Edition**
978-0-7645-9939-2

**Networking For Dummies,
8th Edition**
978-0-470-05620-2

SharePoint 2007 For Dummies
978-0-470-09941-4

**Wireless Home Networking
For Dummies, 2nd Edition**
978-0-471-74940-0

OPERATING SYSTEMS & COMPUTER BASICS

iMac For Dummies, 5th Edition
978-0-7645-8458-9

Laptops For Dummies, 2nd Edition
978-0-470-05432-1

Linux For Dummies, 8th Edition
978-0-470-11649-4

MacBook For Dummies
978-0-470-04859-7

**Mac OS X Leopard All-in-One
Desk Reference For Dummies**
978-0-470-05434-5

Mac OS X Leopard For Dummies
978-0-470-05433-8

Macs For Dummies, 9th Edition
978-0-470-04849-8

PCs For Dummies, 11th Edition
978-0-470-13728-4

Windows® Home Server For Dummies
978-0-470-18592-6

Windows Server 2008 For Dummies
978-0-470-18043-3

**Windows Vista All-in-One
Desk Reference For Dummies**
978-0-471-74941-7

Windows Vista For Dummies
978-0-471-75421-3

Windows Vista Security For Dummies
978-0-470-11805-4

SPORTS, FITNESS & MUSIC

Coaching Hockey For Dummies
978-0-470-83685-9

Coaching Soccer For Dummies
978-0-471-77381-8

Fitness For Dummies, 3rd Edition
978-0-7645-7851-9

Football For Dummies, 3rd Edition
978-0-470-12536-6

GarageBand For Dummies
978-0-7645-7323-1

Golf For Dummies, 3rd Edition
978-0-471-76871-5

Guitar For Dummies, 2nd Edition
978-0-7645-9904-0

**Home Recording For Musicians
For Dummies, 2nd Edition**
978-0-7645-8884-6

**iPod & iTunes For Dummies,
5th Edition**
978-0-470-17474-6

Music Theory For Dummies
978-0-7645-7838-0

Stretching For Dummies
978-0-470-06741-3

Get smart @ dummies.com®

- **Find a full list of Dummies titles**
- **Look into loads of FREE on-site articles**
- **Sign up for FREE eTips e-mailed to you weekly**
- **See what other products carry the Dummies name**
- **Shop directly from the Dummies bookstore**
- **Enter to win new prizes every month!**